The Sun in the Church

The Sun in the Church

Cathedrals as Solar Observatories

J. L. Heilbron

HARVARD UNIVERSITY PRESS

CAMBRIDGE, MASSACHUSETTS

LONDON, ENGLAND

Copyright © 1999 by the President and Fellows of Harvard College

All rights reserved

Printed in the United States of America

Second printing, 1999

Publication of this book has been supported through the generous provisions of
the Maurice and Lula Bradley Smith Memorial Fund

Book design by Dean Bornstein

Library of Congress Cataloging-in-Publication Data

Heilbron, J. L.

The sun in the church: cathedrals as solar observatories / J. L. Heilbron

p.   cm.

Includes bibliographical references and index.

ISBN 0-674-85433-0 (alk. paper)

1. Astronomical observatories — Italy.   2. Astronomy, Renaissance.
3. Religion and science — Italy — History — 17th century.   4. Catholic Church —
Italy — History — 17th century.   I. Title.

QB29.H33   1999

520'.94 — dc21   99-23123

For Alison

Contents

Color illustrations follow page 180.

Acknowledgments

I am indebted to so many librarians and archivists for help in procuring unusual materials that I can only thank them generically, but as sincerely as if the name of each, and also the names of their predecessors, who acquired and catalogued the books now in their charge, were separately set down. They include the present custodians of collections in out-of-the-way learned societies, like the Fisiocritici of Siena and the Zelanti of Acireale, and of major cathedrals, like the duomos of Florence and Milan, as well as the staff of my principal source of books, the Bodleian Library in Oxford.

Individuals who have kindly given me advice, documents, information, or illustrations, or have listened patiently to my enthusiasms, include my wife, Alison Browning; Jim Bennett, Tony Simcock, and Gerard Turner (Oxford); Gerard Chevalier and Paul de Divonne (Paris); Claus Jensen (Copenhagen); Tore Frängsmyr and Sven Widmalm (Uppsala); George Coyne (Rome); Paolo Galluzzi and Marco Beretta (Florence); Giuliano Pancaldi (Bologna); Arturo Russo (Palermo); Gaetano Gravagno (Aci); Owen Gingerich and Elizabeth Knoll (Cambridge); and Diana Wear (Berkeley), who also presided over making the final manuscript. Elizabeth Gilbert expertly prepared the manuscript for the printer. Dean Bornstein designed the book and made the technical drawings. I thank them all.

The Sun in the Church

Introduction

The Roman Catholic Church gave more financial and social support to the study of astronomy for over six centuries, from the recovery of ancient learning during the late Middle Ages into the Enlightenment, than any other, and, probably, all other, institutions. Those who infer the Church's attitude from its persecution of Galileo may be reassured to know that the basis of its generosity to astronomy was not a love of science but a problem in administration.[1] The problem was establishing and promulgating the date of Easter.

The old theologians decreed that Easter should be celebrated on the Sunday after the first full moon after the vernal equinox — that spring day on which the hours of daylight and darkness are equal.[2] This special full moon can be observed easily in principle and also often in practice. One needs only to recognize the equinox, wait until the next full moon, and declare the following Sunday to be Easter. That would give the right day for Easter, but not enough time to prepare for it. There lies the administrative problem. In addition, the equinox and the full moon occur at different times at different places on the earth, as, of course, does

Sunday. Even if all observations were correct, Easter might be celebrated on different days in different places. That was unacceptable to an organization struggling to make good its claims to unity and universality.

To avoid these inconveniences, the Church calculated the dates of Easter several years in advance and required observance on the Sundays specified in its tables. Since neither the lunar nor the solar year contains an even number of days, and, moreover, the year does not contain an even number of full moons, and, again, Sundays do not recur regularly on the same calendar dates, the computation of the Easter canon was neither easy nor accurate. Everything depended on exact average values of the periods between successive vernal equinoxes and between successive full moons. Administrators frequently have to make decisions on insufficient data. The bishops accepted the Julian year of 365.25 days as the interval between spring equinoxes and, as an approximation to the average number of full moons, or lunations in a Julian year, a value previously used by the Greeks.

The popes forced consensus on these numbers and procedures during the sixth century. By the twelfth century, however, they could see that the parameters of their predecessors no longer gave Easters in harmony with the heavens. In this emergency, popes encouraged the close study of the apparent motions of the sun and moon. The experts consulted ancient Greek mathematical texts that were then, luckily for them, just being turned into Latin from Arabic intermediaries. The most important of these texts was Ptolemy's "Great Compilation," which showed how to represent the motions of the stars, the luminaries (the sun and the moon), and the planets as seen from a stationary earth. Ptolemy's mathematical hypothesis, that the earth stands still at the center of the universe, seemed the most obvious and satisfactory basis for an exact astronomy. Not only did it conform to the evidence of the senses, it fit perfectly with the physical books of Aristotle, then, by 1200, also newly available.

The key parameter in the Easter calculation was the time of return of the sun to the same equinox. The most powerful way of measuring this cycle was to lay out a "meridian line" from south to north in a large dark building with a hole in its roof and observe how long the sun's noon image took to return to the same spot on the line. The most convenient such buildings were cathedrals; they came large and dark and needed only a hole in the roof and a rod in the floor to serve as solar observatories. The accuracy of the results depended on the care taken in installation: correct positioning of the hole, proper orientation of the rod, and exact leveling of the floor.

This book originated in the pleasant viewing of meridian lines installed in four Italian and one French cathedral during early modern times. They have the merits of beauty and utility, which to some observers were one and the same thing. "If the beauty of astronomical instruments depends on the usefulness of the results that can be obtained from them, then meridian lines can be counted among the most beautiful of things."[3] The writer of these lines, reviewing a *meridiana* built in the early eighteenth century in the assembly room of a small academy of sciences in Siena, had in mind not only measurements for correcting the calendar but also a wide range of fundamental studies. The academy's line proved too small to contribute to either purpose. Its model, the *meridiana* in the grand cathedral of San Petronio in Bologna, was another matter. According to the most authoritative compendium of the natural knowledge of the eighteenth century, it "made an epoch in the history of the renewal of the sciences."[4] Together with its counterparts in Florence, Paris, and Rome, "it would be celebrated in ages to come [to quote another eighteenth-century enthusiast] for the immortal glory of the human spirit, which could copy so precisely on the earth the eternal rule-bound movements of the sun and the stars."[5]

The story of the *meridiane* lies at the intersection, or, perhaps more fairly, at the margins, of many fields now usually held apart: architecture, astronomy, ecclesiastical and civil history, mathematics, and philosophy. Two world-historical events shaped the marginal story: the transformations in applied mathematics, with astronomy in the lead, that are supposed to have given Europeans the capacity to conquer the world,[6] and the Reformation, which, among much else, oversensitized the Catholic Church to deviant thinking. The Catholic Reformation met the new astronomy in the deviant thinker Galileo. The ill-advised condemnation of the theory of a moving earth that resulted from the collision made new difficulties for administrators. The Church needed up-to-date astronomers as much after its mistaken martyring of Galileo in 1633 as it had when the popes first puzzled themselves over Easter.

Renaissance and Astronomy

The Europeans read their translations of the Arabic versions of Greek mathematics, together with Arabic commentaries and analyses, over and over again, expecting—for such was the authority of an Aristotle or a Euclid over the unprepared minds of the best-educated Europeans—that the ancient books would have the answers to contemporary questions. Of all the recovered books, Ptolemy's *Almagest* (to give the great compilation the Anglicized Latinized Arabic title by which it is generally known) was technically the most demanding. The standard medieval version, made from the Arabic in 1185, began to circulate just as natural places for its study, the universities of Paris and Oxford, were forming. But the book far exceeded the capacities of almost all the members of both academies. The primer they needed was supplied by a professor in Paris educated at Oxford, Johannes de Sacrobosco, or John of Hollywood, perhaps the most successful writer of textbooks after Moses and Euclid. Sacrobosco's introduction to Ptolemaic astronomy, *On the sphere,* written around 1220, became required reading for the B.A. at Paris and Oxford; it was the first technical work on astronomy ever printed; and it inspired hundreds of reprintings and commentaries in many languages well into the seventeenth century.[7]

Those who advanced from Sacrobosco to the Latin-Arabic *Almagest* and its commentators had the mixed satisfaction of mastering the method and discovering that it did not work. Most of them relied upon tables derived on Ptolemaic principles by the able polyglot astronomers of Alfonso the Wise of Castille in the middle of the thirteenth century.[8] Alfonso's tables provided the raw material for the calculation of planetary positions for three hundred years even though the predictions disagreed systematically with observations. One adept who noticed a discrepancy was Copernicus, who recorded in his copy of the Alfonsine tables published in 1492 that he once saw Saturn almost a month behind its calculated position.[9]

· WORKING OVER PTOLEMY ·

Two paths lay open to European astronomers frustrated by the disparity between their observations and Ptolemy's theories as gathered from Arabic sources and applied in the Alfonsine tables. One way would be to seek another way, to condemn the Ptolemaic system as fundamentally flawed and, as Copernicus began to do, construct an alternative. This *via moderna* did not recommend itself to men strongly gripped by the achievements of the ancients. Supposing therefore that neither they nor Ptolemy could be responsible for the shortfall in their science, they blamed the long train of transmission to and through the Arabs. During the first half of the fifteenth century scholars fleeing the Turkish stranglehold on Constantinople brought west what many astronomers believed would be the key to modern astronomy, Ptolemy in his original Greek.

The knight who carried this grail to Italy was Cardinal Johannes Bessarion, who, it is said, had learned Latin expressly to translate Ptolemy.[10] Ambassadorial assignments for the Holy See interfered with his project, however, until they took him to Vienna in 1460 to negotiate help for Pius II's proposed crusade to reconquer Constantinople. There he met the professor of mathematics Georg Peurbach and his student Johannes Müller, then twenty-four, who became celebrated under the name Regiomontanus. Peurbach had been trying to improve the old Latin version of Ptolemy by sheer brain power, working backward from known errors. Bessarion

arranged for both of his new friends to go to Venice, where he kept his manuscripts; Peurbach was to make an abridgment or *Epitome* of the *Almagest* and Regiomontanus the long-sought translation. Peurbach died just before they were to
leave Vienna. The loss, which saddled Regiomontanus with the *Epitome,* was the
lesser of the impediments to his realization of Bessarion's project. The greater was
that he knew no Greek.

In Italy Regiomontanus busied himself studying Greek and cultivating acquaintance with humanists versed in it. Among them were Paolo del Pozzo
Toscanelli, who was to design the first cathedral *meridiana,* and George of Trebizond, a translator of Greek mathematical texts or, as Regiomontanus styled him,
"a perverse and impudent blabbermouth." That might have been a mistake. After
spending a few years back in central Europe, Regiomontanus returned to Italy in
1476 to advise the Pope, Sixtus IV, about correcting the calendar. He died suddenly
in Rome, some say of the plague, others of an adverse comet, still others of poison
administered by the vengeful sons of George of Trebizond.[11] Regiomontanus died
before he could produce an *Almagest* purged of the errors of the ages. The Greek
text he worked on appeared in print in 1538, just as Copernicus was completing
the book that would make it obsolete. The true Ptolemy proved little better than
the transmitted one for controlling what Regiomontanus, in his indulgent way,
called the "worthless" tables of Alfonso.[12]

"Just as Hercules held up the heavens when Atlas became tired, so when his fellow countryman Johannes Regiomontanus passed away, Copernicus revived the
science of heavenly motions."[13] This assessment, wrong only in all the details (Regiomontanus was German, Copernicus Polish, and the new Hercules was only
three when his Atlas died), is right, where it counts. Copernicus began where Regiomontanus had been thirty-five years before, in Italy, studying Greek and working on Peurbach's *Epitome* (in Copernicus' case, the printed edition of 1496).
Copernicus had been sent south by his uncle, the Bishop of Varmia, to prepare
himself for a canonry in the bishop's cathedral. The young man studied medicine,
law, and the classic literatures as well as astronomy, spending seven years soaking
up the sun and the Renaissance. He returned home in 1503 to help his uncle run
the diocese and face down the Teutonic Knights. To improve his Greek and to establish himself as a humanist, he published as his first book a Latin translation of
an obscure Byzantine poet. The choicer Greek authors had long since found modern editors, luckily for them. Copernicus' Greek was as accurate as the Alfonsine
tables.[14]

Among the classical writers Copernicus read was Plutarch, who recorded the
Pythagorean opinion that the earth revolves about a central fire and also spins like
a wheel.[15] The basic qualitative conception of a planetary system centered on the
sun was not far to seek in Renaissance Italy. But no one before Copernicus had had

the conviction and energy to develop this classical commonplace into a quantitative astronomy. A crude idea of the magnitude of the technical task of rewriting Ptolemaic theory for a moving earth, taking into account the observations made by star gazers of unequal capacities over a thousand years, may be obtained by counting. In its standard English edition, Copernicus' *De revolutionibus orbium coelestium* has 330 folio pages, 143 diagrams, a hundred pages of tables, and over 20,000 tabulated numbers. It took Copernicus most of the forty years from his return from Italy until his death in 1543 to work it all out. What made him do it is not known.

For the practical astronomer or astrologer concerned to predict the positions of the planets, the merit of the new system would lie in the accuracy of the tables deduced from it. A Protestant professor at the University of Wittemberg, Erasmus Reinhold, undertook the task. His *Tabulae prutenicae*, so named after his patron, the Duke of Prussia, appeared in 1551. They did not make an epoch in astronomy. Since, technically, Copernicus' system was only Ptolemy's with the sun and earth interchanged, and Copernicus' celestial parameters were not always better, and sometimes worse, than his predecessors', Reinhold's tables could not have been much more accurate than Alfonso's.[16] From a purely quantitative point of view, Copernicus left the field much as Regiomontanus had found it a century earlier.

Qualitatively, however, Copernicus' scheme had striking advantages over Ptolemy's. Merely by assigning the earth the third orbit from the center, it explained why the apparent motions of the so-called inferior planets, Mercury and Venus, differ so greatly from those of the superior ones, Mars, Jupiter, and Saturn. Since the inferior planets lie within the earth's orbit, they never appear far from the sun; owing to this confinement, Venus serves as the morning star when west of the sun and the evening star when east of it. The superior planets can appear at any angular distance from the sun as seen from the earth. They are brightest when opposite the sun because they are then closest to the earth. Ptolemy's system needed special and implausible assumptions to account for these and other appearances that followed naturally on Copernicus'.[17]

Against these advantages weighed the obnoxious assumption of the continuous, complicated, and rapid revolutions of a huge heavy body apparently at rest. Aristotelian physics disallowed the earth an enduring natural motion around the center of the universe. Common sense observed that the earth's spinning, which Copernicus proposed as an explanation of the alternation of day and night and the motions of the stars, would cause great easterly winds and leave the birds behind. Theologians objected that Scripture expressly stated that the sun moves, for otherwise Joshua would not have commanded it to stand still.

Copernicus spoke to all these objections in a lengthy dedication of *De revolutionibus* to Pope Paul III. He told the Pope that he had almost resolved not to make

his system public for fear that its novelty and peculiarity would arouse the ignorant. But two of his friends, a bishop and a cardinal, pressed him to publish. He
yielded and chose the Pope as judge and dedicatee, "for even in this very remote
corner of the earth where I live you are considered the highest authority by virtue
of the loftiness of your office and your love for all literature and astronomy [the
pope was a devoted astrologer]."[18]

As for the babblers who might bend Scripture into censure, "I despis[e] their
criticism." They know nothing. Great theologians have written childishly about
the earth's shape. They should stick to their business. "Astronomy is written for astronomers," that is, for mathematicians. They will recognize the theoretical merits of the new approach and also the utility to the church in its values for the
parameters needed to reform the calendar. "But what I have accomplished in this
regard, I leave to the judgment of Your Holiness in particular and of all other
learned astronomers."[19]

Copernicus was too ill and too distant from his publisher to see his masterpiece
through the press. The supervision was entrusted to a bellicose Lutheran theologian, Andreas Osiander, who cultivated astronomy primarily for its usefulness in
dating the Apocalypse. He thought to strengthen the protection afforded by the
dedication to the Pope, which carried negative weight with his co-religionists, by
a foreword of his own, which he did not sign, probably so as not to add to the
book's burden an association with a controversial theologian. The truth, according to Osiander, is that there is no truth in astronomy. Various hypotheses can account for the appearances. In this predicament, the astronomer should pick the
one easiest to grasp. "Therefore alongside the ancient hypotheses, which are no
more probable, let us permit these new hypotheses also to become known, especially since they are admirable as well as simple and bring with them a huge treasure of very skillful observations. So far as hypotheses are concerned, let no one
expect anything certain from astronomy, which cannot furnish it, lest he accept as
the truth ideas conceived for another purpose, and depart this study a greater fool
than when he entered it."[20]

· WORKING OUT COPERNICUS ·

In 1563 a Danish noble boy (he was then sixteen) attending the University of
Leipzig observed a grand conjunction (an alignment of Jupiter and Saturn) a
month away from the date given in the tables. He thought the discrepancy intolerable. He later wrote that the experience inspired his resolution to depart from
the usual practice of astronomers, who based their hypotheses on a few measurements of the planets made at theoretically significant times, and to renew astron-

omy by a regular course of frequent observations.[21] The name of this presumptuous Lutheran adolescent was Tycho Brahe.

Tycho realized his ambition with the help of the King of Denmark, who gave him feudal tenure of the small island of Hven in the Baltic and other sources of income, making him, it is supposed, the richest scholar in Europe. The money went into building and equipping a small palace on Hven that Tycho named Uraniborg. With assistants recruited mainly from Scandinavian universities, Tycho maintained a schedule of some 185 observing sessions a year and amassed a treasure of accurate, systematic data about the luminaries and the planets. To achieve his higher goal of correcting planetary theory, he needed calculators as well as observers.[22] In 1597, as a new King of Denmark forced the Lord of Uraniborg from his fiefdom, the perfect calculator appeared on the scene, a man as driven and compulsive as himself, the greatest mathematical astronomer the world had yet produced, Johannes Kepler.

The son of a "vicious, inflexible, [and] quarrelsome" mercenary and a woman later tried as a witch, Kepler would have led the life of the snarling little dog he took himself to be were it not for a system of state education in the Duchy of Würtemberg that took promising impecunious students from the rudiments of Latin grammar through the University of Tübingen and on to the Protestant ministry. Kepler so distinguished himself in the study of mathematics that the university sent him to teach it at the Protestant seminary in Graz before he had completed his theology course. Thus, as Kepler later explained it, did God call him from the ministry to another form of divine service.[23]

As a student Kepler had come to believe in Copernicus' world picture. It presented him with a deep problem, however. Why did God choose to make six and only six planets and to put them at their respective distances from the sun? On 19 July 1595, a year after arriving in Graz, Kepler had his answer. God had used the Platonic solids (geometrical figures like the cube and regular pyramid that have all their faces, sides, and angles equal) as spacers between the planets. Since even He could not change Euclid's finding that only five such solids can exist, He could make six and only six planets. Voilà! The matter of the distances took a little longer to work out, since there are many different ways—120, to be exact—in which the solids can be nestled inside one another. After days and nights spent in calculation, Kepler discovered an order that worked, well enough. "It will never be possible for me to describe with words the enjoyment I have drawn from my discovery." The ecstatic young teacher developed his demonstration into a book, the *Mysterium cosmographicum*, which appeared in 1597. Later in life he judged it with his usual understatedness: "No one ever produced a first work more deserving of admiration, more auspicious, and, as far as its subject is concerned, more worthy."[24]

The next step for Kepler was to obtain data that would refine and confirm his solution to the cosmographic mystery. He sent Tycho a copy of his book. Tycho did not like it, for two reasons. First, if it were possible to deduce the detailed structure of the universe by reason alone, Tycho's years of painstaking measurement would have been a waste of time. Second, Tycho knew that Kepler could not be right. The hasty young mathematician had begun with the wrong system. Tycho had worked out a world picture of his own, which he regarded as his most significant achievement and as close to truth as fallen mankind was likely to come. This privileged system had many of the advantages of Copernicus' without violating sound physics, common sense, and scriptural authority.[25]

Tycho admired Copernicus as the foremost mathematician of his century. In public lectures on *De revolutionibus* given in Copenhagen in 1574, Tycho had discussed ways by which its offensive features might be blunted and its results employed in calculation. Several geometrical hybrids were under consideration: for example, the inferior planets might circle the sun, which would carry them around the earth, or, perhaps, all the planets might do so, thus inverting Copernicus' geometry while preserving all the relative motions. Tycho put forward this latter scheme as his own in 1585, in a book on comets. It was not for him a mere transformation of heliocentric into geo-heliocentric geometry. In working it out, he realized that it required several of the solid spheres usually supposed to carry the planets to collide. His subsequent calculation that the comet of 1577 would have had to penetrate several of these spheres confirmed the inference that their solidity had been exaggerated.[26] About the only thing that would have pleased Tycho in Kepler's *Mysterium* was the evanescent character of the Platonic "solids" supposed to fill the spaces between the planets.

Tycho had already left Hven to set up as the Holy Roman Emperor's mathematician in Prague when he received Kepler's book. He noted that its author knew how to calculate and invited Kepler to join him. Their collaboration, which was not pleasant, lasted less than two years until Tycho died in 1601. Kepler succeeded to his treasure of observations and to his title of Imperial Mathematician. After many tortuous attempts to fit Tycho's data to Copernicus' theory, Kepler gave up. He could not make do with the two principles that had dominated, indeed, defined, planetary astronomy since Ptolemy: that the apparent paths of the planets be compounded from motions along circles and that these motions be performed at constant velocities.[27]

Kepler replaced these hoary principles with the products of exasperation, exhaustion, and genius. Instead of the compound circles, he put a single ellipse; in place of constant circular motion, an obscure and elaborate rule involving areas under elliptical arcs. These two prescriptions, now known as Kepler's laws, were buried within a narrative, reported in the detail of military history, of what Kepler called his war on Mars. The subjugation of Tycho's data on Mars to Kepler's laws made an epoch in astronomy. The year was 1609. Two decades later, in 1627, Kepler published tables of planetary motions calculated according to his laws. These Rudolphine tables (after Kepler's employer, the Holy Roman Emperor) saved the phenomena with unprecedented accuracy, thus realizing the goal of Bessarion at the trifling cost of destroying the system he had hoped to perfect.[28]

Counter-Reformation and Cosmology

In the year that *De revolutionibus* issued from the press with its dedication to Paul III, the Pope put in train the great reforming Council of Trent. This meeting, which lasted off and on for almost twenty years, created an atmosphere inimical to the development of Copernican astronomy in Catholic countries. Despite his participation in the usual vices of Renaissance popes — mistresses (he kept one for many years while a cardinal and bishop but not yet a priest), nepotism (he made his teenage grandsons cardinals), and parties (he threw many, with dancing girls, at the Vatican) — Paul understood the need for spiritual revival and stricter discipline for the Church if not for himself. He set in motion administrative reforms that, when completed by his immediate successors, established governance of the Roman church by standing committees of competent cardinals. Before sponsoring the Council, he had created several powerful instruments for maintaining the rules and dogmas it would lay down: the Society of Jesus (1540); the Roman Inquisition (1542); and the Index of Prohibited Books (1543). In a perfect gesture of his split personality, he commissioned the greatest artist of the age to paint the *Last Judgment* in the fun-loving Vatican palace.[29]

· THE SPIRIT OF TRENT ·

The tremendous business of Paul's Council was to forge consensus on dogma and discipline throughout what remained of the Catholic Church. In both departments it let loose a spirit that informed and reformed policy and practice for a hundred years. The disciplinary decrees prohibited the sort of recreation Paul enjoyed, ordered bishops to reside in their dioceses, and provided means to educate and civilize the clergy. Agreement about standards of behavior came easily; but, as usual, the question of education brought the various groups that wanted to control it into collision. The ignorance of the lower clergy was a standing scandal and the exposure of youth to reforming ideas a constant risk. To correct the ignorance, the Council decreed that a seminary be established in every diocese for the instruction of future priests, who were to learn, in addition to grammar and singing and liturgy, "computi ecclesiastici aliarumque bonarum artium doctrinam," the computation of Easter and other useful arts. To defeat the reformers, hard-liners like the Dominicans stuck close to Aristotle in philosophy and to Saint Thomas in theology. The Jesuits, who swiftly became the Church's most influential teaching order, devised a curriculum appealing to the lay as well as to the clerical leadership of Catholic Europe. They did not exclude humanistic texts or eschew "aristocratic sciences" like fencing and dancing; and they emphasized mathematics as a subject pleasant and useful for their more worldly students to know.[30]

In setting forth dogma, the Council strove to define the chief differences between the Roman Catholic and the Protestant ways to salvation. The characteristic definitions turned on the question of authority. Thus, where the reformers gave the Holy Word the last word and individuals the right, in principle, to interpret it, the Council stipulated that Scripture be understood in the light of tradition. The opinions of the fathers, councils, popes, and, above all, of popes acting with councils were determinative. Where the authorities agreed that the Word should be interpreted literally, as when Joshua commanded the sun to stand still in midheaven, Catholics must so accept it; but in other cases, as when the Bible speaks of God as having arms and legs, a figurative reading was allowed and even preferred.[31] In principle, the teaching left loopholes and opportunities for diverse interpretations; but the establishment after the Council of the Latin Vulgate as the correct form of Scripture, the outlawing of vernacular translations, and the censorship of deviant opinion largely diminished the scope and incentive for exegetical innovation.[32] Galileo made the mistake of relying on the possibility of reinterpretation while overlooking the restrictions on freelance theologizing.

Two other central dogmas aired or settled at Trent helped to sharpen the disciplinary machinery available for policing dissident thought. One, which was to intersect with the materialistic natural philosophies associated with Pierre Gassendi and René Descartes, concerned the Eucharist. The Council insisted on the real presence of God in the consecrated wine and wafer and on the scholastic philosophy with which the theologians affected to explain the transformation.[33] With this mixing of faith and science the Council asserted the unity of knowledge. Much of the subsequent intellectual history of the Church turned on finding practical ways to elude this ideal; for unity had the double disadvantage of making an attack on one part of the body of knowledge a threat to other parts and of tying the teachings of the Church to progressively outmoded natural philosophies.

The other dogma aimed at the central sour teaching of Luther and Calvin, that divine caprice, not human purpose, determines salvation. The Council had trouble hammering home a counterposition since theologians had struck the balance among faith, hope, and charity differently over the centuries. The Jesuits preferred a version that stressed good works and free will, the Dominicans and Augustinians one that stressed faith and fate. The Council neither endorsed nor condemned the extreme Augustinian position: endorsement would have brought it too close to Calvin, condemnation would have put it at odds with an authoritative saint.[34]

That was to store up trouble. People concerned about the state of their souls demanded to know whether Jesus' death saved all who accepted Him and showed it by pious acts and good works; or whether only those justified or made right by a special grace from God could overcome the evil impulses that have directed all human behavior since the original sin. The popes did not want to pronounce. For a decade, from 1595 to 1605, Clement VIII examined without deciding a fundamental dispute between the Dominicans and the Jesuits over grace and free will; in 1607, his less patient successor, Paul V, after receiving much conflicting advice from theologians and universities, decided not to decide; and, in 1611, the Holy Office resolved the business administratively by prohibiting publication of another word on the subject without special and prior approval.[35]

The matter was reopened in 1640 by the posthumous publication of an immense rendition of the extreme Augustinian position by a theologian of Louvain and onetime bishop of Ypres, Johannes Jansen. Two years later the Pope, Urban VIII, condemned Jansen's book. Anathemas came easy to Urban. A decade earlier he had presided over the trial and condemnation of Galileo. As will appear, the operation of the machinery against Jansenism, which eventually succeeded, helps in understanding the application of the censorship to astronomy, which did not.[36]

The Council of Trent did not concern itself with astronomy, apart from its directive that computing as applied to calendrics be taught in the seminaries; and its records make no reference to Copernicus, although concern about the conformity of his system with the usual interpretations of Scripture had been raised by two Dominicans immediately after the publication of *De revolutionibus*. One of these whistle-blowers was a competent astronomer; the other, the Master of the Sacred Palace, was the chief theologian in Rome. They died before they could take any action beyond recording their opinion.[37] No doubt, construed realistically, Copernican theory violated the obvious sense of bits of Scripture. So? Questions of planetary geometry occupied too few people to worry a church fighting soul-threatening schisms. When, however, Galileo raised this private molehill into a public mountain, the authorities took an interest in the philosophical and theological status of Copernican theory.

The bureaucratic inquiry was opened around 1612 by Robert Bellarmine, "the brightest star of the Jesuit school and the systematizer of controversial theology." A cardinal who breathed the spirit of Trent, a scholar with a steel will, a diplomat, inquisitor, and philanthropist, Bellarmine fought the Vatican's wars from Venice to England and ended up a saint. He began his training as a Jesuit during the last sessions of the Council of Trent. He developed his skill as a controversialist against the Protestants when teaching in their neighborhood at the University of Louvain. Unfortunately, he was to transfer his refined sensitivity to errors of reformers to assertions by cosmologists. Recalled to Rome, he became a bishop, rector of the Jesuits' Roman College, cardinal, and member of the congregations of the Inquisition and the Index. He did not forget the poor, lived largely on water and garlic, and died literally in the odor of sanctity.[38]

Bellarmine had taught astronomy at the level of Sacrobosco's *Sphere*. He developed an independent line. His heavens were corruptible. He regarded the planet-bearing spheres as useless fictions and Ptolemy's machinery of compound circles as computing gimmicks. He took his astronomy from the Bible. The seventh step in his *Ascent of the mind to God* plays with a passage in Psalm 19, which compares the sun running in its course to a bridegroom bounding from his chamber; "God wished that the sky itself be the palace of the sun in which it might roam freely and do its work." And roam quickly. Observing that the earth's circumference exceeds 20,000 miles, Bellarmine marveled at the speed with which the sun drops through its own diameter into the sea. Why, it had to cover much more than 7,000 miles (because it is much bigger than the earth) in less than the time it takes to read through Psalm 51 twice. "Who would believe that, unless it had been demonstrated by certain argument?"[39] Bellarmine could be touched by quantitative reasoning.

So could Bellarmine's fellow student Christoph Clavius, who had been received into the Jesuit novitiate in 1555 by Saint Ignatius himself. Nine years later, at the age of twenty-two, Clavius published the first of his several editions of Sacrobosco, the grandest of all versions of the old book, from which Galileo was to learn and teach astronomy. Clavius' sphere rolled him into a seat on Gregory XIII's committee on the calendar. Like updating Sacrobosco, explaining and defending the Gregorian calendar gave Clavius steady employment. Appointed to teach mathematics at the Roman College, he urged the merits of his subject against colleagues in philosophy and theology who knew from Aristotle and personal dislike that computing was a lesser art. In his efforts to upgrade mathematics in the Jesuit curriculum he had Bellarmine's support.[40] The two old friends did not agree about the status of mathematical knowledge, however.

Against Bellarmine's skepticism Clavius argued that mathematics could give certain information about the true structure of the world. No doubt he held this position irrespective of the consideration that, if accepted, it would assist his project to improve the status of mathematicians, as well as of mathematics, in the Society. He did not shrink from the logical conclusion that, since Ptolemaic astronomy worked, more or less, and did not conflict with accepted truths in philosophy and received interpretations of Scripture, the solid spheres supposed to carry the planets objectively existed, down to the last epicycle of the mathematical machinery. Could not the same be said for Copernicus? Indeed it could, Clavius replied, since Copernicus used the same mathematical scheme as Ptolemy, were it not that heliocentrism violated physics and opposed Scripture. Copernicus' calculations might give correct predictions. Nothing could be inferred from that. We see every day that erroneous premises can yield true conclusions. Clavius cited the famous syllogism, "all stones are animals, all men are stones, all men are animals," to illustrate the fallacy. But the argument was too strong for his purpose, being as fatal to Ptolemy as to Copernicus.[41]

Clavius' realistic understanding of astronomical constructs agreed perfectly with the views of his younger contemporaries, Kepler and Galileo. To be sure, there was this small difference, that they were sure the earth moved, and he that it stood still.[42] Galileo hoped to persuade Clavius to change his mind.

· THE MATTER OF GALILEO ·

Galileo first directed his newly invented telescope at the heavens in 1609. The result was flabbergasting. He saw lunar mountains, making the moon earthlike; he saw satellites around Jupiter, making the earth planetlike; and he saw a great swarm of stars no one had ever seen before, indicating the magnitude of the ignorance he was called to dispel. He immediately took three steps to secure himself and his career. He baptized the satellites of Jupiter the "Medici stars," as an unprecedented gift to the Grand Duke of Florence, with whom, as luck would have it, he was then negotiating for a position at court; he presented telescopes where they might be used to confirm his findings; and he published the most spectacular of his discoveries in a brilliant pamphlet entitled *Sidereus nuncius,* or *Starry messenger.* His strategy worked perfectly, in the short run. The Grand Duke made him the grand ducal Mathematician and Philosopher; Kepler and Clavius confirmed his findings; the Jesuit astronomers in Rome invited him down for collaborative viewing and friendly discussions; and the Pope, Paul V, granted him an interview.[43]

Galileo's discoveries took the discussion of the merits of Copernican theory from the merely mathematical to the physical or philosophical, where it necessarily conflicted with better-established ideas about the nature of things. A few zealous or jealous clerics insinuated that he held views contrary to the newly codified teachings of the Church. He met these insinuations with a lengthy letter to the Grand Duchess, in which he developed a form of exegesis recommended by Saint Augustine. According to it, everything in the Bible is true, but not necessarily literally. The inspired writers addressed their words to the common people, who naturally supposed that the sun they saw to move, did move; it would have been pedantic and puzzling for Joshua to have called out, "Earth, stand thou still."

Yet, by the light of advancing science we can see that Joshua spoke more accurately than a commonsensical or Ptolemaic interpretation of his words might suggest. For by commanding the sun to stop "in the midst of heaven" he may not have meant it to freeze at noon, as most commentators had assumed, but rather, for the benefit of later sophisticates, he had alluded to the sun's true place in the universe.[44]

Galileo's able and clever excursion into theology appeared to silence his enemies. That tempted him onto the offensive.[45] In a privately circulated manuscript, he castigated as equally intolerable the assertions that the sun-centered system was physically foolish and that Copernicus had regarded it as a mathematical fiction. Reading Copernicus as Osiander recommended made no sense to Galileo, for whom "hypothetical" meant "false." His powerful literal mind had no room for the proposition that systems might be equally hypothetical without being equally plausible. Since the Copernicans were essentially right, the opposing philosophers and theologians had to be wrong, absolutely, utterly, in every point and part. "Those who hold the false side cannot have in their favor any valid reason or experiment, whereas it is necessary that all things agree and correspond with the true side."[46] Copernican theory is true down even to its machinery of compound circles. Theologians must reinterpret Scripture as astronomers direct if the Holy Catholic Church is to avoid the scandal of opposing manifest truth.[47] Some good-natured people regarded these opinions as products of an arrogant and doctrinaire personality.

In 1615 Galileo went to Rome to justify his uncompromising stand and ward off further attacks. He hoped to convince Bellarmine, who already knew about his exercises in exegesis.[48] By the time Galileo reached Rome at the end of 1615, accusations against him had been referred to the Holy Office. Its consultants advised, and it ruled, that the proposition that the sun stands motionless in the center of the world is "foolish and absurd in philosophy, and formally heretical, since it explicitly contradicts in many places the sense of Holy Scripture, according to the literal meaning of the words and according to the common understanding of the Holy Fathers and doctors of theology." The companion proposition, that the earth is not at the center, but rotates and also translates, is likewise false and absurd in philosophy and "in regard to theological truth it is at least erroneous in faith."[49] Thus quixotically did the consultors of the Inquisition reassert the compatibility of science and religion.

Galileo was summoned to Bellarmine's chambers to hear the ruling and to receive a special injunction from the Holy Office: the inquisitors ordered him to abandon the obnoxious propositions and to abstain from holding, teaching, or defending them in any way whatsoever. According to the minute of the meeting, which took place on 26 February 1616, Galileo acquiesced. That was not all. The

Congregation of the Index decreed a week later that *De revolutionibus* be prohibited until corrected and that all other books that taught Copernicanism also be suspended or, if uncorrectable, banned.[50]

Galileo still did not get the point. He wrote the Tuscan Secretary of State that the attack against Copernicus had failed. The Church had not declared the opinion that the earth moves to be heretical; "[it] has only decided that that opinion does not agree with Holy Scripture, and thus only those books are prohibited which have explicitly maintained that it does not conflict with Scripture. . . . As for the book of Copernicus himself, ten lines will be removed from the Preface to Paul III, where he mentions that he does not think such a doctrine is repugnant to Scripture."[51]

In 1620 the Holy Office issued the promised corrections to *De revolutionibus*. These touched only thirteen passages, which offended by not qualifying the system as hypothetical. We may doubt the effectiveness of this bowdlerization; over 90 percent of extant copies, including the one in the Roman College, have no trace of the required corrections.[52] The censor who suggested them had considered three possible ways of dealing with the book: letting it pass, since the Church needed access to the work of all astronomers; banning it outright, as contrary to Scripture; or modifying it in the few places where Copernicus asserted the truth of heliocentrism. Here are his reasons for proposing the last course of action:

> I say that this emendation can be made without prejudice to the truth or holy scripture, since the science Copernicus treats is astronomy, whose most proper method is to use false and imaginary principles to save the celestial phenomena, as the ancients' use of epicycles, equants, apogees, and perigees confirms. If the non-hypothetical passages of Copernicus' *de motu terrae* are made hypothetical, they will not be opposed to holy scripture but rather, in a certain way, will agree with them because of the nature of false supposition, which the science of astronomy is accustomed to use by a certain special rule.[53]

At the same time that it issued the emendations required for *De revolutionibus*, the Congregation of the Index repeated its prohibition against all books asserting the truth of the moving earth. This general condemnation recurred in the updatings of the Roman Index published after 1620.

It did not stop Galileo. Holding fast to his opinions, he developed what he considered to be a demonstration of the motions of the earth from the phenomena of the tides. He bided his time, jousting with the Jesuits over comets and sunspots and lampooning them in a little book entitled *The Assayer*. The book appeared in 1623, just as Galileo's onetime friend Maffeo Barberini became Pope Urban VIII.

The Pope had the *Assayer* read to him over dinner. He was amused. Did that presage a relaxation of the restrictions on Copernicanism? Urban VIII was difficult to read. A pope partly of the Renaissance type (he was an accomplished literary man, an energetic nepotist, and a spendthrift builder) and partly of the Tridentine type (he sponsored missions, sanctioned new religious orders, and enforced residence on bishops), Urban was authoritarian, secretive, and resentful. Ever bullish, Galileo believed that the climate had changed.[54]

He composed a "Dialogue on the tides," which, on the demand of the Master of the Sacred Palace, who licensed it, appeared in 1632 under the apparently neutral title *Dialogue on the two chief systems of the world*. This *Dialogo* pretends to discuss the physical principles and observational evidence for an earth-centered and a sun-centered system fairly and hypothetically. But in fact it is one long attack on Aristotle and his modern followers and a continuous demonstration of the merits of a literal interpretation of Copernicus. The omission of the Tychonic system made Galileo's task easier, although it belied his title, and it slapped the Jesuits, who had made Tycho's scheme their own. The *Dialogo*, witty, caustic, unfair, brilliant, irritating, and overly clever, hit hard. Its victims immediately understood its message and attacked its author for violating the injunction laid upon him in 1616.

At the age of seventy Galileo was brought to Rome to answer for his behavior. Consultants advised that the *Dialogo* taught the philosophically absurd and formally heretical propositions condemned in 1616 and 1620. The book was banned. As for its author, the tribunal found that his behavior had made him "vehemently suspected of heresy." The inquisitors forced him to recant his opinions in public, forbade him to teach or write further about them, and sentenced him to imprisonment for life (soon alleviated to restriction to his home in Arcetri near Florence). Urban VIII played a prominent part in the proceedings. He presided over the trial at the Holy Office and ordered that its reaffirmed condemnation of the propositions of the moving earth and stationary sun be printed up and disseminated, together with Galileo's recantation, through the widely dispersed network of inquisitors and nuncios. These agents were to take particular care to alert all professors of philosophy and of mathematics in the universities, "so that knowing how Galileo was treated, they will understand the seriousness of the error he made, in order to avoid it as well as the punishment."[55] The Index of Prohibited Books as revised under Alexander VII and issued in 1664 took over all preceding sanctions against heliocentrism: the general prohibition of 1620, Copernicus' great work, two books banned in 1616, and Galileo's *Dialogo*.[56]

The shameful treatment of Galileo has served many critics of the Catholic Church. The philosophes of the Enlightenment exhibited it as behavior typical of the loathsome beast they intended to crush. The nineteenth-century historian of the warfare between science and theology represented that the Church as a whole delighted in the condemnation of Galileo as a proper assertion of its prerogatives against upstart astronomers. "Having gained their victory over Galileo, living and dead, having used it to scare into submission the professors of astronomy throughout Europe, conscientious churchmen exulted. Loud was their rejoicing that the 'heresy,' the 'infidelity,' the 'atheism' involved in believing that the earth revolves on its axis and moves around the sun had been crushed by the great tribunal of the Church."[57]

The charge that the Catholic Church contributed nothing to the advance of natural knowledge after the trial of Galileo but unreasoning opposition has life in our time. A sustained recent version is Georges Minois' *L'église et la science.* Minois dates the onset of modern mathematical science precisely. "Freeing itself from myths, fables, and imaginings, from qualities, deductions, and Aristotelian elements to rely on observation, experiment, and mathematics, [modern science] appears precisely during the 1620s, celebrated by historians of science as the birth of 'mechanism.' "[58] Since the Church opposed materialism in all forms, it had to attack mechanism and therewith the infant that, in time and despite its persecutions, would mature into quantum mechanics and the computer. "Ruling out the new hypothesis, the Church diverged from the evolution of science and forced it to develop outside a religious framework. . . . It condemn[ed] as necessarily erroneous, even heretical, all scientific results capable of challenging the traditional interpretation of the Bible as well as the dogmas based on an Aristotelian-Thomistic synthesis." It follows that the hierarchy of the Church had to be hostile to science and that the cultivation of science was incompatible with a clerical career.[59]

This thesis cannot be sustained. Many learned clerics rated, or affected to rate, mixed mathematics, including astronomy, as having no ontological value. It therefore could have no fundamental connection to the rest of the body of knowledge, including the truths of faith. This consideration explains Minois' own finding, of which he makes no sense, that most of the papers published by clerics in the *Mémoires* of the Paris Academy of Sciences up to 1720 concerned mathematics and astronomy.[60] The Church judged these studies to be neutral and also useful, and supported them. The work of the meridian makers shows that men whose careers were underwritten in whole or part by the Church could contribute importantly to the development of astronomy, that is, to the leading sector of natural knowledge during the seventeenth century. This proposition is not intended as an apology for the Catholic Church, but as a correction to the view, found even in the best modern historians, that the Church's action in the matter of Galileo made "Copernican astronomy a forbidden topic among faithful Catholics for . . . two centuries."[61]

Saint Peter sits in his basilica in Rome between the statues of two popes. We know them both: Paul III and Urban VIII. No doubt they stand there because Paul began the construction, and Urban accomplished the consecration, of the basilica. But that is only the literal and figurative meaning of the statues. They also offer an epitome of the Church's relationship to science. Paul III represents encouragement of probably useful technical innovation; Urban VIII, repression of possibly subversive cosmological ideas. There are other arresting reminders of astronomy in and around Saint Peter's. The sarcophagus of Gregory XIII portrays Clavius and others presenting the new calendar to the Pope. Shafts from the sun fall through the dome and windows of the cathedral to make puddles of light on the marble floor. Shadows cast by the obelisk in the square outside serve as a gigantic solar clock. At sunset, rays shining through the stained-glass window over the western altar dramatize the presence, and indicate the aptness, of the sun in the church.

Wider Uses of *Meridiane*

The church observatories ceased to have astronomical significance around the middle of the eighteenth century. That did not stop the drawing of meridian lines in cathedrals. They were intended not to advance science or purify liturgy but to tell time. Citizens in towns lucky enough to have good ones, like Milan, Bergamo, Palermo, and Catania, used them to correct their mechanical clocks. The latest of the type were built in the 1830s to set the time for the Belgian state railroad. With the introduction of better mechanical watches and electrical clocks, however, these noon marks went the way of the old *meridiane*.

They have not outlived their usefulness, however. They are valuable to that consumer of last resort, the historian, as probes or soundings of historical developments too big or too poorly known for synthetic treatment. Three such developments are explored in this book. One is what might be called the Catholic Merton thesis. According to Merton, natural philosophy and applied mathematics prospered in England during the second half of the eighteenth century owing to a harmony between the values of science and Protestant ethics.[62] This thesis was developed further by Joseph Ben David, who attributed the success of science in England to the fashioning of sociological niches in which it could be cultivated. In this more general and commonsensical version, science flourishes where the wider society respects and encourages it by giving its students a positive social role.[63]

During the sixteenth and seventeenth centuries the Catholic Church supported a great many cultivators of science. A systematic study of the niches they occupied, the infrastructure they moved through, and the roles they fulfilled does not exist. The courts and households of the big ecclesiastical patrons, the popes and cardinals, afforded many openings for the learned. The great orders, especially the Jesuits, supported some of their brethren as writers, mathematicians, architects, and engineers. Lay patrons would often share the burden with ecclesiastical ones. The careers of the builders of meridian lines in Italian cathedrals provide a sampling of these niches. The sample discloses the existence of a widely based support for a "scientist's role" (to use Ben David's anachronistic phrase) in Catholic countries during the Scientific Revolution.

The second development probed with the help of the meridian lines is the working of the censorship of astronomy books after the condemnation of Galileo. The soundings suggest that upper administrators of the Church, and even many of the cardinals who ran the Index of Prohibited Books and the Holy Office, recognized that Urban VIII had made a bad mistake and that the best policy would be to ignore Copernican writings unless pushed to intervene. Even the Jesuits were teaching heliocentrism before the end of the seventeenth century, using the convenient fiction that it was a convenient fiction. Those willing to call a theory a hypothesis could publish any astronomy they wanted. So flexible was the system that in 1741 the Church licensed a reprinting of Galileo's *Dialogo* although it still stood on the Index of Prohibited Books and was to remain there for another eighty years.

A useful symbol of the effective relationships between the leading astronomers of the church, the system of clerical patronage, the legacy of Galileo, and the progressive institutions of early modern science is Francesco Bianchini's *Hesperi ac phosphori nova phaenomena* of 1728. The protégé of five popes, for one of whom he built the meridian line in Santa Maria degli Angeli, Bianchini was the first to make a reconnaissance of the surface features of Venus. He succeeded through patience, astounding eyesight (Venus' constant cloud cover is opaque in the visible spectrum), and costly instruments paid for from income from sinecures given him by his popes and from gifts from the Catholic king of Portugal. In keeping with the style used for the moon, Bianchini named the features he saw on Venus after people, among whom Portuguese dignitaries figure abundantly. But we also find a straight named after Gian Domenico Cassini, the builder of the meridian line in San Petronio, equally observant as a Catholic and as an astronomer; a large sea to honor Galileo, "the prince of all [astronomers]"; and promontories named after the academies of science of Bologna and Paris, the one in a papal province, the other in a secular capital, in return for their vigorous and complementary promotion of observational astronomy.[64] The naming loses nothing of its symbolic value by the fact that the named features do not exist.

The third development sampled via *meridiane* is the bearing of astronomical discoveries on practical observation and data reduction. During the heyday of the cathedral observatory, from 1650 to 1750, astronomers found out how to correct their observations for atmospheric refraction and for the displacement of their instruments from the center of the earth, to which their theories referred their measurements. The decisive steps in the evaluation of these corrections were taken using data obtained at the *meridiana* of San Petronio. The discovery during the first half of the eighteenth century of the aberration of starlight and the periodic nodding or nutation of the earth's axis indicated the need for further refinements in the reduction of data. When corrected for all the effects discovered and quantified since 1650, observations made in the 1750s at the last *meridiana* to serve science yielded up the very delicate long-term change in the inclination of the earth's axis to the plane of its motion around the sun.

We are not finished with the merits of *meridiane*. Their study provides an opportunity for people unacquainted with the elements of astronomy to learn them easily and pleasantly. Most of what is needed lies at the level of that hardy medieval primer, the *Sphere* of Sacrobosco, and is supplied in this book from the edition of Sacrobosco made for his pupils by an early builder of *meridiane*, Egnatio Danti. Further information comes in connection with the niceties of the calendar and corrections to astronomical observations. The treatment follows the principle laid down by an eighteenth-century Italian calendrist in explaining his business to his countrymen. "Although to facilitate matters I put in theories and tables not found in other authors, nonetheless they do not depart in the least from the usual ways of computing, and I add nothing that goes beyond the ordinary, for, as the poet says, a plinth adds nothing to a statue of the colossus, or a boot to the stature of a dwarf."[65]

Finally, there are the instruments themselves, constructions both beautiful and useful, conduits of light through vast dark spaces, defunct sites of science, living objects of wonder. Visit Bologna when the sun shines. Charles Dickens did. He liked nothing in the city except "the great meridian on the pavement of the Church of San Petronio, where the sun beams mark the time among the kneeling people." There is something romantic, even sublime, in witnessing the faithful rendezvous of sun and rod arranged centuries ago. Among the old meanings of "matematizzare," to mathematize, was "to cast a spell."[66]

1: The Science of Easter

When God made everything according to measure and number, He chose hard ones for astronomy. He assigned 29.53059 days for the moon to fulfill its phases and 365.2422 days for the sun to run from one vernal equinox to the next. Calendar makers must do what they can to adapt one or both of these cranky numbers to the uses of humankind. By far the easier expedient is to employ either the sun or the moon, but not both, to specify the times of civil and religious transactions. The Hebrews picked the moon, the Romans, after Caesar, the sun, and the Christians, compromising as usual between Jewish and Gentile thinking, both.

To fix a calendar date in the seasonal year, the solar cycle must be constructed to make 0.2422 into a whole number. Adding one day every four years made the year 0.0078 days too long. To compensate, 0.0078 might be set equal to 0.0075 or to 0.0080. In the first case, three days would be dropped every 400 years; in the second, four days every 500 years; in both cases leaving a remainder. The lunar cycle offers similar challenges and uncertainties. Discovering, synchronizing, and disseminating the two cycles was a work of supererogation that enrolled several saints.

The Luminaries and the Calendar

The Hebrew month began with the first sighting of the new moon at dusk, when, in fact, the moon was not perfectly new (for it was then in front of the sun and invisible) but a day or two old. The official sighting, by a committee appointed by the Sanhedrin of Jerusalem, was promulgated by the chief rabbis. The astronomical moon, the object in the sky, thus determined the onset of the month. Since the moon requires a little more than 29.5 days on average to run from one new or full moon to the next, the lengths of the months, which must be an integral number of days, could not be predicted far in advance. The Arabs adopted the same lunar calendar. To this day the beginning of the religious month and, in some Islamic countries, the civil month as well, is fixed by detection of the first sliver of moon after its conjunction with the sun.[1]

A pure lunar calendar does not respect the seasons. The Hebrews did not carry their calendrical reckoning to the extreme of ignoring the year, however. They had a year, consisting of twelve or, as required, thirteen named months, and they had seasonal holidays with fixed dates in the various months. To take the case that was to give rise to the computing industry of the Dark Ages, the Hebrews always began their feast of Passover, or Pesach, on the day of the full moon of the month of Nisan, and they insisted that the holiday occur during the spring. Now, twelve lunations fall short of a seasonal year by a little less than eleven days. If in one year the fourteenth day of Nisan (the day of the Paschal full moon) falls on the spring equinox, the following year, twelve lunar months later, it will occur eleven days before the equinox. In the third year it will anticipate the equinox by twenty-two days, and in the fourth by over a month; if that were allowed to continue for six more years, the spring holiday of Passover would take place in midwinter.

The Hebrews retained Pesach as a spring feast by introducing or intercalating a thirteenth month just before Nisan whenever the accumulated excess of the seasonal over the lunar year required it. In the example just given, the intercalation would have occurred in the fourth year. But this is an easy example, with an excess of thirty-three days, well over the length of an average lunation. When, however, the accumulated excess amounted to twenty-eight or twenty-nine days, the keeper of the calendar might not know whether to intercalate or not. The ancient Hebrews did not have a fixed rule for these cases. A year was embolistic (of thirteen months) when the head rabbi so decreed.[2] The practice worked for a people concentrated in an area small enough to permit timely dissemination of the decision.

Eventually, however, it proved to be troublesome. In a decision of great consequence for the history of science, the Jews resolved to adopt rules for intercalation and to accept tabulated moons rather than sightings of the true moon for fixing the date of Passover. The effect of the intercalations can be seen in any modern calendar that records the civil date of the Jewish new year, Rosh Hashanah. We find, for the years 1992–1995, 28, 16, 6, and 25 September. The first two intervals, –12, –10, correspond to the eleven-day difference between the solar and lunar years; the third, +19 = 30 – 11, is the consequence of the insertion of a second Nisan.

The Romans began with a calendar that tried to respect both sun and moon. But the arithmetic and the politics of intercalation became so entangled that Julius Caesar had to make the year we now call 44 B.C. 445 days long to restore the spring equinox to its traditional calendar date of 25 March. He decided to keep it there by detaching the lengths of the months from the motions of the moon. Tradition and caprice produced the arbitrary months that have passed into our calendar and are learned by rhyme rather than by reason. But however the days are distributed among the months, they will never sum to a year. To keep the vernal equinox on or close to 25 March, Caesar had to have a precise value for the length of the year and a scheme of intercalation to fix it in the calendar. To advise him on this nice matter he brought an astronomer named Sosigenes to Rome from Alexandria, the center of mathematical learning in the Hellenistic world. Sosigenes gave the year 365.25 days and observed that, if one year in four had 366 days and the other three 365, the vernal equinox would stay put in the calendar. The advice was good, but not good enough, since Sosigenes made the year too long by about eleven minutes.[3]

Caesar took Sosigenes' advice and promulgated it as the Julian calendar. After some neglect in introducing leap days, his successors imposed the new calendar throughout their empire. They placed a leap day between the 24th and 25th of February. Hence it received the name "bissextile," not because the day was devoted to sexual extravagances but because of a peculiarity of the Roman system of specifying dates. Instead of counting forward from the beginning of a month, the Romans counted backward from the first day of the following month or from special days within the month. The counting was inclusive. What we would call 24 February a Roman would name "ante diem vi Kalendas Martias," or "vi. Kal. Mar.," the sixth day before the first day of March. Hence "bissextilis," the "second sixth day." The special days within the months that served as date markers were the Ides (which fell around the middle of the month) and the Nones (which came at the beginning of the second week). The Middle Ages rang various changes on this system, the Bolognese style being perhaps the most attractive: in Bologna one counted days in the first half of the month forward from its beginning and days in the second half backward from the kalends of the succeeding month.

Sosigenes' reckoning and the faithful use of the bissextile kept the equinox on 25 March for many years. But little errors often repeated can sum to big mistakes. As we know, the difference between the Julian year and the time between equinoxes is 0.0078 days, which, in 400 years, amounts to three days and a little more. So, in 400 years, the vernal equinox was falling three days earlier in the calendar than it had in Caesar's time. At that rate, it would come earlier by a month in 4,000 years, should the world last so long; and in 12,000 years, if the slip were uncorrected, the spring equinox, and therefore spring, would come in December. The human race might easily have adjusted to this slow separation of the months from the seasons, which would have had no discernible calendrical consequence in a single lifetime. This was not the view that developed in the Catholic Church.

The earliest Christians felt no need to trouble with the niceties of the calendar. The evangelist Mark advertised the second coming within a lifetime of Christ's death. "The sun will be darkened, the moon will not give her light, [and] the stars will come falling from the sky," he wrote, a prophecy that could not have encouraged astronomers.[4] When it became clear that the Lord's return would be delayed indefinitely, the Church shouldered the burden of planning for centuries and even millennia. To its everlasting glory, it raised the little Julian excess of eleven minutes or so a year into a major problem. If uncorrected, the excess might imperil the souls of those who, by Sosigenes' error, were led to celebrate Easter on the wrong day; and it would certainly cause discord between segments of the Church that tried to correct for the error in their own ways.

The crucifixion occurred on the day of Christ's Last Supper, which was a Passover feast. Christ did not die before he ate; the Jewish day, like the Jewish month, began at sundown, so that dinner was the first meal of the day. The Christians decided to observe the anniversary of the Passion at the Jewish season of Passover. Like Pesach, the Christian holiday would celebrate rebirth and redemption, not suffering and death. Hence it would celebrate not the crucifixion but the resurrection, which took place one or perhaps two days after the entombment. The uncertainty over timing was calendrical, not doctrinal. On the Jewish reckoning, the events of the Passion all occurred on the same day, the sixth day of the week, and Christ was buried in the evening, that is, the beginning of the seventh day, or Sabbath. He rose on the eighth day and spent therefore one day (but two nights) entombed. On the Roman reckoning, the Passover feast came on Thursday evening, the day before the crucifixion. The deposition occurred on Friday and Christ lay two "days," Friday and Saturday, in his tomb. Consequently, the Christians first thought to set their commemoration of the resurrection one or two days after 14 Nisan.[5]

Unfortunately, only the rabbis could say when Nisan began. Early Christian communities had to apply to the leaders of a rival church to learn when to celebrate their principal feast. The ignominy of this procedure, and the difficulty of a timely dissemination of the result as the church spread, forced the bishops into arithmetic. They sought a way to determine their own Nisan and to compute the dates of Easter far in advance. Let us call the Christian Nisan "Luna": it is the lunar month that contains the full moon that determines the holiday. The bishops' solution was to give the dates of 14 Luna in the Julian calendar for several years. They based their computations on "luni-solar" cycles that returned 14 Luna to the same calendar dates after a certain number of Julian years. Their line of thought may be expressed easily with the help of algebra they did not possess. God knows how they did it.

Suppose that the cycle consists of J Julian years and K lunar months. The condition of the problem is that the J years and K months are to contain the same number of days. Hence if a tabulated full moon falls on 1 January of the first year of the cycle, then, J Julian years later, there will again be a full moon on 1 January and the sequence of moons will repeat itself on the same dates. The calendar maker would need to specify only the sequence of calendar dates (not days of the week!) on which the full (or new) moon fell during J years to have, and broadcast, the sequence forever.

Since the tropical year (the interval between successive spring equinoxes) exceeds the lunar year, we know that $K > 12J$. Call the excess p. Observation teaches that the mean lunation is longer than 29.5 days. Therefore, most or perhaps all of the p additional tabulated months should be 30 days long. Let's assume all. Also, there are $J/4$ leap days in J Julian years. We can further increase the average length of a lunation by assigning 30 days to the tabulated month that contains the bissextile. In leap years therefore the lunar year will have 5 months of 29 days and 7 of 30 days or, if it is embolistic, 5 months of 29 days and 8 of 30 days.

With these stipulations, $365.25J = 354.25J + 30p$, that is, the number of days in J Julian years equals the number of days in J lunar years and in p intercalated months of 30 days each. We must find whole numbers J and p that satisfy as nearly as possible the relation $11J = 30p$, as required by the previous equation, and that also give, to a close approximation, the mean length of the month $\Lambda = (354.25J + 30p)/(12J + p) = 29.53059$ days. The attractive possibilities appear in Table 1.1.

All the mean lunations Λ are a little too large. Apparently we have put in too many thirty-day months. But the table gives cause for hope: there is one too many days in the lunar count in the nineteen-year cycle. Subtracting this day makes $M = N$ and $\Lambda = 29.53085$, agreeing with the modern value to better than one part in a hundred thousand. A Greek mathematician named Meton is credited with the discovery of the nineteen-year cycle. But since ancient peoples did not know Λ to seven figures, the Metonic cycle competed for some time with its constituent parts, the eight- and eleven-year cycles, each adjusted to the total of Julian days by

Table 1.1 Solar-lunar cycles

Julian years, J	Embolistic years, p	Total months, K	Days in J years, M	Days in K months, N^a	Mean lunation, Λ
3	1	37	1095.75	1092.75	29.5338
8	3	99	2922.00	2924.00	29.5353
11	4	136	4017.75	4016.75	29.5349
19	7	235	6939.75	6940.75	29.5351
30	11	371	10957.50	10957.50	29.5350

a. That is, in $(6J-J/4)$ months of 29 days each and $(6J+J/4) + p$ months of 30 days each.

taking the appropriate mix of twenty-nine- and thirty-day months. In the eight- and nineteen-year cycles, $N > M$, and, consequently, day(s) had to be dropped from the lunar count. Calendar makers nonetheless maintained the fiction that all embolistic months had thirty days by allowing the moon's age to skip a day (in the nineteen-year case) or two days (in the eight-year case) somewhere in the cycle. Since the jump (the *saltus lunae*) could come in principle in any thirty-day month, the practice gave cyclists the opportunity to personalize their tables and to multiply confusion.

· THE MODEST AND THE VENERABLE MONK ·

The cycles specify where 14 Luna falls in the Julian calendar. It remained to decide how to go from these dates to Easter. Two schools of thought had developed by the middle of the second century of our era (to use a concept then not yet invented). One, later declared heretical, did as the Jews did, ignored Sunday and the resurrection, and celebrated on 14 Nisan as specified by the rabbis. This easy solution, embraced primarily by the Asian Church, bore the heavy label "Quartodeciman," for "fourteenth day." The practice was despised by Christians who wanted greater distance from their roots. The Roman and African churches, including Alexandria, insisted that Christians celebrate on the Sunday nearest 15 Luna (Alexandrian style) or 16 Luna (Roman). Each party went its own way despite calls for a common observance until Emperor Constantine consented to tolerate Christianity. That was in A.D. 313. Twelve years later, in 325, the first ecumenical council of the Church convened in Nicaea in Asia Minor to discuss the suppression of heresy. It declared Quartodeciman practice heretical, affirmed that Easter should be celebrated at the same time by all Christian communities, and assigned to the wise men of Alexandria the task of computing Easter and informing the rest of the world.[6]

The Alexandrians used the Metonic or nineteen-year cycle, with the *saltus* in the last year, and took as 14 Luna the *tabulated* full moon that fell on or next after the vernal equinox. They placed the equinox four days earlier than Sosigenes' out of respect for the astronomical facts and for the 318 bishops of the Nicene Council, who had observed the beginning of spring on 21 March. And as ordered, they disseminated their results, sometimes in tables covering periods far longer than the administrative lives of their recipients. For example, early in the fifth century the Romans received a table from Bishop Cyril of Alexandria that covered five cycles, that is, ninety-five years, from 437 to 531. Cyrillan tables included specification of the day of the week for a date near the equinox, usually 24 March.[7] A nonmathematical bishop distant intellectually and geographically from Alexandria thus had everything he needed to announce Easter Sunday to his flock for as long as he remained in office.

Unfortunately, about a century before the meeting in Nicaea the Romans had devised a system of their own, written down in *The little tablet (Laterculus)* of one Augustalis. This Augustalis preferred a cycle that does not fit the series so far discussed. He took $J = 84$, which, with 6 jumping moons and 30 embolistic years, agrees to within 1.3 days with 1,039 lunations. Why 84? Probably because it is the smallest multiple of 28 that gives a tolerable fit. Why 28? Because it is the "circulus solaris," the period after which the same weekdays came back to the same calendar dates in the Julian calendar. (Since $28 = 4 \times 7$, in 28 years every possible sequence of weekdays and leap years will occur.) If 84 years were a good solar-lunar cycle, it would support a perfect Easter canon since at its conclusion the sequence of new (or full) moons would return to the same calendar dates *and days*. Although Augustalis' cycle soon would be superseded, it carried a technical improvement that has endured. This is the practice of stating the "epact," or age of the moon in days on some reference date, say 1 January or, as became common, 22 March, in the calendar. If in one year of the cycle the epact is 3, in the following year it would be $3 + 11 = 14$. From the epact the computer could find the date of the Paschal moon by easy addition.[8]

Since the Romans adhered more or less to Augustalis and kept the vernal equinox on 25 March, Easter computed *à la Romana* differed from the Alexandrian reckoning after Nicaea. Around 455 Pope Saint Leo I, who had talked Attila the Hun out of savaging Italy, took up the harder problem of reconciling the computations of Alexandria and Rome. He commissioned Victorius of Aquitaine to look into the matter. Victorius reported, correctly, that the discrepancies arose from a difference in basis (the cycles and the *saltus*), in the date of the equinox, and in the difference in allowed terminal dates for Easter Sunday (the Roman 16–22 Luna against the Alexandrian 15–21). He suggested that Rome set the equinox earlier than 25 March and adopt the nineteen-year cycle. He himself worked out a table on these principles beginning with the Passion, which compulsive chronologists had placed in the year 5229 after the creation of the world. Taking the first year of creation also to be the first year of a nineteen-year cycle, he calculated that Christ died during the fourth year of the 275th cycle; thence he computed merrily onward, setting out Easter for 430 years down to his own time and a hundred and more years beyond, until he stumbled over a great fact. After 532 years everything repeated itself, the moons, the dates, the days, the Sundays, the Easters. Victorius was a true computer. Instead of multiplying together the nineteen-year luni-solar cycle and the twenty-eight-year *circulus solaris* to uncover the great Paschal cycle of 532 years, he had calculated without thinking until he noticed that the entries in his table began to recur.[9]

Victorius' tables, though a great step toward the unification of practice, did not clear up the trouble since he waffled about the date of the equinox and disallowed Easter on 15 Luna. He did not insist, however. A model bureaucratic computist, he listed the many cases where the Greek reckoning of Easter gave dates different from his and left the problem for the Supreme Pontiff. "It is not my business to fix anything, but to provide a choice . . . for the leader of the universal church to decide under the circumstances what day should chiefly be considered for this special feast." The popes usually chose the Greek dates in the interest of unity. [10]

Around 525 the Romans decided to get to the bottom of things. A new pope, John I, consulted a new expert, Dionysius Exiguus, Denis the Lowly, a compiler of canon law and a good computer. Dionysius replied in 526. He insisted on the nineteen-year cycle and the vernal equinox on the day the Nicene fathers observed it. He supplied a table for ninety-five years, beginning where Cyril's ended, in the year then denoted the 247th of the era of Diocletian.[11] The idea of keying a table of Easter celebrations to the reign of a persecutor of Christians revolted Dionysius. A more reasonable style would be to count from the Incarnation or the Passion. He chose the Incarnation and renamed the 248th year of the era of Diocletian the 532nd year of the reign of Christ. Why he chose 532 is not known. If he knew Victorius' discovery, he did not say so.[12] If he did know it, he might have been struck by the fact that in the twelfth year of his second nineteen-year cycle, Easter fell on 25 March. Exactly 532 years earlier, it would have fallen on the same date. According to a received tradition, Christ died at the age of thirty on the day of the vernal equinox, which, in Roman times, was 25 March. If we put Christ's birth in A.D. 1 and his Passion in A.D. 31, the twelfth year of Dionysius' second cycle would be A.D. 563. Thence followed the first year of his first cycle, (563 − 12 − 19) = A.D. 532.[13]

A transcription of Dionysius' first cycle may clarify the principles of Easter calculations and luni-solar calendars (Table 1.2). Columns C–G are interconnected. For example, the epact on 22 March 535 was 3 (column C); that made the date of 14 Luna, when the moon was eleven days older (3 + 11 = 14), 2 April (column E).

Table 1.2 Excerpt from the Easter table of Dionysius

A	B	C	D	E		F		G
532b	1	0	4	5	Apr.	11	Apr.	20
533	2	11	5	25	Mar.	27	Mar.	16
534	3	22	6	13	Apr.	16	Apr.	17
535	4	3	7	2	Apr.	8	Apr.	20
536b	5	14	2	22	Mar.	23	Mar.	15
537	6	25	3	10	Apr.	12	Apr.	16
538	7	6	4	30	Mar.	4	Apr.	18
539	8	17	5	18	Apr.	24	Apr.	20
540b	9	28	7	7	Apr.	8	Apr.	15
541	10	9	1	27	Mar.	31	Mar.	18
542	11	20	2	15	Apr.	20	Apr.	19
543	12	1	3	4	Apr.	5	Apr.	15
544b	13	12	5	24	Mar.	27	Mar.	17
545	14	23	6	12	Apr.	14	Apr.	18
546	15	4	7	1	Apr.	8	Apr.	21
547	16	15	1	21	Mar.	24	Mar.	17
548b	17	26	3	9	Apr.	12	Apr.	17
549	18	7	4	29	Mar.	4	Apr.	20
550	19	18	5	17	Apr.	24	Apr.	21

Source: PL, 67, 495–8.

Note: The table covers only the first cycle, A.D. 532–51; omits two columns (the indiction and the lunar cycle); and adds the current year of the cycle beginning in 532. Column A is the date A.D., in which "b" signifies a leap year; B, the current year; C, the epact (the moon's age) on 22 March; D, the "ferial number," where 1 is Sunday, 2 is Monday, etc., of 24 March; E, the day of 14 Luna; F, Easter Sunday; G, the age of the moon on Easter day.

Now 24 March was a Saturday (column D); hence 2 April was a Monday and the following Sunday, 8 April, Easter (column F). The moon by then had waned six days beyond full, and so had attained the age of 20 (column G). Is it not all beautifully clear?[14]

With a few innocent tricks, like forging the last cycle of Cyril and claiming the authority of Nicaea, Dionysius carried the day on the Continent.[15] The Irish and the British, however, stuck to Roman traditions, including the eighty-four-year cycle. And worse: they allowed Easter on 14 Luna if the day was a Sunday. They thereby not only celebrated differently from Rome but also, occasionally, in unison with the Jews. In 664 a showdown took place at the Synod of Whitby, where a young gladiator educated at Rome, one Wilfrid, who would be a bishop and, what is as incredible as the Easter canon, a rich man and a saint, opposed an Irish clergy "befogged with mental blindness." The point at issue was not computistic detail but Christian unity. Wilfrid demanded to know whether "a handful of people in one corner of the remotest of islands is to be preferred to the universal church of Christ spread throughout the world." The president of the synod was not a cleric but a king. Although he had been educated by the Irish, this king, Oswy of Northumberland, appreciated Wilfrid's argument. Fringe groups menace authority. Oswy ordered agreement. That brought the British communities apart from the Picts, and most Irish ones apart from Iona, into communion with Rome.[16]

Rome was encouraged but not convinced. Most of the British clergy had been trained in the Irish error. To be safe, Archbishop Theodor of Tarsus, who began his long service in England in 669, and who made Wilfrid bishop of all Northumbria, excommunicated the lot. "Whoever have been ordained by the bishops of the Scots [the Irish] or Britons, who are not Catholic as regards Easter and the tonsure, are not deemed to be in communion with the Church, but must be [re]confirmed by a fresh laying on of hands by a Catholic bishop."[17] There is no better indication of the importance of the correct reckoning of Easter to the medieval Church.

Wilfrid's words at the Synod of Whitby come from *The ecclesiastical history of the English people* completed by the Venerable Bede in his monastery in Northumbria in 731. Bede had a special interest in the dating of Easter. That was thought a peculiar taste. Wilfrid's biographer had omitted his subject's contributions to computing "in order not to insert . . . anything disgusting to my readers."[18] Bede was made of sterner stuff. In 703, after the ukase of King Oswy had been confirmed and reconfirmed, Bede drew up an outline of best computistical practice for his students at the monastery. He had before him the treatise and tables of Dionysius. He explained the nineteen-year cycle and its relation to eight- and eleven-year cycles, and also the twenty-eight-year *circulus solaris;* but he did not mention the great Paschal cycle of Victorius.[19]

A few years later, around 710, Nechtan, king of the Picts, applied for instruction in the proper way of observing Easter. Bede no doubt had a hand in the reply, which he later set out at great length in his *Ecclesiastical history.* The recipe offered Nechtan: begin with the vernal equinox, which occurs on 21 March as recorded by the Nicene fathers and "as we can also prove by inspecting a sundial"; take the full moon that falls on or next after the equinox as the Paschal moon; and you have Easter as the Sunday of the third week of the moon, that is, the Sunday that falls on any of the days 15–21 Luna. Fortunately, the Picts did not have to perform any computations themselves since the ninety-five-year calendar of Dionysius contained everything necessary. When the ninety-five years ended, new tables would be supplied. "There are so many mathematicians [*calculatorum copia*] today that even in our churches here in Britain there are several who have committed to memory these ancient rules . . . and can easily continue the Easter cycles." How far? "For an indefinite number of years, even up to 532 years if they wish; after this period, all that concerns the succession of the sun, the moon, the month, and the week returns in the same order as before."[20]

The great Paschal cycle also appears at the end of Bede's *De temporum ratione*, an expansion of his student handbook finished in 725. This version contains a useful description of the apparent motions of the sun and the moon, including an original but irrelevant discussion of the connection between the tides and the moon's position; an explanation of time and time telling, both civil and ecclesiastical; and, of course, a full account of the Dionysian *computus*.[21] The manual became a standard reference and pattern for medieval computists. An abundant pseudepigraphical literature grew up under Bede's name. There is extant an Easter table in the Dionysian style (one cycle of which is exhibited as Table 1.2) attributed to him that runs up to 1595.[22]

Other sorts of tables by would-be Bedes circulated widely. The form illustrated in Table 1.3 for the month of January was easy to use and left wide margins for annotations. Column A gives the day of the month counting forward; column B, the date Roman style; column C, the *numerus aureus,* the golden number, that is, the year in the nineteen-year cycle in which 14 Luna falls on the indicated date; column D, a letter identifying the day of the week, that corresponding to Sunday being "dominical." For example, if we were in the fifth year of the cycle and 1 January was a Tuesday, we would know from the ninth line of the table that a full moon would occur on Wednesday, 9 January, and that the dominical letter is *f.* One such table is needed for each month and two dominical letters, one valid to 24 February and one valid after, in leap years.[23]

When suitably simplified, the *computi* descended from Bede's *De temporum ratione* became a text for schoolchildren. In a dialogue from one of the spurious works of Bede a master tells his student that God's church is built from four subjects: the divine canon, grammar, history, and "numerus, in quo facta futurorum et solemnitates divinae dinumerantur," "number, by which the events of the future and the Lord's feasts are reckoned." Similar ingredients, for example, grammar, music, law, and *computus*, recur in many didactic texts, usually attributed to Saint Augustine. The manuals on computing kept mathematics alive in the Latin West during the Dark Ages and also conveyed a little exact information about the physical world. "Take away number from everything and everything will perish. Deprive our time of *computus* and blind ignorance will seize everything. Those who do not know how to calculate can not be distinguished from animals."[24] Thus our schoolmaster. Had not Saint Augustine declared that a man who could not compute was not worthy to be a priest? The celebration of Easter became a cause and source of numeracy among the Catholic clergy and an incidental carrier of natural knowledge.[25] We shall encounter many other examples of the promotion of what looks to us like science as a by-product of advancing the core interests of the Church.

Another useful practice traceable to the Easter problem is the universal dating by *Anno Domini*. Dionysius did not use his innovation to date events in civil or ecclesiastical history. For him the reckoning from the Incarnation served solely to locate his Easter cycles in time. Owing to the custom of recording significant events in the blank spaces in calendars, however, the Paschal dating gradually took on a wider role. Bede was perhaps the first influential writer to apply Dionysius' reckoning to the general purposes of chronology. He gives the dates in his *Ecclesiastical history* in the style *Anno Domini*.[26]

A Scandal in the Church

The Easter canon as set forth by Bede and his imitators would have established the date of the feast of the resurrection for all time had the numbers or the Church been less difficult. For, on the one hand, the lengths of the tropical year and the synodic month, on which the recipe depended, were not known with a precision that supported calculations for a millennium, let alone an eternity; and, on the other hand, the church insisted on celebrating Easter on a Sunday close to the first full moon of spring in accordance with tables prepared long in advance. Already in Charlemagne's time, around 800, discrepancies between the tabulated Paschal moons and the observed ones were noticeable. By then the annual difference $\Delta = 0.0078$ days between the Julian year of 365.25 days and the truer value of 365.2422 days had accumulated to $(800 - 325)\Delta = 3.70$ days since the Council of Nicaea; the vernal equinox was then coming on 17 or 18 March rather than on 21 March as assumed by the tables. By the same time the difference $\varepsilon = 0.00026$ days between the average lunation of 29.53085 days as assumed in the nineteen-year cycle and the synodic month of 29.53059 days had accumulated to $(235\varepsilon)(475/19) = 1.53$ days; the real moons were coming around a day and a half earlier than the tabulated ones.

At first no one knew whether the numbers, the rules, or the implementation of the rules were at fault or how to find the error or, once found, how to correct it. Charlemagne, who had a table in Bede's style done up in gold and silver, applied to his advisor in intellectual matters, the very learned Alcuin of York, in whom the tradition of Bede culminated. Alcuin could give no practical suggestion for im-

Table 1.3   A standard form of medieval calendar

A	B	C	D
1	Kal. Jan.	3	A
2	iv Non. Jan.		b
3	iii Non. Jan.	11	c
4	pridie[a] Non. Jan.		d
5	Non. Jan.	19	e
6	viii Idus Jan.	8	f
7	vii Idus Jan.		g
8	vi Idus Jan.	16	A
9	v Idus Jan.	5	b
10	iv Idus Jan.		c
11	iii Idus Jan.	13	d
12	pridie Idus Jan.	2	e
13	Idus Jan.		f
14	xix Kal. Feb.	10	g
15	xviii Kal. Feb.		A
16	xvii Kal. Feb.	18	b
17	xvi Kal. Feb.	7	c
18	xv Kal. Feb.		d
19	xiv Kal. Feb.	15	e
20	xiii Kal. Feb.	4	f
21	xii Kal. Feb.		g
22	xi Kal. Feb.	12	A
23	x Kal. Feb.	1	b
24	ix Kal. Feb.		c
25	viii Kal. Feb.	9	d
26	vii Kal. Feb.		e
27	vi Kal. Feb.	17	f
28	v Kal. Feb.	6	g
29	iv Kal. Feb.		A
30	iii Kal. Feb.	14	b
31	pridie Kal. Feb.	5	c

Source: PL, 90, 759–60; the full table is in ibid., 759–88. Cf. Bickerman, *Chronology* (1980), 125.
a. "The day before."

proving upon Dionysius.[27] After Alcuin, the computists lost even the rudiments of astronomy and could only watch helplessly as errors in the reckoning of Easter compounded. The first computist of whom we have record to identify the cause of the difficulty was a certain Conrad, who entered into the subject in 1200, "lest we be like beasts, similar to chimeras, ignorant of the very idea of time." Conrad made Δ = one day in 120 years (= 12 minutes per year). As for ε, the discrepancy between the observed and calculated average lunation, which had accumulated to three days in his time, Conrad set it equal to zero and blamed the trouble on Adam. In his ingenious explanation, the Dionysian cycle went back to the beginning of time; but Adam mistakenly judged the first sliver of moon he ever saw to be brand new, as he was, whereas, as every reader of Genesis knew, it was then three days old.[28]

When Conrad wrote, the Latin West was just coming into possession of its lost intellectual patrimony. As the new mathematicians waxed in knowledge and competence, they supposed that they not only could do better than the old computists in tabulating average moons but also, what was unheard of, that they could furnish reliable Easter dates from their theories of the motions of the real moon.

· THREE HUNDRED YEARS IN COMMITTEE ·

They could begin in a familiar place. Their cicerone in astronomy, Sacrobosco, also wrote a *computus* ("the science that considers time from the motions of the sun and moon") that had many versions down into the sixteenth century. Sacrobosco observed that Ptolemy had known that neither the Julian year nor the Metonic cycle exactly fit the facts and suggested adopting Ptolemy's values, Δ = 1 day in 300 years, ε = 1 day in 310 years. To correct the calendar and the canon, Sacrobosco suggested dropping a few days to restore the equinox to the Nicene date and throwing out a day from time to time to keep it there.[29] Sacrobosco's suggestions — returning the equinox to 21 March, retaining the tabular approach, and canceling a day from the year and the month every few centuries — were adopted after the problem had been discussed by academics, prelates, committees, and commissions for a quarter of a millennium. One cause of delay was a well-taken objection to Ptolemy's values for Δ and ε. Campanus of Navarre, a thirteenth-century astronomer newly acquainted with Arab learning, observed that the great Islamic authority al-Battani had made Δ one day in a hundred years, three times Ptolemy's value.[30] How could the perplexed Westerner choose among such great and irreconcilable experts?

An obvious way was for Western astronomers to declare independence of Greeks and Arabs and measure Δ and ε for themselves. So advised Roger Bacon, who went partway by combining his own observations of the vernal equinox with one reported by Ptolemy. The result, $\Delta = 1$ day in 125 years $= 11.52$ minutes/year, hit remarkably close to the value later used in the Gregorian reform. Bacon also anticipated later practices by urging the popes to support the study of astronomy to find the correct values of Δ and ε. Something had to be done. The computists had shown their incompetence to all the world in 1276, when they erred by an entire month in setting the date of Easter; "men fasted when they should have been jubilant, and ate meat when they should have fasted." By the end of the thirteenth century, Western astronomers had come as close to Δ as their purposes required. The Alphonsine tables made Δ one day in 134 years (10.75 minutes/year), exactly the value that would be used in the Gregorian calendar three hundred years later.[31]

The popes began to play a part in the middle of the fourteenth century, when Clement VI asked astronomers at the University of Paris and elsewhere for advice. They responded that all would be well if days were dropped from the cycles and the golden numbers recomputed; but they differed about the number of superfluous days. Some took the Alfonsine year as basis, others the shift in the day of the equinox since Christ's time (making $\Delta = 12$ minutes, a little less than twice as far off as the Alfonsine value). Although they differed over the details of the solution, they agreed that the problem was urgent; the calculated Easter differed from the true date by eight days in 1345 and by a month in 1356. They warned that a miscomputed Easter could easily coincide with the new moon, which would destroy the miracle of the solar eclipse at the crucifixion. The Pope worried too, but could scarcely act in so important a matter when his experts disagreed about the measures he should take.[32]

The Church councils of the early fifteenth century proved no more able to initiate reform. The Council of Constance (1415) heard from Pierre d'Ailly, a cardinal competent in astronomy, whose geographical errors were to encourage Columbus. D'Ailly recommended dropping one leap day in 134 years (according to Alfonso's reckoning) and moving the golden numbers to catch up with the moon. The council did not follow his advice. At the Council of Basel (1434), another cardinal astronomer, Nicholas of Cusa, suggested omitting one leap day in 150 years, as a compromise between Alfonso and Ptolemy. He further proposed that days be dropped from the calendar to restore the date of the vernal equinox to 21 March, so that, in a Pickwickian sense, the Nicene injunction against monkeying with the calendar would be obeyed; also that, to bring the golden numbers into line with the new calendar and the true moons, the year 1439 be declared the twelfth year rather than (as it was) the fifteenth year of the current Metonic cycle. Nothing happened. The business was too technical for the councils, and the Pope, Eugene IV, threatened by schism, an antipope, and uprisings in the Papal States, had trouble more serious than calendar reform. Easter was celebrated five weeks late in 1424 and one week off in 1433 and 1437.[33]

The professional astronomers then took the matter in hand, calculating not from average but from real moons. Peurbach and Regiomontanus showed how to find the moon's position at any time on Ptolemaic principles. Regiomontanus calculated his moons to minutes and seconds, and published a table comparing his Easter dates with those prescribed by the golden numbers (Figure 1.1). He did not say how to reform Church practice but protested that its continuance was an insult to the newly won competence of Western astronomers. "Most of all [he wrote] we should be ashamed at the scandalous words that the obduracy of the Jews throws at us, since they continually snipe that we have not understood how to follow the simplest rules of God's law." Regiomontanus' calendars circulated widely. The then new printing press made it plain that the Church was perpetuating bad astronomy.[34]

In an attempt to square the Church with astronomy and Easter with the moon, Pope Leo X and the Holy Roman Emperor Maximilian I asked the principal universities to send their suggestions for repairs to the Lateran Council scheduled to convene in 1511. As usual, the expert differed. The Viennese view, as expounded by former students of Regiomontanus, was that Easter should be computed in the manly manner, from true moons, not from cycles; that the vernal equinox should be fixed in the calendar, at either 10 March or 25 March, by dropping one leap year in 134 years (the Alfonsine Δ); that the golden numbers, the triumph of medieval computing, should be scrapped; and that Easter should be celebrated throughout expanding Christendom on the date computed for Rome according to exact Dionysian definitions. (As the Viennese professors observed, if a Paschal full moon occurred in India just after midnight on a Sunday, Easter would be celebrated there a week later; but the same moon would be timed on Saturday afternoon in Lisbon and Easter would be observed in Western Europe the next day.) Pope Leo recommended to the Lateran Council that it follow the astronomers' advice to compute the new moons astronomically, but retain the golden numbers for each 134-year period. The Council did not act.[35]

A consensus soon developed against the professorial or Viennese solution: golden numbers and average values should be retained for convenience and tradition and to ensure that everyone everywhere celebrated Easter on the same day.[36] Meanwhile, the Protestant Reformation ensured that the problem so stated had no prospect of a solution. Luther decided that in his church dates had nothing to do with faith. He recommended dissolving the problem by nailing Easter in the calendar like Christmas. The Council of Trent decided not to grapple with the calendar but instead referred it to the Pope. That was in December 1563. It took another twenty years to bring the matter to a conclusion.[37]

.	.	.MARTIVS.		SOLIS PISCES.		LVNAE S. G.		S. G.	
1	d	KL		20	37	2	11	2	4
2	e	6 non	Simplicis papę	21	36	2	24	2	17
3	f	5 non		22	36	3	7	3	0
4	g	4 non	Adriani martyris	23	35	3	20	3	13
5	A	3 non		24	35	4	3	3	26
6	b	2 non	Victoris martyris	25	34	4	16	4	9
7	c	Non	Perpetuę & Fœlicitatis	26	33	5	0	4	22
8	d	8 id⁹		27	32	5	13	5	5
9	e	7 id⁹		28	31	5	26	5	18
10	f	6 id⁹		29	30	6	9	6	1
11	g	5 id⁹	Clauis pascę · ARIES	0	29	6	22	6	15
12	A	4 id⁹	Gregorij papę	1	28	7	6	6	28
13	b	3 id⁹		2	27	7	19	7	11
14	c	2 id⁹		3	26	8	2	7	24
15	d	Idus		4	25	8	15	8	7
16	e	17 kal	APRILIS	5	24	8	28	8	20
17	f	16 kal	Gerdrudis uirginis	6	23	9	11	9	3
18	g	15 kal		7	22	9	25	9	16
19	A	14 kal		8	21	10	8	9	29
20	b	13 kal		9	20	10	21	10	12
21	c	12 kal	Benedicti abbatis	10	18	11	4	10	25
22	d	11 kal		11	17	11	17	11	8
23	e	10 kal		12	16	0	0	11	21
24	f	9 kal		13	15	0	14	0	4
25	g	8 kal	Annūciatio Marię	14	13	0	27	0	17
26	A	7 kal		15	12	1	10	1	1
27	b	6 kal		16	10	1	23	1	14
28	c	5 kal		17	9	2	6	1	27
29	d	4 kal		18	8	2	20	2	10
30	e	3 kal		19	7	3	3	2	23
31	f	2 kal		20	6	3	16	3	6

FIG. 1.1. March, from Regiomontanus' Latin calendar of 1474. The table has the days in the first column; the domenical letter in the second; the day's date in Roman style in the third ("kal." = Kalends, "noñ." = Nones, "id[us]" = Ides, "3 noñ" = three days before the Nones); notable saints' days in the fourth; the degree and minute of the sun's zodiacal position in the fifth and sixth; and, in the remaining columns, the moon's longitude (in sign, "S," and degree, "G") on two different reckonings. From Regiomontanus, [Kalendarium] (1476).

The first step was the production of a revised prayer book, a Breviary, published in 1568 under Pius V, which put back the golden numbers by three days to take into account the accumulated error in the nineteen-year cycles. By this ingenious stopgap, a new moon scheduled for, say, 10 September would be expected on 7 September, and so on. Perhaps inspired by this demonstration that change was possible, Aloisius Lilius, or Luigi Giglio, a physician from southern Italy, whose opinion had not been requested, managed to give Pope Gregory XIII a full plan for calendrical reform. It appears that the conduit was another man from the South, Cardinal Guglielmo Sirleto, who had headed the reform of the Breviary in 1568.[38]

Sirleto had arrived at his station by a route followed, not always with the same success, by several clerics who will play major parts in our story. From a modest background, Sirleto left Sicily for Rome with nothing but facility in the three biblical languages and a knowledge of mathematics. That was enough to recommend him to a cardinal who, luckily for both, soon rose to pope as Paul IV. Sirleto immediately became tutor to the papal nephews and advisor to His Holiness. Paul IV raised him to the cardinalate in 1565. Appointed Vatican librarian in 1570, he waxed in learning until he could write dissertations in his head and dream in Greek.[39] When Sirleto recommended Giglio's plan to Gregory, the Pope referred it back to him as president of a new commission on the calendar. It was to have as its technical members Antonio Giglio, brother of the reformer, the rising expert Christoph Clavius, and Egnatio Danti, a Dominican astronomer and geographer.[40] The committee judged Giglio's plan meritorious and, in 1578, sent out a synopsis of it for comment to the universities of Catholic Europe.

Giglio proposed to secure the vernal equinox on 21 March by dropping ten days from the calendar. He did not say how he would do it; in the event, when a modified version of his plan went into operation in Italy, 5 October 1582 was followed immediately by 15 October.[41] (The time was chosen so as to eliminate as few saints' days as possible.) To keep the equinox settled, Giglio altered the length of the year. Again he hesitated over details, in this case between the Alfonsine value for the tropical year (365 days 5 hours 49 minutes 16 seconds) and a value some 4 seconds shorter derived from the calculations of Copernicus. Either choice led him to the same procedure for correcting the Julian year and the Metonic cycle. He proposed to omit three leap days every 400 years and one day in the lunar cycle every 312.5 years. All corrections were to take place in "century years," that is, years divisible by 100. Century years not also divisible by 4, such as 1700, 1800, and 1900, became ordinary years to repair the solar cycle. To fix up the lunar cycle, to remove the excess of a day or so that accumulates in 300 years on Dionysius' scheme, Giglio made the January lunation, which ordinarily would have thirty days, twenty-nine days once every 300 or 400 years. It takes 2,500 years to accomplish the adjustment: a day is dropped seven times on century dates at intervals of 300 years and once after another 400 years, for an average interval of 2500/8 = 312.50 years.[42]

All this was child's play compared with the redoing of the golden numbers. Dionysius had fixed them permanently; his new moons return to the same calendar dates and in the same sequence every nineteen years forever. Giglio knocked them loose, in two ways. The first was a onetime measure to bring the moons to their proper places. Since the error in the Metonic cycle amounted, by his calculation, to eight days in 2500 years, the golden numbers in his time specified moons that came 8(1580 − 325)/2500 ≈ 4 days later than they had during the Council of Nicaea, when they agreed with the true moons. To regain synchronization, he raised the golden numbers by four lines in the calendar. But his plan to omit ten days altogether to restore the vernal equinox to the Nicene date required lowering the golden numbers by ten lines. In the net, therefore, he proposed dropping down 10 − 4 = 6 lines; a new moon computed for, say, 11 February 1583, was rescheduled for 17 February, and so on. A drop of seven lines gave a better fit, however. Gregory's committee accomplished it by switching the epoch from the Nicene Council to the time of Dionysius Exiguus, which reduced the lunar correction from four days to three.[43]

Even great computers make mistakes. Clavius made a subtle one that threw off the moons in century dates not also leap years (like 1700, 1800, and 1900) and thereby made a mess of the Easters of the eighteenth century. The error was disclosed at the end of the seventeenth century by the "celebrated astronomer and mathematician of the Holy See and of his Most Christian Majesty [Louis XIV], Giovanni Domenico Cassini." Cassini designed the first church *meridiana* built to modern notions of precision. The second, erected in Rome under papal auspices just after 1700, had the job of resolving the discrepancies between the Gregorian Easters and nature's foreseen for the eighteenth century.[44]

The Gregorian lowering of the golden numbers by seven throughout the nineteen-year cycle produced a new distribution of what we may call platinum numbers good through the year 1699. The year 1600 was a leap year on both the Julian and the Gregorian calendar, and, by convention, 1500 was taken as the last year in which the lunar correction of one day in three hundred years had been imposed; hence the Dionysian scheme using platinum numbers sufficed for the seventeenth century.[45] The year 1700 brought something new. The loss of the leap-year day drove the new moons one day later in the calendar, requiring a new set of precious-metal numbers. These held sway until 1900, since in 1800 the omission of the leap day, which moves the new moons back in the calendar, was compensated by the omission of the day in the lunar cycle, which brings the moons forward. In 1900 an-

Table 1.4 Epacts through the third millennium

Cycle year (Golden number)	Epacts valid for 100 years beginning in						
	1500, 1600	1700, 1800	1900, 2000, 2100	2200, 2400	2300, 2500	2600, 2700, 2800	2900, 3000
1	1	0 (30)	29	28	27	26	25
2	12	11	10	9	8	7	6
3	23	22	21	20	19	18	17
4	4	3	2	1	0 (30)	29	28
5	15	14	13	12	11	10	9
6	26	25	24	23	22	21	20
7	7	6	5	4	3	2	1
8	18	17	16	15	14	13	12
9	29	28	27	26	25	24	23
10	10	9	8	7	6	5	4
11	21	20	19	18	17	16	15
12	2	1	0 (30)	29	28	27	26
13	13	12	11	10	9	8	7
14	24	23	22	21	20	19	18
15	5	4	3	2	1	0 (30)	29
16	16	15	14	13	12	11	10
17	27	26	25	24	23	22	21
18	8	7	6	5	4	3	2
19	19	18	17	16	15	14	13

other new set of numbers came in, which will be valid until 2200 (2000 is a normal leap year and in 2100 the solar and lunar corrections again compensate). Sirleto's commission proposed a way to introduce all this into the calendar in a perpetual Easter canon.

The principle may be clear from Table 1.4, which converts the serial year of a nineteen-year cycle into the epact for that year. (The epact here is the moon's age on 1 January.) On Dionysius' scheme, the epact can take only 19 values. On the first year of each cycle, year I, it is 0. In year II, it is eleven because the lunar year is 11 days shorter than the solar. The epact of year III is 22; it is embolistic according to the Metonic cycle, which makes the epact of year IV 33 − 30 = 3. The epacts proceed through the sequence 0, 11, 22, 3, 14, 25, 6, 17, 28, 9, 20, 1, 12, 23, 4, 15, 26, 7, 18, and (at the very end of year XIX), 29. These twenty-nine days make the last intercalary month of the cycle to reproduce the old *saltus lunae*. Hence the epact of year XX, or year I of the next cycle, is 0. Intoxicated with these calculations, Clavius gave tables for determining golden numbers and their corresponding epacts for an optimistic 800 million years, and the cumulative difference in days between the Julian and the Gregorian calendar out to the year A.D. 303,300, "atque ita in infinitum."[46]

The suppression of the three bissextiles every four hundred years has the consequence that the epact can take on any value from 1 to 30 (= 0). The second column of Table 1.4 reproduces the cycle of epacts given in the preceding paragraph. The third column gives the cycle for the case that the epact of year 1 is not 30 (= 0) days, but 29 days; then year I is embolistic and the epact of year II is 29 + 11 − 30 = 10. The construction of the rest of the table will be evident to anyone who knows that the first column held for Giglio's time. This is but the first step into Giglio's labyrinth. The next ones derive the dates of the new moons from the epacts and golden numbers, contrive conventions to avoid the occurrence of more than one new moon on any calendar date within the same nineteen-year cycle, and dispose of the very rare situation with golden number XIX and epact 19, which imply that the last lunation in the cycle begins on 2 December. Since the last lunation is hollow (the *saltus lunae!*), a new moon occurs also on 31 December. Clavius took forty-eight large pages to exhibit these peculiarities. They contain, altogether, 12,000 entries.[47]

Gregory's commissioners consulted widely among those able to wind through Giglio's labyrinth. The opinions received, which came from universities from Poland to Portugal, destroyed the earlier consensus over tables. The report from Vienna rejected cycles as usual and augured from the recent blooming of astronomical studies in the West that the true parameters could be found, if only the pope would build some observatories. The Viennese also observed that the printing press made practicable the periodic updating of the Church calendar. The old concern, that changes could not be copied faithfully and distributed efficiently, no longer carried any weight. The theologians of the Sorbonne saw through this precocious grantsmanship. They perceived that astronomers were "despicable, dangerous, and stupid people," who had invented the discrepancies in the Easter canon for their own purposes.[48]

Among the respondents who preferred average moons and cycles, a plurality approved Giglio's scheme as the best or accepted it in principle but preferred updated golden numbers to epacts, or a shift of the vernal equinox to 25 March rather than to 21 March. They divided over whether to take Alfonso's or Copernicus' values of the year and synodic month as the base of calculation. Giglio preferred Alfonso; several respondents preferred Copernicus, without, however, thereby endorsing heliocentrism. The opinion from the University of Louvain, which called Copernicus himself a "gift from Heaven" and Copernicus' theory sheer "fantasy," was typical of the schizophrenia of those who advocated using his results. The commission decided in favor of Copernicus' numbers despite Clavius' condemnation of their foundations in "uncertain and absurd hypotheses abhorrent to the common opinion of mankind, and repellent to all natural philosophers." Also, despite Clavius' preference for the astronomical way ("to restore astronomy and keep it in good repute"), the commission remained cyclist, primarily on the ground that nothing but confusion would be served if astronomers, who could not agree about the facts, tried independently to set the Easter parameters. By the autumn of 1580 the commissioners had decided for Giglio's scheme, including fixing the vernal equinox on 21 March. They proceeded to its adoption.[49]

The Pope and the relevant rulers promulgated the new calendar in most Catholic lands in 1582. The principal exceptions were the territories of the Holy Roman Empire, where the certainty of upsetting the Protestants prompted some princes to request the Pope to withdraw his handiwork. Still the Protestants protested. Michael Maestlin, Kepler's teacher, exposed the Gregorian reform as a plot to reestablish papal authority in disaffected regions. His insight was widely shared. Again, according to Maestlin, the pompous perpetuity of the Easter calculation was both fatuous (because, in his opinion, the length of the year is not constant) and futile (because the world will not last until the first dropped leap day). Others perceived that the Pope's intention was to offset the loss of traffic in indulgences by the sale of almanacs and calendars, or to steal ten days from time, tricks for which, it was devoutly hoped, he would go to the Last Judgment ten days before everyone else.[50]

The Catholics adduced in response that a nut tree in a village in Germany had elected to bloom on the same day in the new calendar as in the old. Pieces of the pious tree were sent to unreformed princes as signs that the reform agreed with the nature of things. Clavius undertook a more systematic and sensible response in a lengthy *Explanatio* of the new calendar and several polemical essays; published together under an apt bit of Scripture, "God made me to know the course of the year and the dispositions of the stars," they made up one of the five large volumes of his mathematical works. In these several pieces he turned the variability of the year, about which, and little else, he agreed with Maestlin, into an argument for using cycles and averages. The other peculiarities of the Gregorian reform, he said, had been adopted in an effort to depart as little as possible from the old scheme of golden numbers. The new calendar rested on the redundant authorities of the Pope and the mathematician, neither of whom could teach anything "that is false or even [only] probable." If Protestants rejected the reform, it was not because it was defective, but because it was papal.[51]

The Protestants did reject it, and for the reason Clavius had foreseen: the reluctance of the Protestants to accept any innovation, however useful, sponsored by a pope. The great English magus John Dee argued for its utility and convenience, and was supported by fellow-mathematicians, who, to show some independence from Rome, proposed that the superfluous ten days be discarded three each from May, July, and August, and one from June; and that an ephemeris be compiled, showing the dates of Easter for a century or two, "and so easily renewed, as we see yearly Almanachs are, if the Sins of the World do not hasten a Dissolution."[52] Their recommendation, which Queen Elizabeth and her council accepted, was felled by the copious doctrine of Anglican bishops, that "all changes are dangerous." To this they added that the Bible enjoined true Christians to have no commerce with the Antichrist, that is, the Bishop of Rome, and that the business was not worth the bother. "Because that the latter day [is] approaching . . . we doe think that the pope might very well have spared his labour."[53]

Neither side had any trouble finding astronomers willing to lend the prestige of their special knowledge to arguments that were no longer technical, but theological and political. The spectacle of his fellow astronomers selling out to special interests disgusted the greatest of them all, Johannes Kepler, who, though a Protestant, approved of the Gregorian reform. He railed against the precociously modern sin of prostituting expertise. "The parties [in the dispute] should no longer say, our mathematicians think thus, and the others so, they should rather say, we can make our mathematicians say what we please, for they are our slaves."[54]

2: A Sosigenes and His Caesars

Florence

· PRINCELY PROJECTS ·

For two decades before his death in 1574, Cosimo I dei Medici was immersed in immense construction projects. Foremost among them was the elevation of himself from the lowly Duke of Florence to the Grand Duke of Tuscany. The preliminary work in this construction, completed with the help of a friendly pope, Pius V, in 1569, included the conquest of Siena and other neighboring territories, the transformation of the ducal palace into a residence fit for a first-class sovereign, and the renovation (some say, and said, wanton destruction) of two important churches. These churches, Santa Maria Novella and Santa Croce, respectively the strongholds of the Dominicans and the Franciscans, retained the medieval construction that shut off the monks during mass from the laity on whose behalf they prayed. The Council of Trent advised that greater participation by Catholics in the mass and frequent, easy-access communion would protect them from Protestantism.

Cosimo, who needed the help of the papacy to achieve his political aims, thought to demonstrate his religious fervor by rearranging the interiors of the great Dominican and Franciscan churches to suit.[1]

Cosimo's principal agent in the opening of the churches and the refurbishing of the palace was the artist, architect, and historian Giorgio Vasari. Among the many artists and artisans engaged in the work, a sculptor named Vincenzo Danti figured prominently. Vincenzo, born in Perugia in 1530, was working under Vasari by 1560, for in that year he finished a bronze door to a cupboard in which Cosimo kept his important papers. A good courtier as well as an elegant sculptor, Vincenzo placed at the center of his door an allusion to Cosimo as Augustus Caesar. He soon became head of the ducal foundry. A contemporary described him in the understated rhetoric of the time as "an unusual young man of acute and sublime genius, gracious and kind, whose power and tremendous mastery of sculpting is worthy of immortal honor." The unusual youth, who died, still young, in 1576, assisted in the reconstruction of S. M. Novella. Vasari tore out its barrier or rood screen in 1565 and then removed a welter of old tombs and chapels. Not even the blessed Giovanni da Salerno, the founder of the Dominican order in Florence, was safe from this evacuation; and the relocation of his bones provided a commission to Vincenzo Danti to embellish their new resting place.[2]

A few years before translating the blessed Giovanni, Vasari began an ambitious project of great interest to Cosimo: the decoration of a new room to house a set of cabinets to contain the Grand Duke's prize possessions. The decoration was to consist of over fifty maps depicting various regimes around the world, drawn according to the best available knowledge and projected on sound geometrical principles.[3] Mural maps had decorated important walls elsewhere in Italy, notably the maritime republic of Venice; but the series designed for Cosimo, some fifty-seven maps in all, was a great novelty, very probably the first extensive atlas of the world, and certainly one of the least convenient.[4] To create it required a man who combined the skills of cosmographer, designer, and painter. Vincenzo Danti happened to know such a man. It was his younger brother Egnatio.

Egnatio Danti possessed in even greater measure than Vincenzo the peculiar *virtù* of his family. Its patriarch, their grandfather Pier Vincenzo Rinaldi, was by trade a goldsmith and by inclination a poet, architect, and astronomer. He was called Dante by his friends because, according to Egnatio, "his cleverness seemed to approach the acuteness of the great poet." Pier Vincenzo obligingly styled himself "Dante de Rinaldi," which shortened, eventually, to the more modest "Danti."[5] He was followed in this affectation by his younger brother, who showed the family flair for invention and display by flying to a wedding on his own wings.[6] Or so it is said. The Dantis knew how to advertise themselves.

Most of the very little known about the patriarch Per Vincenzo Danti comes from the autobiographical fragment that he prefaced to a translation he made, in his omniscient way, of Sacrobosco's *Sphere*. According to this preface, the occasion of the translation was an outbreak of plague in Perugia in the early 1490s. Pier Vincenzo fled to a remote villa with his family and passed his time observing the stars. It kept him alive. "Undoubtedly [he wrote] the long physical indisposition I have suffered for so many years would have put me under the ground if my mind, continually sweetened by the contemplation of astronomy, had not attenuated the [weakness] of my body." So strengthened, he imposed his tastes upon his children, Teodora and Giulio. Both mastered the sphere. They also cultivated the artistic gifts of the family. Teodora studied with Pietro Perugino and acquired a local reputation as a painter. Giulio became an architect and the father of Vincenzo (born 1530) and Egnatio (born 1536).[7]

Egnatio learned drawing and savoir-faire from his father and mathematics from his aunt. He later wrote, in his grandfather's style, of "the great delight and contentment the mind experiences in the contemplation and exercise of these most pleasant arts, which . . . have always charmed the minds not only of ordinary men, but also of great seigneurs, of kings, and emperors."[8] Further to his education in the Danti style, he had the example of his elder brother Vincenzo, who at an early age made himself famous and popular in Perugia by restoring water to a plugged-up fountain before leaving home to study with Michelangelo.[9] Vincenzo became a leading proponent of the study of human anatomy by artists. He outlined a treatise on the elements of all the knowledge a practicing artist would need, but managed to see only a small part of it through the press. Egnatio followed Vincenzo's work and later published a synopsis of it.[10]

In 1555, at the age of nineteen, having attended the university in his home town, Egnatio joined the Dominicans. He continued his studies of mathematics along with philosophy and theology. In the few places in his subsequent mathematical writings in which he expresses opinions about the structure of the world, he followed Dominican teaching: Aristotelian in physics, Ptolemaic in astronomy, Thomistic in astrology ("with this science, we can know our bad inclinations, and keep ourselves from many sins and errors").[11] He had just finished preparing himself for a life of teaching and preaching when he received the call to cosmology from Cosimo. He thus entered into the well-oiled machinery of joint patronage, by Church and State, through which talented clerics advanced in Italy. His superiors transferred him to their convent at Santa Maria Novella to make it easier for him to commute to court.

In October 1563 Vasari announced to Cosimo the arrival of the "friar for the maps of Ptolemy." The friar proved prolific. Maps and paintings poured forth from his cell; in one instance, five porters were required to transport the current issue to the palace. Cosimo took notice and began to patronize his cosmographer. Eventually the inconvenience of the transport and the desire to have Danti "continually available for service in cosmological matters" prompted Cosimo to request permission from the Dominican General for Danti to live in the palace. He moved in in 1571. Danti's preferment excited the jealousy of some of his brethren, who were filled with the uncharitable spirit of counterreformation that Cosimo had encouraged in the convent, and who complained that Danti lived in a manner unbecoming a monk. Cosimo protected him and added to his dignity, and to the envy of others, an appointment to the chair of mathematics at the University of Pisa.[12] Danti's congenial duty was to give one lecture a month; he replaced a theologian who had prepared for early retirement by giving none at all.[13]

While Danti resided in the palace, he and Cosimo discussed two new projects that, had they brought them to completion, would have covered them with glory. One was a canal through Tuscany linking the Mediterranean and the Adriatic, which Danti said (sounding like his uncle the flyer) he could accomplish "with less difficulty than people thought."[14] This was bravado indeed. Danti had had little experience with large construction apart from a commission in the mid-1560s to design a church and convent for the Dominican Pope (later Saint) Pius V. The second big project was the reform of the calendar. Here Danti the courtier exploited Cosimo's conceit of a resemblance between Tuscany and Rome, and himself and Julius Caesar. The prospect of improving the calendar promulgated by his ancient model thrilled Cosimo. Thus Danti, writing four years after his Caesar's death:

> Because the Grand Duke always had a heroic mind inclined to the greatest projects, and having always been an emulator of outstanding deeds of the ancients and knowing (in addition to its general usefulness) how much glory Julius Caesar had obtained from his reform of the year, which though good is not perfect; he got the idea of devoting himself with all his power to this most honourable task, which (having proposed it to the Pope) he would doubtless have completed if first a long and troublesome illness, and then death, had not prevented him.[15]

In order to play the part of Sosigenes to Cosimo's Caesar, Danti needed to make accurate measurements of the length of the tropical year. Or so he claimed: in fact, as we know, the length adopted by Gregory XIII, in the reform that superseded Caesar's, had been recommended by experts since the thirteenth century. To advertise the great project, and perhaps also in the hope of obtaining useful data, Danti proposed to mount instruments on the beautiful face of S. M. Novella. Cosimo favored the plan. With the same highhandedness that had demolished the inside of the church, he caused the officials in charge of the building to permit the cementing of two small excrescences onto the façade some seven meters above the ground. They are still there (Figures 2.23 and 2.26).[16]

The instruments consist of two hoops of wire intersecting one another perpendicularly (an armillary sphere) and a slab of marble bearing a quarter of a circle marked off in degrees (a quadrant). They work on the same principles as *meridiane*. These principles — the elements of naked-eye astronomy — may be presented most aptly by a paraphrase of Danti's grandfather's annotated translation of Sacrobosco's treatise, which Danti had printed to ease the access of his own students to the basic geometry of the heavens.

· DOCTRINE OF THE SPHERE ·

Sacrobosco's sphere is the bowl of the heavens, studded with stars, that every observer imagines above his head on a clear night. The untutored solipsist thinks that he stands at the center of this sphere (Figure 2.1); medieval astronomers, like Sacrobosco, knew that the earth is round and that its center coincides with that of the great celestial sphere (Figure 2.2). (Sacrobosco gives Aristotle's reasons for believing in a spherical earth: the stars do not rise and set everywhere simultaneously, the elevation of the polestar increases as one goes north, ships sailing away disappear hull first.)[17] A third sphere, concentric with the other two, marks the closest approach of the moon to earth; another, that of the sun; between them come spheres for Mercury and Venus; beyond the sun's, those for Mars, Jupiter, and Saturn (Figure 2.3, Table 2.1). The moon's sphere separates the sublunary region, where all is change, corruption, and generation, from the celestial, where the stars and planets continue in the same state, invariably, for ever and ever. Why? We might answer, because Aristotle, whom Sacrobosco and the Dantis followed

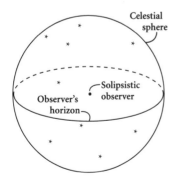

FIG. 2.1. Horizon and celestial sphere.

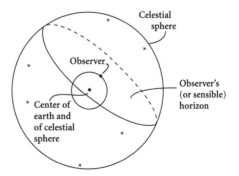

FIG. 2.2. Observer, horizon, earth, and celestial sphere.

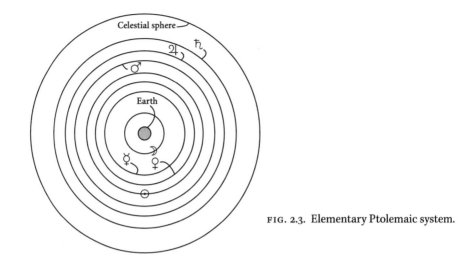

FIG. 2.3. Elementary Ptolemaic system.

faithfully, said so. They answered: "because thus God, glorious and sublime, created and disposed [the world]."[18]

The stars appear to be fixed inside the celestial sphere as if (to borrow an image from grandfather Danti) they were nails driven into a plank.[19] Of course, only the stars nailed into one half of the sphere are visible to a single observer at any time. The other half lies hidden beneath the horizon plane, the "sensible horizon," which just touches the earth at the observer's station. To divide the sphere exactly, the plane would have to pass through the earth's center. The hypothetical plane through the earth's center parallel to the sensible horizon is called the "rational horizon." Because the radius of the earth is so small in comparison with the indefinitely large radius of the sphere of the fixed stars, the sensible and rational

Table 2.1   The planetary signs

○	Earth
☽	Moon
☿	Mercury
♀	Venus
☉	Sun
♂	Mars
♃	Jupiter
♄	Saturn

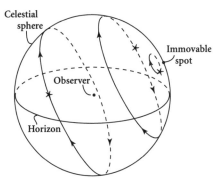

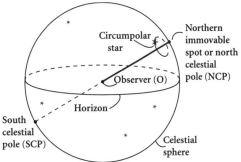

FIG. 2.4. Apparent diurnal paths of stars and the "immovable spot." The star on the left has just risen; that in the near right is close to setting; that on the far right neither rises nor sets.

FIG. 2.5. Poles and circumpolar stars.

horizons coincide so closely that in most geometrical demonstrations the sensible horizon can be supposed to pass through the center of the earth. For the sorts of solar observations made in cathedral observatories, however, the difference between the sensible and the rational is material and, as will appear, troublesome.

An observer who looks at the hemisphere visible to him or her every two hours or so during a clear evening will see the supposedly fixed stars move with respect to the horizon. Some low in the sky rise, some high up decline, some disappear entirely, and others come into view. A particularly acute observer will note that not all stars set and that some, in a certain region of the sky, appear to move in circles around an unmarked spot in the heavens (Figure 2.4). Such an observer is an astronomer.

To represent the apparent motions of all the stars — those that rise and set and those that stay above the horizon — it is enough to assume that the heavenly sphere turns once a day around an axis that runs from the unmarked spot through the earth's center. This spot is called the north celestial pole (NCP). The south celestial pole lies where the axis of rotation, extended beyond the earth, again meets the starry sphere. The stars that do not rise or set are called "circumpolar" (Figure 2.5). The North Star, though not the most conspicuous of them, is the most useful, since it describes a circle very close to the NCP and thus gives a rough indication of the pole's location.

The angular height, or "elevation," of the pole above the horizon depends on the observer's latitude. A little geometry shows that the elevation just equals the latitude. In Figure 2.6, C is the center of the earth, whose diameter is much exaggerated in size, O is the position of the observer, and ON runs to the point on the horizon directly under the NCP. Then α is the elevation. Now enlarge the central

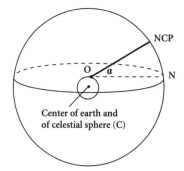

FIG. 2.6. Altitude of north celestial pole.

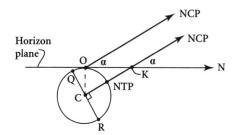

FIG. 2.7. Horizon, equator, and north celestial pole. The altitudes at O and K are the same because the earth's radius QC = CO is negligibly small compared with the distance C·NCP.

part of Figure 2.6 into Figure 2.7, to which a few lines have been added: one through the earth's center C parallel to the line from the observer to the NCP and intersecting the horizon plane at K; the radius CO perpendicular to the horizon; and the diameter QR perpendicular to O·NCP and C·NCP. (Because the distance to the NCP is so large compared with the radius OC, O and C virtually coincide on the scale of figure 2.6; that is the reason that the lines from O and C to the same point, the NCP, can be considered parallel.) The intersection of the line C·NCP with the earth's surface is the geographical or terrestrial north pole, the NTP; hence C·NTP is the northern half of the earth's axis, and the diameter QR, perpendicular to the axis, is in the plane of the earth's equator.

Now comes the geometry. Since O·NCP and C·NCP are parallel, $\angle OKC = \alpha$ (by Euclid I.29, as Danti would say). Note that $\angle COK$ and $\angle KCQ$ are right angles. Therefore α and $\angle QCO$ are complementary to the same angle, $\angle OCK$, and therefore equal to each other. But $\angle OCQ$ is just the latitude ϕ of the observer's station O, as will be clear from Figure 2.8. The circle NTP·OQ is the intersection of the earth with a plane through the north geographical pole and the surface point O; by definition, the arc NTP·OQ is the meridian of longitude through O. Similarly, the intersection of a plane through O parallel to the equator with the earth is O's parallel of latitude. Latitude runs from 0° at the equator to 90° at the NTP; in between it is given by $\phi = \angle OCQ$. Understanding might be eased by redrawing the earth in its customary position with the NTP on top (Figure 2.9). To define the longitude, the reference meridian NTP·GS, which should be imagined to go through Greenwich, has been added to the diagram. The longitude of O is defined as $\lambda = \angle SCQ$ in the plane of the equator. The observer O appears to be a little north and far west of Greenwich, probably somewhere in Alaska.

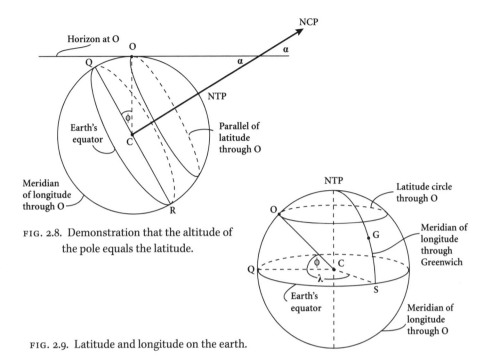

FIG. 2.8. Demonstration that the altitude of
the pole equals the latitude.

FIG. 2.9. Latitude and longitude on the earth.

We rise again to the celestial sphere. Sacrobosco defined ten imaginary reference circles on it to make easy and possible an exact description of the apparent
motions of celestial objects. (The Dantis thought it necessary to emphasize that
the circles do not exist in the sky; "mathematicians [use them] . . . to enable us to
understand celestial motions with greater facility.")[20] We are already familiar with
the first of these circles, the horizon plane, or, rather, its intersection with the celestial sphere. The second circle, which plays the principal part in the theory of
meridian lines in churches, is perpendicular to the horizon plane and passes
through the NCP and the SCP (south celestial pole). This capital circle is called the
prime meridian. Its intersections with the horizon mark the directions of north
(N) and south (S) for the observer O; its intersection with the line CO extended,
which runs from the earth's center through O's head and thus perpendicular to the
horizon, is the observer's zenith; the diametrically opposite point is the nadir (Figure 2.10). Like the horizon, the prime meridian is fixed to the observer; stars rise
to it and descend from it as the earth turns. All stars that rise and set attain their
greatest angular height, or altitude, for a given observer as they cross the observer's meridian. Circumpolar stars cross the meridian twice each day, once at
their highest and once at their lowest altitudes (Figure 2.11). Because they reach
their maximum height at the meridian, they are said to "culminate" there.

The third circle, the equinoctial, is the intersection with the celestial sphere of

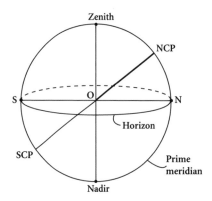

FIG. 2.10. Zenith, prime meridian, north, and south.

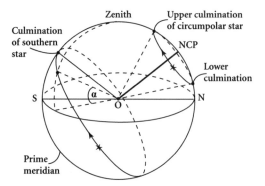

FIG. 2.11. Culminations (meridian crossings) of southern and circumpolar stars.

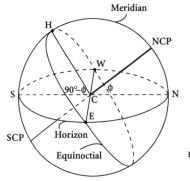

FIG. 2.12. Meridian, horizon, equinoctial, east, west.

the plane through C perpendicular to the axis of the world NCP·SCP. It is to the great sphere what the equator is to the earth. The intersections of the equinoctial with the horizon circle mark the observer's east (E) and west (W) (Figure 2.12). The equinoctial is best pictured as a line painted on the celestial sphere. Every point on it revolves around the observer once a day, tracing the equinoctial in its course. Since the equinoctial, being perpendicular to the axis of rotation, revolves in place for every observer, it always occupies the position indicated in the figure: its culminating point H always being 90° − ϕ above the southern horizon in northern latitudes. The sense of rotation is from east to west: every point on the equinoctial rises due east and sets due west. Because the horizon divides it into equal halves, every point on the equinoctial spends the same amount of time above as below the horizon. Therefore, when the sun appears to be on the equinoctial, days and nights are equal.

The sun appears on the equinoctial twice a year, on the days of the vernal and autumnal equinoxes, around 21 March and 21 September. At other times it is either north (toward the NCP) or south (toward the SCP) of the equinoctial. When

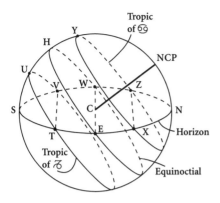

FIG. 2.13. Equinoctial, tropic of Cancer, and tropic of Capricorn.

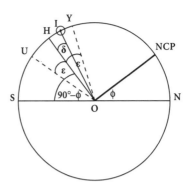

FIG. 2.14. Sun's noon positions at midsummer and midwinter.

north, it describes a longer arc above the horizon than below it; when south, a shorter arc. The first situation corresponds to summer, the second to winter, in the Northern Hemisphere. These arcs are parts of circles parallel to the equinoctial. Two of these circles have special names. The circle described by the sun on the day on which it is farthest north of the equinoctial, that is, on midsummer day, is called the summer tropic or tropic of Cancer. That described by the sun on midwinter day is the winter tropic or tropic of Capricorn (Figure 2.13). Both these circles, being perpendicular to the axis of the world, rotate in place, that is, keep their orientation with respect to the observer's horizon and meridian.

Every point on the winter tropic rises at T, south of east, culminates due south at U, and sets at V, south of west. Similarly, every point on the summer tropic rises at X, north of east, culminates due south at Y, and sets at Z, north of west. The angular distance measured along the meridian of the culminating sun above or below the equinoctial is called its declination, δ, taken as positive to the north (when the sun crosses the meridian between H and Y) and negative to the south (culmination between H and U). Maximum (or minimum) declination occurs when the sun is at the summer (or winter) tropic. The situation is symmetrical: the absolute values of the maximum and minimum declination, ε, which is called the obliquity of the ecliptic, are equal. We have therefore on the tropic of Cancer $\delta = +\varepsilon$, and on the tropic of Capricorn, $\delta = -\varepsilon$. One of the primary objectives of measurements at meridian lines was to find an exact value for ε. The altitudes of points U and Y in Figure 2.13 are $90° - \phi - \varepsilon$ and $90° - \phi + \varepsilon$, respectively, as appears from Figure 2.14, where, for good measure, the sun is shown at noon (that is, culminating) at a point I of positive declination δ. The maximum declination ε is approximately 23.5°, a little more than a fourth of a right angle. This ε, which is a protagonist in this book, is of more than theoretical interest. By taking the sun's noon altitude $\alpha = \angle SOI$,

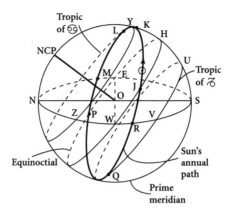

FIG. 2.15. Tropics, equinoctial, and sun's annual path (ecliptic).

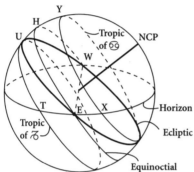

FIG. 2.16. Ecliptic, tropics, and horizon with vernal equinox (VE) rising.

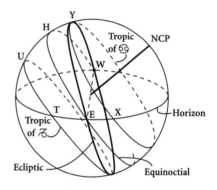

FIG. 2.17. Ecliptic, tropics, and horizon with vernal equinox setting.

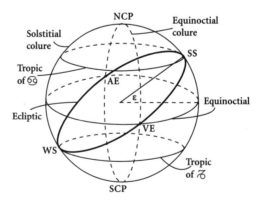

FIG. 2.18. Tropics and colures; SS signifies the summer solstice, WS the winter.

navigators could find their latitude if they knew what day it was and had accurate tables of the solar declination for every day in the year.[21] As appears from Figure 2.14, $\phi = 90° - \alpha + \delta$.

Let us now enjoy the privilege of Joshua, for whom, to complete the slaughter of the Amorites in the light, God stayed the sun on the meridian for an entire day.[22] That means that God arrested not only the rotation of the sphere of the fixed stars, which gives us day and night, but also the rotation whereby the sun moves from one tropic to the other, which determines the seasons. Now, if God had stayed only the rotation of the sphere that bloody afternoon, the amazed warriors might have seen the sun move back in the sky, from west to east, against the usually prevailing diurnal revolution from east to west. If the battle had begun on the day of the vernal equinox about 3:00 P.M., the sun would have been at J in its diurnal course (Figure 2.15). It would then have gone back toward the east, rising slowly toward the north, to culminate at K between H and Y about six weeks later. In another six weeks, still proceeding east and north, it would have met the northern tropic at L and turned south. ("Tropic" signifies turning.) If the battle had continued another six weeks, the sun would have set to the north of east at M on its course toward the autumnal equinox, which, lying in the equinoctial diametrically opposite J, lay beneath the horizon at P. In another six months the sun would have risen south of west at R (truly a miracle) to regain its original position at J, having touched meanwhile the winter tropic at Q in the dead of night.

The circle JKLMPQRJ, which the sun would appear to trace out against the celestial sphere without the diurnal motion, is called the ecliptic. (The name expresses the fact that eclipses can occur only when the moon appears on or near the ecliptic, since only then can earth, moon, and sun be aligned.) Contrary to all the other circles Sacrobosco defines, the ecliptic does not remain in place with respect to the horizon and meridian without a miracle of Joshua's type. Figure 2.15 shows the position of the ecliptic at a single instant, namely, 3:00 P.M. on the day of the vernal equinox. Figures 2.16 and 2.17 show it at two instants particularly easy to draw, with the vernal equinox rising and setting, respectively.

Although it is constantly changing its relationship to the horizon and meridian of any given observer, the ecliptic circle—the sun's annual course against the fixed stars—does not move around the celestial sphere. Like the equinoctial, therefore, it can be imagined as painted on the sphere. The two circles intersect in two points, which are, of course, the equinoxes. The circle that goes through the celestial poles and the equinoxes is called the equinoctial colure. The points where the ecliptic touches the tropics are called "solstices," that is, places where the sun appears to stand still as it switches directions from north to south, or vice versa; the circle passing through the solstices and the celestial poles is the solstitial colure (Figure 2.18).

Like the sun, the moon has its own apparent path around the earth from west to east against the diurnal motion of the heavens. However, the moon goes much faster on this private path than the sun does on the ecliptic. Let us suppose that Joshua's battle began on the day of a new moon and that, as before, God arrested only the diurnal motion. (In fact He stopped the moon altogether.) Then the moon would have been seen in the direction of the vernal equinox at 3:00 P.M. on that extraordinary spring day. In about 84 hours it would have culminated; in another seven days it would have set on the eastern horizon; and two weeks later it would have returned to where it was when the slaughter commenced (J in Figure 2.15), long before the sun reached the meridian. Before the sun returned to J, the moon would have made thirteen revolutions.

In its motion to the east, the moon would have gone off the ecliptic, to which its apparent path is inclined by about 7°. The intersections of this lunar circuit (as we may call the moon's apparent path) with the ecliptic are known as the moon's nodes. Unlike all the circles so far presented, the lunar circuit does not remain fixed on the celestial sphere and so cannot be represented by a painted line. There is something constant, however, in its dance. While the moon's nodes slip along the ecliptic, the circuit maintains the same inclination, of 7° or so, to the ecliptic (Figure 2.19). Consequently, the moon remains within a band of around 14° centered on the ecliptic. So do the planets.

Astronomers have found it useful to give this region a name. It is the zodiac, so called because the stars located in it seemed to some imaginative observers to outline animals ("zodiac" = "circle of animal figures"). In astronomical practice, already ancient by the time of Sacrobosco, the twelve original asterisms gave way to twelve conventional signs, each indicating a piece of the ecliptic 30° long together

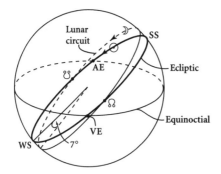

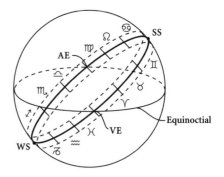

FIG. 2.19. Ecliptic, equinoctial, and lunar circuit showing nodes and equinoxes; ♌ and ☊ signify the ascending and descending nodes (the lunar analogies to the vernal and autumnal equinoxes), respectively.

FIG. 2.20. Equinoctial, ecliptic, and zodiac.

Table 2.2   The zodiacal signs

♈	Aries, the ram	♎	Libra, the balance
♉	Taurus, the bull	♏	Scorpio, the scorpion
♊	Gemini, the twins	♐	Sagittarius, the archer
♋	Cancer, the crab	♑	Capricornus, the goat
♌	Leo, the lion	♒	Aquarius, the water bearer
♍	Virgo, the maiden	♓	Pisces, the fish

with its accompanying band 7° to either side. The signs begin with Aries at the vernal equinox and continue to the east in the direction of the sun's motion through them (Figure 2.20). Table 2.2 lists the signs, their meanings, and their symbols.

The vernal equinox (VE) is the first point in Aries. The summer solstice (SS) comes 90° later, at the first point in Cancer; the autumnal equinox (AE) at 180°, at the beginning of the seventh sign, the first point in Libra; and the winter solstice (WS) at the tenth sign, the first point in Capricorn. Hence the tropics of Cancer and Capricorn.

Sacrobosco defined two more circles, making ten in all. The circumpolar stars that just manage not to set touch the north point N of the horizon during their diurnal course. The angular distance of these stars from the NCP must equal the pole's height, that is, the latitude. At sufficiently high latitudes the sun too can be circumpolar. The minimum latitude at which it can achieve this feat is where the tropic of Cancer just touches the north point of the horizon. Figure 2.21 shows that, since $\angle\text{NCP·OY} = \angle\text{NCP·ON}$ and $\angle\text{NCP·ON} + \varepsilon = 90°$, this latitude must be

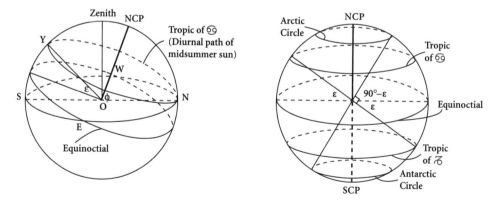

FIG. 2.21. The tropic of Cancer as seen from the latitude of the Arctic Circle.

FIG. 2.22. Tropics, equinoctial, and Arctic and Antarctic Circles.

90° – ε, or about 66.5°. That parallel is called the Arctic Circle; the Antarctic Circle is the corresponding parallel in the Southern Hemisphere. The Arctic Circle may also be interpreted as the circle on the celestial sphere made by the zeniths of all points on the earth at latitude 90° – ε, or, what is the same thing, the diurnal circles traced by the poles of the ecliptic. In the same vein, the latitudes of 23.5° north and south are the highest at which the sun can appear in an observer's zenith (Figure 2.22). The sun can never stand overhead in continental Europe.

It will be useful to group together Sacrobosco's circles and their definitions (Table 2.3). In addition, the tropics and arctic circles can be defined on the earth, the first as the latitudes on which the sun stands overhead, the second as the latitudes at which the sun just does not set, on midsummer day.

The parameters that regulate the system, like ε and the tropical year, cannot be deduced a priori. As grandfather Danti observed, they are what they are because God so ordained; to learn their values, one must measure, carefully. The Dantis report five values for ε (Table 2.4).

The table raised the reasonable conjecture that ε decreases steadily in time. Did it and does it? Will the ecliptic and the equinoctial eventually coincide, and all days equal all nights? That is the sort of question that might interest even people who are not astronomers. As for the tropical year, the Dantis observed that Sosigenes made it too long, but they did not specify by how much; and they made the time of the lunar circuit 27.1333 days.[23] Neither indication sufficed for calendar reform.

· THE INSTRUMENTS OF SANTA MARIA NOVELLA ·

In the spring of 1574, Danti cemented into place on the façade of his convent church his version of the instrument with which Sosigenes, and after him, Ptolemy, detected the arrival of the sun at the vernal equinox. He chose the location, Danti explained, not because S. M. Novella belonged to the Dominicans, or because Cosimo had lately been doing as he pleased with it, but because "it was the most convenient and stable in Florence, being strong enough to stand immobile as long as the world lasts, and being freely exposed to the south so as to receive the rays of the sun at the times of the equinoxes from morning until evening."[24] Danti required a south-facing façade so that his instruments could catch the noon sun throughout the year. He would have had to use the nave wall had S. M. Novella had the conventional orientation with front door to the west, which would have been a poorer venue for his wares. Luckily for him, the Dominicans had rotated the church when they rebuilt it in 1279.[25]

Figure 2.23a shows the instrument as Danti drew it, Figure 2.23b as it stands to-

Table 2.3   Sacrobosco's celestial circles

Name	Characteristics
1. Horizon	Tangent to the earth at observer's station; at rest with respect to observer.
2. Meridian	Perpendicular to the horizon; passes through NCP; at rest with respect to observer.
3. Equinoctial	Perpendicular to the axis of the world; passes through east and west points of the horizon; rotates in place.
4. Ecliptic	Sun's apparent path during its annual motion; inclined at an angle ε to the equinoctial; changes orientation with respect to observer.
5. Equinoctial Colure	Passes through the equinoxes and celestial poles; changes orientation with respect to the horizon.
6. Solstitial Colure	Passes through the solstices and celestial poles; changes orientation with respect to the horizon.
7. Tropic of Cancer	Parallel to the equator and tangent to the ecliptic at the summer solstice; rotates in place.
8. Tropic of Capricorn	Parallel to the equator and tangent to the ecliptic at the winter solstice; rotates in place.
9. Arctic Circle	Locus of zeniths of points on the earth where the sun is circumpolar only on midsummer day in the northern hemisphere; rotates in place.
10. Antarctic Circle	Locus of zeniths of points on the earth where the sun is circumpolar only on midsummer day in the southern hemisphere; rotates in place.

Table 2.4   Values of ε delivered by Danti, 1571

Measurer	Date	Value
Ptolemy	ca. 150	23°51'20"
Albategni	880	23°35'00"
Arabel	1070	23°34'00"
Almeone	1140	23°33'00"
Danti et al.	1570	23°29'00"

Source: Danti, *Sacrobosco* (1571), 28.

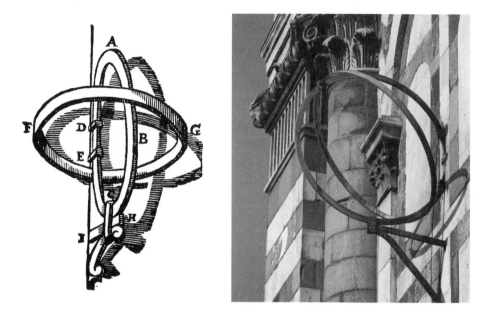

FIG. 2.23. The armillary sphere on the façade of S. M. Novella, Florence. (a) Danti's drawing; (b) what remains. From Danti, *Primo volume* (1578), 317, and Comune di Firenze.

day. The ring ABED lies in the plane of the meridian looking south; the observer may be supposed to sit at the ring's center O. When the sun comes into the plane of this meridian, it is noon for the observer and for everyone else standing in the plane. The ring FGE lies in the plane of the equinoctial. Should the sun come to an equinox exactly at noon for the observer at O, its rays would not penetrate to O, being blocked by both the rings. Even on a bright day O would have a dark noon. This is the principle on which the armillary of Sosigenes, Ptolemy, and Danti rests. To make it work, one must fix it so that a diameter of the equinoctial hoop makes the proper angle ϕ with the vertical diameter of the meridianal hoop (Figure 2.24).

If the sun were a point, as indicated in Figure 2.25a, then, if the equinox occurred at noon for O, the entire concave surface of the ring EFG would be in shadow. The diurnal motion would carry the sun in the plane of the equinoctial during the afternoon and the interior of the ring would remain dark. However, the sun's annual motion would be taking it north, above the plane of the ecliptic, so that, toward evening, the top half of the back concave surface of the ring around E would be illuminated (Figure 2.25b). Since the sun is not a point — it subtends an arc of about 30' at the observer's eye — and very rarely presents itself at the equinoctial very close to noon, the appearances are more interesting than those just described. When the center is on the equinoctial, two dim bands of light appear at the top and bottom of the hoop separated by a zone of darkness. As the sun

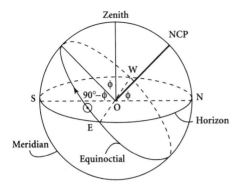

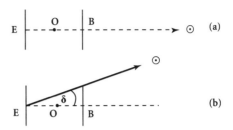

FIG. 2.24. Sun's apparent diurnal path at the equinoxes.

FIG. 2.25. The sun rising out of the plane of the equinoctial. (a) The sun's center in the equinoctial plane, represented for simplicity as occurring at noon; (b) the sun's center a little later at positive declination δ. B is the center of the front of the equatorial ring.

rises toward the north, the light vanishes from the bottom of the hoop and occupies more and more of the top (details in Appendix A).

The day of the equinox of 1574 was perfectly clear and bright. Danti brought "many gentlemen" to the Dominicans' convent church to see the play of light on the brass hoops some twenty feet above their heads. They saw the sun's rays vanish from the bottom of the equinoctial hoop and advance slowly down from the top. "And at that instant, and a little before and a little after, the [center of the] sun was in the equinoctial, the sun's rays appeared as a subtle thread on the edges of the concave part of the equatorial armillary, above and below, and then, in an instant, the southern thread vanished and the northern began to grow." That happened on 11 March at 22 hours and 24 minutes, according to the astronomer's method of reckoning time.[26]

Astronomical time ran from noon to noon on a 24-hour clock. Hence Danti observed the vernal equinox of 1574 at what would be 10:24 A.M. modern style. (He erred slightly in placing his hoop in the plane of the equinoctial and in fact misplaced the equinox by almost two hours and a half.)[27] The modern style of time telling did not prevail in Italy during the sixteenth century, however, although it was well established then in Northern Europe. Northern Italians counted days as 24 hours beginning half an hour after sunset; a system more cumbersome and peculiar than the Bohemian, which, according to Danti, counted the 24 hours from sunrise. To assist in reading the time of equinoxes, Danti cemented another modernized ancient instrument onto the patient façade of S. M. Novella. This quadrant, like the vertical hoop of the armillary that followed it two years later, had to

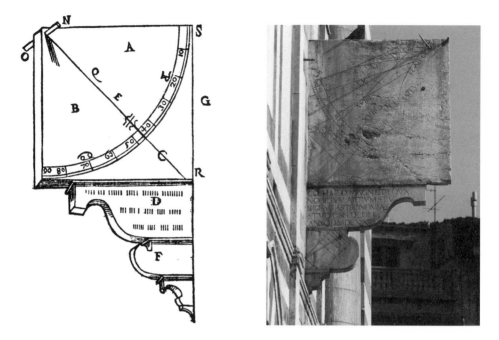

FIG. 2.26. The astronomical quadrant on the façade of S. M. Novella, Florence. (a) Danti's drawing (b) what remains. From Danti, *Primo volume* (1578), 282, and Comune di Firenze.

be placed exactly in the plane of the meridian. So situated, the narrow brass cylinder at its center threw shadows on sundials that told time in the astronomical, transalpine, Italian, and Bohemian manners. In Figure 2.26a, the quadrant, about 1.5 meters square, bears an Italian dial at A, a Bohemian at B, an astronomical at C, and a transalpine at F. Conversion of Italian to transalpine time became an important function of meridian lines.[28]

The main purposes of the quadrant were to observe the equinox, obtain the length of the year, and check the value of ε. The quadrant is shown at noon on the day of an equinox, when the shadow of the cylinder fell along the line QE, the intersection of the quadrant with the equinoctial. Then (Figure 2.27), $\angle$SP♎ = $\angle$Eq·PQ = the sun's noon height at the equinox = 90° − ϕ = 43°40′ using Danti's value for the latitude of S. M. Novella. (That the equinoctial intersects the horizon at an angle 90° − ϕ will be clear from Figure 2.24.) As with the armillary, repetition of the determination of the equinox would give a value for the length of the year. The magnitude of ε followed from measuring the angles SP♑ = WS·PQ and SP♋ = SS·PQ made by the shadow cast by the cylinder at the longest and shortest days. In an inscription on the quadrant, Danti gave the overprecise result 2ε = $\angle$SS·P·WS = 46°53′39″50‴, or ε = 23°26′49″55‴.[29] This was not a good value. Nor did Danti come very close to the length of the tropical year, which he set at

365^{d}5^{h}45^{m}36^s, 3^{m}36^s shorter than the Gregorian value. His instruments were more artistic than precise, and his observations of little use in comparison with later ones.[30]

Danti's instruments were in place in the spring of 1574. Cosimo had by then cleared with the Pope his intention of perfecting the calendar and Danti had acquired useful, and potentially important, additional patronage by cultivating Cosimo's brother, Cardinal Ferdinando. The instrument of this cultivation was a large and beautiful astrolabe, said to be the best ever fashioned in Italy, which Danti made, or caused to be made, as a gift for Ferdinando.[31] Earlier, he had procured a fine Mercator astrolabe for Cosimo. These princely objects figure occasionally in the exchange of gifts between an astronomer and his patron; for two other examples, Regiomontanus gave one to Bessarion and Pier Vincenzo Danti, né Rinaldi, gave another, said to have been his own handiwork, to Alfano Alfani, the pontifical treasurer of Perugia.[32] The princes had as much trouble understanding how an astrolabe works as ordinary people do; Danti tried to smooth their way with a treatise on the instrument, which he was allowed to dedicate to Cardinal Ferdinando.[33] In short, all signs presaged that the mobilization of the Medici patronage system would produce great works in astronomy—all signs, that is, except Cosimo's deteriorating health.

Cosimo's death in 1574 brought to power his son Francesco, a Caesar who did not want a Sosigenes or anything else that reminded him of his father. Despite the protection of Cardinal Ferdinando, Danti's position at court was irreparably undermined. The spiteful and envious, who had muted their chorus during Cosimo's

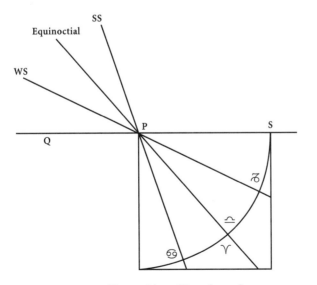

FIG. 2.27. The working of Danti's quadrant.

reign, renewed their song of scandal. Bowing to his and their desires, Francesco pressured the Dominican General into ordering Danti to repair to a convent outside of Tuscany within 24 hours. The General's order has the date 23 September 1575, about two weeks after the autumnal equinox.³⁴ The interval between Francesco's accession and Danti's ejection may indicate an agreement that Danti could remain in Florence to observe the equinoxes of 1575. The spring's was as clear and beautiful as that of 1574.³⁵ The fall's appeared to mark the beginning of an eclipse. But it is hard to read the heavens. Danti's superiors reassigned him to perhaps their noblest house, San Domenico in Bologna, containing a rich library, works of Michelangelo, and the bones of the founder of the order.³⁶

Bologna

· THE LINE ·

Even if Danti had taken great pains in the work, the instruments mounted on the face of S. M. Novella to advertise Cosimo's coming amendment of the calendar could not have been placed or read with sufficient accuracy to give a close estimate of the length of the tropical year.³⁷ Another approach was required. Shortly before Cosimo died, Danti knocked a hole or two in the great circular window in the upper story of S. M. Novella, 21.35 meters above the ground (Figure 2.28). On the floor of the church he laid out a line running to the north from a point immediately under one of the holes. The line and the hole therefore defined a meridianal plane: the sun's rays, entering the church at noon, threw an image of the solar disk onto the line—as if the church were a gigantic camera obscura. As the seasons changed, the sun's noon image would have run up and down the line, falling closest to the façade on midsummer day, when the sun stands highest in the sky, and furthest from it six months later, at midwinter noon. A line engraved on a paving stone indicted the position of the sun's lower limb at the winter solstice of 1575. Dante expected to determine the length of the tropical year by counting the number of days between reappearances of the sun at the same equinox. But "owing to the death of Grand Duke Cosimo, [the installation] was not finished."³⁸

Danti did not complete his *meridiana* (as the hole and line were called) at S. M. Novella. Had he finished it, he might well have obtained better results than he found with the armillary and the quadrant. Figure 2.29 presents the geometry: the sun's disk, subtending an angle of 30' at the hole H, makes an image centered at C on the line SN; AB indicates the diameter of the intense part of the image along the line. The matters to be measured were the position of the image at times of as-

FIG. 2.28. Sketch of the façade of S. M. Novella, Florence, showing the placements of Danti's armillary, quadrant, and gnomon. From Righini-Bonelli and Settle, IMSS, *Annali, 4:2* (1979).

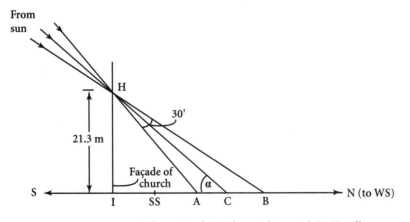

FIG. 2.29. Schematic of Danti's *meridiana* at S. M. Novella; C marks the center of the sun's image.

tronomical interest, like the equinoxes, and its size at different times of the year. A few numbers will give an idea of the scale of observation. Between the solstices, the sun's image traveled 47.6 meters, half the length of an American football field, down the nave of the church. The image of the midwinter sun fell over five times as far from the entrance to the church as that of the midsummer sun. The size of the diameter AB increased even more steeply, from 0.21 meters at midsummer to 1.56 meters six months later, a change by a factor of almost 8.[39] The noon image of the lower limb of the midwinter sun just touched the edge of the choir of the church as rebuilt by Vasari.[40] That no doubt explains the choice of 21.35 meters for the height of the hole in the window. It also provides a principal rationale for using great churches for solar observatories. Their size made possible the installation of very long measuring devices capable of revealing fine details about the size and motion of the sun.

Danti expected that the sun's noon image on the day of the first vernal equinox after he completed his *meridiana* would be centered at a point close to 22.37 meters from the façade ("close to" because the equinox can occur at any time of day). It would then move south, coming to within 10.36 meters of the front wall on midsummer day; recede toward the equinoctial position; continue beyond it a long way, to 57.96 meters on midwinter day; and move once again toward the equinoctial station. On the anniversary of his first spring observation, Danti would have watched to see how closely the sun's image came back to its initial position. Thus, he said, it would be easy to find the year and the true time of Easter, "which is almost impossible to do with any other instrument."[41]

The *meridiana* of S. M. Novella was not the first designed for a church in Florence. In 1475, almost a century to the day before Danti poked a hole in the rose window of his convent church, Toscanelli had done the same in the newly completed lantern of the vast cathedral of S. M. del Fiore (Plate 1).[42] Toscanelli was a friend, advisor, and mathematical tutor of the dome's architect, Filippo Brunelleschi, who died in 1456.[43] No piece of writing from Toscanelli has come down to us to indicate why he or Brunelleschi wanted to make a *meridiana* in the cathedral. But we know their purpose from Toscanelli's friend Regiomontanus. It was to check whether the inclination of the earth's axis changes over time.[44]

The Florentine Duomo, unlike S. M. Novella, is oriented in the manner prescribed for Catholic churches, with its main axis east-west. Toscanelli's *meridiana* therefore had to lie across the church, in the transept. Because the dome is so high (Toscanelli's hole is 277.3 Paris feet or over 90 meters above the pavement), an entire meridian line could not fit within the transept. He contented himself therefore with a short piece running from the main altar (Figure 2.30) into the north wall of the transept and usable only for a few weeks on either side of the summer solstice.[45] Even so, it has a length of almost 10 meters. The fact that the sun's image

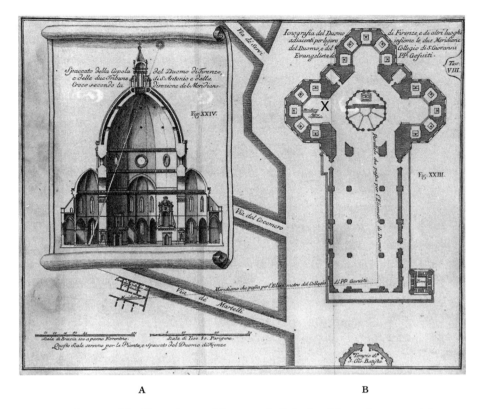

A   B

FIG. 2.30. Layout of the Florentine *meridiana* as rebuilt in the eighteenth century: (A) a ray
through the lantern; (B) plan of the church, showing at X the short section of the line in the
north transept. The dashed line running from left to right across the lower part of the figure
indicates the meridian of longitude passing through the *meridiana* in the Jesuit college north
of the cathedral. From Ximenes, *Gnomone* (1757), plate viii.

could be observed at only one solstice gave rise to an ambiguity: a shift in the location of the solstitial image might indicate not that the ecliptic was changing its
obliquity, but that Brunelleschi's masterpiece was moving. Indeed, the architects
and engineers responsible for the maintenance of the cathedral read the *meridiana* to determine the state of the building, not of the ecliptic, while few astronomers tried to use it for its original purpose.[46] Since Danti wanted to make
unambiguous measurements at the equinoxes as well as at the solstices, he could
not rely on Toscanelli's instrument and turned, though without consummation, to
the conveniently south-facing church of S. M. Novella.

While at leisure in Bologna, Danti built a small *meridiana* (height = 4 m) in the
chamber of the Inquisition in his convent. It runs — it has recently been restored,
although none of Danti's original installation is preserved — some 6.4 meters along
the floor before vanishing up a pillar.[47] Danti needed a larger building with an un-

obstructed view to the south and an orientation sufficiently close to east-west to accommodate an entire line, from one solstice to the other, on a scale big enough to keep alive his hope of determining the year and deciding the obliquity of the ecliptic. He found the church he required in the heart of Bologna, opposite the ancient university: the great basilica of San Petronio, begun at the end of the fourteenth century and not yet furnished with a finished façade, a condition in which it remains today.

The interior of San Petronio offered more than a good technical setting for solar observations. Its huge open nave, paved in the traditional Bolognese style of ruddy and creamy marbles, made it a perfect theater for the daily rendezvous of the sun's image with Danti's line. Plate 2 gives an impression of this extraordinary interior, which, according to architectural historians, had no close precedent, but rather hovered between a dying late Gothic, without the extravagance of the cathedral in Milan, and an emergent humanism. It is a harmonious composition of opposites: light and shadow, to be sure, but also "monumentality and clarity, solemnity and delicacy, a power at once supple and diaphanous."[48] In the wall of a side chapel of this solemn monument, Danti put a hole and, on the floor, a line, all of marble, embellished with plaques marking the sun's entry into all the zodiacal signs.[49]

To penetrate San Petronio Danti required the permission not of clerical authorities, but of five lay custodians of the building (*fabbricieri*). Four of these gentlemen were chosen by the Senate of Bologna from among its membership for two-year terms; the fifth, the *presidente perpetuo,* was a senator appointed for life by the Pope, who, since 1506, had been the ruler of Bologna. The Pope exercised his jurisdiction through a Legate, usually a rising cardinal, who divided authority with the Senate, or Reggimento, which had forty (after 1590, fifty) members drawn from leading families in which Senate membership was hereditary. The Legate had charge of the justice system and public order and the Senate had power over finance and general administration. The Senate also had the rights to send its own ambassador to Rome and to maintain a small standing army.[50]

During Danti's time in Bologna, the president of the fabric (the chairman of the committee responsible for the church as a building) was a strong-willed character named Giovanni Pepoli. He ran roughshod over his fellow *fabbricieri* and made the mistake of treating the Legate in the same way. He refused to hand over a bandit caught on his property, perhaps because he was in league with the felon. The Legate reported to the Pope, the decisive Sixtus V, who decided to let the Senator suffer the bandit's punishment. Pepoli thus immortalized himself as the first Senator of Bologna to be executed. With the bullheadedness that was to cost him his life, he gave Danti permission to make San Petronio into a camera obscura without consulting his colleagues.[51]

There was, alas, one detail about the fabric over which Pepoli had no power. The

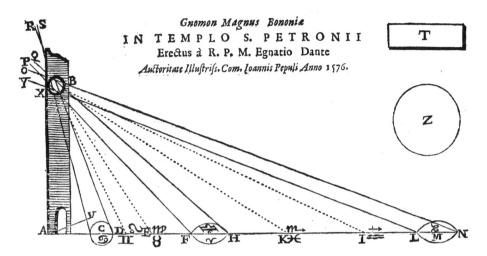

FIG. 2.31. Danti's *meridiana* in San Petronio. T is the cross-section, Z the size and shape of the hole. From Riccioli, *Almagestum novum, 1:1* (1651), 132.

architect had placed the piers that supported the nave where they would block the sun's rays from reaching the pavement around noon at some season of the year no matter where the hole was made in the roof. Unable to make the piers both supple and diaphanous, and unwilling to abandon San Petronio, Danti did the best he could, and ended up with a line that ran more than 9° off true. Figure 2.31, taken from the *Almagestum novum,* the Jesuit guide to astronomy of the mid-seventeenth century, gives a rough idea of Danti's meridian in San Petronio.[52]

We learn from the same source that the hole at B stood 65 feet and 9.25 inches Bolognese measure above the rosy pavement. That would put the image of the equinoctial sun about 65 feet from the point directly under the hole (φ for San Petronio being about 44°30'). Users of the instrument disagreed among themselves by as much as half a foot about where the center of the equinoctial image crossed the line. Now half a foot at 65 feet corresponded to a difference in declination of about 13 minutes.[53] At one minute of declination to one hour of time, the observers of the equinox at Danti's *meridiana* disagreed about the moment of its occurrence by over half a day. Danti was correct in theory when he wrote, in a broadsheet explaining the uses of the meridian line, that its great size made possible observation of fine detail, and the fixing of the solstitial and equinoctial points, "more accurately and more easily than with any other astronomical instrument." It also made possible, again in theory, a determination of the relative distances between the earth and the sun at different times of the year.[54] In practice, only rough results, like the finding that the winter solstice of 1576 fell on 11 December, came from Danti's meridian line at San Petronio. The 9° deviation in itself did not destroy Danti's project, since it scarcely would have affected determination of the length of the year. But he had not leveled the line carefully, the plate holding the hole slipped (a common problem with meridian lines), and the location of the foot of the perpendicular was uncertain.[55]

It is not easy to make an accurate *meridiana.* An apposite illustration of this proposition, and an episode also instructive in other respects, is the line laid down in 1636, in Marseilles, by two people who knew what they were doing. They were the priest Pierre Gassendi, professor of mathematics, restorer of atomism, and, in consequence of his observation of the transit of Mercury across the sun's disk in 1631, one of the leading astronomers of Europe; and his patron, Claude Fabri de Peiresc, likewise an experienced observer, the first man in France to see the moons of Jupiter and the first man anywhere to find, via lunar eclipses, a good value for the length of the Mediterranean Sea.

In 1636 these gentlemen attended to a request from Gottfried Wendelin, another priest-astronomer and onetime protégé of Peiresc's, to measure the sun's altitude at Marseilles on the day of the summer solstice. Wendelin wanted to compare their measurement with one made there by the Greek geographer Pytheas during the reign of Alexander the Great in order to determine whether the obliquity of the ecliptic had changed in two thousand years. Gassendi and Peiresc decided to make a *meridiana* for the purpose, since, Gassendi wrote to Wendelin, in words reminiscent of Danti's, "when it is leveled and adjusted to the perpendicular, there is nothing more exact."[56] After searching Marseilles for a proper site, they settled on the Oratorian College, which was big, dark, unobstructed to the south, and unfinished. Only one intermediate floor had been completed. Peiresc asked that part of it be removed to allow the sun's rays to reach the ground through the hole he proposed to make in the new roof. No request was too great when Peiresc made it. "The excellent fathers agreed to have it [the floor] broken into and the city officials, under whose auspices and at whose expense the college is being built, ordered that it be completely demolished."[57]

The diameter of the hole was about half an inch; its height above the pavement, about 52 feet, as measured by sticks (strings and wires being unreliable because they stretch under their own weight); and the direction of the line, precisely north-south, as determined by a good compass corrected for the variation. The careful measurement of the sun's solstitial altitude at Marseilles in 1636, when compared with Pytheas', showed a decline in the obliquity of around a minute a century. In a shower of learning, Gassendi showed that other determinations of the obliquity, both ancient and modern, allowed the inference that it had not changed at all. It was only necessary to suppose that Pytheas had been mistaken by about a third of a degree. And that, in Gassendi's judgment, was not the least unlikely, the Greeks in general being very unreliable.[58]

It turned out, however, that Pytheas chanced to be nearer the mark than Gassendi and Peiresc. Not knowing that the magnetic variation changes over time, they had not corrected properly for it. Their line had declined from the truth by about 3°, as Gassendi informed Wendelin seven years after the fact.[59] That did not change their value of the obliquity by much. It amply demonstrated, however, that they had not taken sufficient pains, or lacked the skill, to make an accurate *meridiana*. Subsequent meridian builders eschewed the compass and learned from one another's experience.

· OTHER LINES ·

While engaged on the *meridiana*, Danti did not neglect other instruments for the advancement of science and himself. He made a replica of the armillary of Santa Maria Novella for the cardinal-archbishop of Bologna, Gabriele Paleotti, a specialist in, and in implementing, the decrees of the Council of Trent. This worthy disciplinarian, restorer of churches, founder of seminaries, hospitals, and charities, thought Danti's armillary, an instrument of ancient exact astronomy, a fit centerpiece for the courtyard of an episcopal palace. Danti gave anemoscopes (wind gauges), his particular and favorite invention, to Paleotti and other influential people.[60] After he had passed a year in Bologna gathering patrons, the university decided that it should seize the opportunity of adding him to its staff of the "most eminent professors." The Reggimento obtained from its temporal and spiritual lord, Pope Gregory XIII, and the Dominican General permission to establish a second chair of mathematics and to offer it to Danti. He was appointed by letter dated 28 November 1576 to the "afternoon chair," so called because its incumbent would lecture only in the afternoon. Danti taught the usual range of "mathematics": arithmetic, geometry, astronomy, perhaps navigation, and certainly fortification, on which he wrote a treatise explaining how to lay out ground plans in the shape of regular polygons.[61]

Again his brethren desired to clip Danti's wings, this time by insisting that he teach in the Dominican college rather than in the university. But the university was unwilling to surrender so illustrious a scholar, whose lectures, we are told, drew an unprecedented seventy or eighty auditors. The Senate of Bologna mobilized three cardinals and the Pope to defeat the Dominicans. The mobilization resulted in a letter from their General confirming Danti's seconding to the university and an intervention from the Pope's *nipote* (nephew), Jacopo Boncompagni (to whom Danti dedicated his *Scienze matematiche* of 1577), which raised his salary.[62] Professors of mathematics in those days were expected to make themselves useful to the municipalities that hired them. Danti served the Senate by helping with

FIG. 2.32. The radio latino in use. BC is a plumb line, AB and AC adjustable
slats. From Danti, *Radio latino* (1583), fig. vi, p. 27.

works to control the Reno, Bologna's major river, and by making surveys related to
boundary disputes.

In 1577 Danti was in Perugia attending to his sick brother Vincenzo. He left his
usual calling cards: an anemoscope for the Palazzo dei Signori and another for the
pontifical governor. The governor and the municipal council (Signoria) responded
by appointing him to make a survey and cadaster of Perugia, including all natural
and artificial points of interest, mountains, rivers, castles (all 223 of them), and

also the quality of the air, water, land, and local dignitaries. Danti completed the survey in a month, scarcely dismounting from his horse, using his rough-and-ready methods, some earlier work of his father's, and a universal measuring device, the *radio latino,* "an instrument so well and cleverly made by its illustrious inventor [Latino Orsini] that even a mediocre performer will use it perfectly."[63]

Figure 2.32 suggests how Danti used this rare instrument. "Great and marvelous are the excellence of mathematical instruments [thus Danti, in praise of his *radio*] since they make known to us things that appear not only most difficult, but even impossible to believe." He had in mind finding the heights and widths of things at a distance. This lavish praise of instrumental aids to sublime, if practical, knowledge appears in the preface of Danti's edition of Orsini's description of the *radio,* which he published in Rome in 1583. There he signed himself "Maestro in theologia & cosmografo di N.S.P.P. Gregorio XIII," master in theology and the Pope's cosmographer. The *radio* was the instrument of his promotion from professor to pontifical mathematician.

Danti published his report to the Signoria on the situation of Perugia by painting it 15 feet square, in color, on a wall of the governors' palace.[64] It proved a great success. Danti presented a copy or draft of the map to the pontifical nephew, Boncompagni, who arranged a commission for a similar survey of the entire Papal States and supplied the necessary servants and helpers. Danti received this commission in 1578 while still teaching at Bologna. Again he distinguished himself both by his speed and by his virtuosity. By 1580 he had done the Romagna, Bologna, Perugia, and part of Umbria. The Pope decided to bind him more closely to the Vatican. In 1581, by order of Gregory XIII conveyed through Boncompagni, Danti was called to Rome and lodged in the apostolic palace. His agreeable assignment: to advise about the calendar and to oversee the painting of maps of regions of Italy on the walls of the Vatican's Belvedere gallery.[65] This time the Senate of Bologna was on the losing side. It directed its ambassador in Rome to do everything he could to retain Danti as a professor at its university.[66] He could do nothing. The Pope wanted Italy's top cosmographer in Rome.

Rome

Popes had had cosmographers since at least the middle of the fifteenth century, when Aeneus Silvius Piccolomini (Pope Pius II) engaged map makers to help him keep track of the results of Portuguese exploration down the west coast of Africa.[67] The practice of painting maps on walls of palaces has been traced back as far as 1342, to the Palazzo Ducale in Venice; but it did not flourish until the great age of cartography following on the discovery of the New World. There survive a series of six large maps, including a *mappamundo,* at the Palazzo Farnese in Caprarola, begun in 1529 and finished about half a century later; the murals in Cosimo's wardrobe; and several sets in the Vatican, of which those designed by Danti beginning in 1580 were the last. A proposal of 1507, that Bramante should paint a map of Italy in one of the rooms of Julius II, unfortunately came to nothing. The Medici Pope Pius IV made a good start in 1559, with a dozen maps of Europe based largely on Mercator's latest publications, painted on the walls of the west wing of the Terza Loggia above the Belvedere, the gallery that connects the Pinacoteca with the Sistine Chapel. They were finished about the time that Danti began to work for Cosimo.[68]

The rationale of papal interest in world geography is not far to seek. Maps indicated wide open spaces for the propagation of the faith; popes favoring missions to foreign places, like Gregory XIII and Urban VIII, embellished their palace with cosmographical indications of the church's claim to universality.[69] Danti was Gregory XIII's chief instrument in achieving a major cartographical project in three parts. One concerned the scenes and targets of missionary work. It comprised a huge *mappamundo* and separate large representations, according to the best available information, of Africa, China, the Indies, and the New World. Danti not only designed these maps, but probably also did the murals. They deteriorated as the missions advanced. Little or nothing of Danti's work survived restoration except parts of the *mappamundo,* which in its time was the "most important of all the geographical pictures in the Vatican."[70]

The second project was to decorate the walls of the Belvedere with forty maps of regions and towns in Italy. Danti made all the designs, most of which he had completed by the end of the year, as appears from a letter of his dated Christmas Eve, 1580, to the Dutch dean of European cartographers, Abraham Ortelius. The letter describes the layout of the corridor, asks whether Ortelius would be interested in publishing a book of the maps, and includes, as a gift and a specimen, a copy of Danti's map of Perugia. Ortelius did not engage for the book, but he published Danti's "Perugino agri, exactissima novissimaque descriptio" in his great atlas, the *Theatrum mundi,* which also has a map of Latium done after one of Danti's murals in the Belvedere.[71]

The gallery is 150 meters long. Danti imagined it divided longitudinally by the Appenines and placed depictions of regions toward the east on one wall and toward the west on the other. Most of the maps are large, about 3.3 m × 2.5 m; all are marked with degrees of latitude and longitude; many are beautifully drawn and colored, with castles, churches, ships, and sea creatures as appropriate. They represent the transition between pictorial maps and the sterner cartography of the seventeenth and eighteenth centuries. Danti used astronomically determined coordinates whenever available and drew accurate coastlines where he could rely on the Venetian and Genoese charts or on his own measurements in the Papal States; but his interiors, especially the mountains, woods, and roads, are often only rough, though decorative, approximations.[72] They were not published in their entirety until 1952. Danti reproduced his design for at least one of them, however, the town of Viterbo, as a gift for one of its leading citizens. The provident courtier, like a building contractor, plans for the next patron before finishing with the current one.[73]

Although Danti used the best material he could find, the Belvedere murals naturally were not free from errors. Some arose not from faulty data but from congenital haste. Danti could not have devoted more than a week on average to the design of any one map if he finished all the cartoons by Christmas of 1580. The murals excited the admiration of everyone who saw them, except rival cartographers. One of these was Giovanni Antonio Magini, who had succeeded Danti as professor of mathematics at the University of Bologna. But, although he worked on his atlas of Italy for decades, Magini could not always do better than his predecessor, whose map of Perugia he adopted, and on whose fallible sources he had sometimes to rely. Substantial corrections and additions to the murals were made during extensive restorations commissioned a half-century later by Urban VIII.[74] Today, after several later restorations, the murals of the Belvedere make one of the unforgettable images of a visit to the Vatican.

The third project connected Gregory's maps with his calendar reform. He had had built in connection with the Belvedere a Torre dei Venti, a "tower of the winds," on the roof of which Danti constructed an anemoscope.[75] A wind rose (a compass showing directions of the prevailing winds) was a common embellishment on early maps. On the floor of the top room of the tower, beneath the wind gauge, Danti inscribed a meridian line. Plate 3 shows the line and the hole through which the sun's rays enter; the frescoes surrounding it, which depict biblical scenes, were designed by Danti according to an iconography he explained in detail in an addition to his book on astrolabes.[76] The line itself indicates the position of the sun's image at its entry into the zodiacal signs from Cancer only down to Scorpio/Pisces, the room being too small to admit the images for Sagittarius/Aquarius and Capricornus. To fix the position of Cancer, Danti used the value $\varepsilon = 23°29'$. The line declines from the north by only $1°10'$, which suggests that Danti could have done much better than he did at San Petronio had he not had to dodge the pillars.[77]

The main point on and about the *meridiana* in the tower was the equinoctial position of the solar image. According to an old story, Danti's demonstration to Gregory in the Torre dei Venti that the equinox occurred on 11 March rather than on the canonical 21st convinced the Pope of the need to reform the calendar.[78] In fact Gregory had decided on the reform in principle in 1577 and had received the report of his special commission on the calendar in 1580, before Danti's line went down in the Torre dei Venti. Although Danti signed the commission's report, he played only a minor part, and his *meridiane* none at all, in the creation of the Gregorian calendar.[79]

After the promulgation of the calendar and the completion of the mural maps, Gregory rewarded his computer and cosmographer with a bishopric, that of Alatri, not far from Rome. The appointment came on 14 November 1583, immediately after the death of the incumbent. Danti did not enjoy his office long. Gregory's successor, the redoubtable Sixtus V, called him back in 1586 to help correct a serious error of the ancient Romans. They had planted a huge obelisk that they had stolen from the Egyptians to the side of the then newly rebuilt Saint Peter's. Sixtus decided to move it to the center of the unfinished square in front of the great basilica.[80]

The engineer in charge of the project, Domenico Fontana, required seventy-four horses and nine hundred men to raise the huge monolith, lower it to a cradle, move it to its new location, and raise it again, without cracking it. To place all the capstans properly, he had to tear down some of the church. Danti's primary assignment was to install instruments at the obelisk's base that would indicate the solstices and equinoxes, and also the winds. They may never have been set up; no drafts or drawings of them have been found and nothing remains of them now. The slip between design and implementation derived from an interruption in Danti's usual method. He could not oversee the work. Obedient to the Tridentine decree that bishops reside in the dioceses, he returned to Alatri soon after the obelisk went up on 10 September 1586. He came down with pneumonia on the home journey and died in his hard-won see on 19 October.[81]

Danti's career demonstrates that competence in mathematics, including astronomy, could take its possessor far in the early modern Catholic Church, especially if combined with an ability to draw maps and to get along with princes. Danti had the support of a range of powerful people within and outside the Church: within, several cardinals, the Dominican General, the Pope's *nipote*, and the Pope himself; outside, a grand duke, governors of Perugia, senators of Bologna, and university officials. He changed patrons, from lay to clerical and back again, as opportunity and his superiors allowed. That his superiors understood his utility to his Order appears implicitly in his career and explicitly from a letter addressed to the Duke of Urbino in September 1576 concerning a treatise on the sphere Danti was then writing. "If I can finish it before All Saints I'll bring it on the way to Rome, where the General has sent me again to serve some signore."[82] From this scrap we see that the Duke of Urbino should be counted among Danti's potential or active supporters and that, once again, a piece of his work was a suitable gift to acknowledge or create an obligation.

This is not to say that Danti did not have a program of his own. From a letter of 1578 to a senior colleague in the Church, it appears that he had wanted to undertake the survey of the Papal States because, "if I succeed in this, it will be easy later to finish the entire chorography of Tuscany done in the same detail as I've done with the survey of Perugia." He enjoyed teaching and argued successfully with his General that he should be allowed to stay at the University of Bologna for the sake of himself, his students, and the Order: "Such teaching is most desirable and necessary; it is a marvelous ornament, and useful to the Studio."[83] The Dominican life offered a way to realize his objectives. "Thus defended and protected . . . from the dangers of the time, and from the hubbub of a world always an industrious inventor of new dissipations . . . it is no wonder that he made such easy progress in his favorite studies in the placid quiet of the cloister."[84] There may be some truth in that, except for the bit about the cloister.

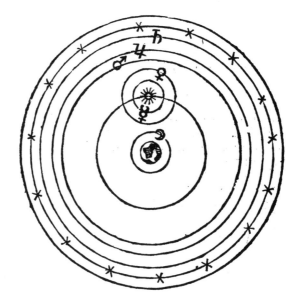

3: Bononia Docet

A New Oracle of Apollo

· IN TEMPLO ·

The Grand Cassini

During his education at the hands of the Jesuits of Genoa in the 1640s, Giovanni Domenico Cassini conceived so strong a religiosity that, but for the scruple that he had no calling for it, he would have entered the Church. Had he done so, he would have risen as high as Danti did, and perhaps higher: the Church needed the services of widely respected and demonstrably Catholic mathematicians all the more after its misguided silencing of Galileo than before. "It was a delicate matter," so judged Cassini's obituarist Bernard le Bovier de Fontenelle, the secretary of the Paris Academy of Sciences. "In Italy a learned ecclesiastic can reach so high a position he thinks nothing is above him; there is no other situation likely to pro-

FIG. 3.1. The Paris Observatory around 1675. Wolf, *Histoire* (1702), plate X.

vide such great rewards; but Cassini did not feel a calling, and the same piety that made him worthy of entering the Church kept him from it."[1]

Instead of finishing up an Italian archbishop, Cassini ended his days in Paris, where he had served as the Sun King's astronomer for over forty years. "He arrived in Paris at the beginning of 1669, brought from Italy by the King, just as Julius Caesar brought Sosigenes to Rome."[2] Established in Paris, he sired a dynasty of astronomers royal, who, beginning with himself, ran the Paris Observatory for over a century. His appointment gave him many opportunities to indulge that penchant for display first realized in his redoing of Danti's *meridiana* in Bologna in 1655. He had gigantic telescopes built in the courtyard of the Paris Observatory (Figure 3.1). He projected a trigonometrial survey of the kingdom of France, which his grandson and great-grandson completed. Like most great men, he was both admired and disliked. "That was because he had in him much to admire, because he worked hard, because he constantly held the attention of the public, [and] because he usually used extraordinary methods, like his gnomon [*meridiana*] and his long telescopes."[3] Despite his old-fashioned piety, he was modern in pursuing his profession: "He served science twice over, with great discoveries and the ability to advance them."[4]

Cassini owed his discoveries to the ability to see, and the power to recognize, the tell-tale detail. He owed the opportunity to deploy these qualities in the service of astronomy to the Jesuits. They deflected his youthful preoccupation with theology to mathematics, which then included astronomy and astrology. Attracted by the exactness of the one and the complexity of the other, he mastered both with the help of an abbot who taught him how to calculate planetary tables (ephemerides). These skills made a useful stock in trade. Many wealthy Italians then cultivated astrology and supported mathematicians able to improve the art; and many of them, like Cassini, shared their employers' convictions.[5] For a time. Eventually the passion for accurate prediction, realized in astronomy, could not survive in the same mind with tolerance for the fuzzy forecasts of astrology.

Cassini was losing his faith in astrology when he arrived at the villa of the Marchese Cornelio Malvasia, Senator of Bologna, generalissimo of the armed forces of Modena, and onetime commander of the galleys of Urban VIII. In these hazardous employments, Malvasia welcomed any reliable indication of the future, and he used to draw up an astrological almanac every year for the guidance of himself and his friends. In his retirement he had undertaken a more serious project. As he described it, in a military metaphor later popular among physicists, "I rather changed arms than abandoned them when I began my celestial expedition: I pitched a new camp, aimed new machines, not loaded with iron balls but fitted with glass lenses, at the heavens; and I set out new patrols, to seize the roving and fleeing planets, to chain them with perpetual laws and, at last, to subjugate them to human knowledge."[6] To speak plainly, he determined to set up an observatory in his villa, improve astronomical tables, and apply the results to astrological forecasts.[7] He brought Cassini to help him. One day—the day predicted in Malvasia's forecast for the year—a terrible storm hit the villa. Cassini suggested that they redo the calculations. The upshot: the predicted storm had been made to coincide with the real one by a printer's error. According to Cassini's autobiography, the realistic old soldier thereupon surrendered his conviction and proclaimed that "only astronomy deserved attention."[8]

Among Malvasia's visitors were the Jesuit astronomers Giambattista Riccioli, who then, at the age of fifty, was seeing his monumental encyclopedia of astronomy, the *Almagestum novum,* through the press (it took three years to print), and his collaborator Francesco Maria Grimaldi, then thirty-two, who has enduring fame in the history of science as the discoverer of optical diffraction. Both despised astrology, but, reasoning Jesuitically, did not attack it lest they weaken support for astronomy; and, acting encyclopedically, inserted into the *Almagestum novum* an account of its technical apparatus, a consideration of Copernicus' horoscope, and a warning that prognosticating had been condemned by the Pope. "If the public did not believe in astrology," they said, "books on astronomy would not sell."[9]

Through Riccioli and Grimaldi, Cassini came to the attention of the mathematicians at the University of Bologna. Danti's old chair had not been reassigned since the death of its last incumbent in 1647.[10] That was Bonaventura Cavalieri, one of Galileo's most able disciples, a creative mathematician and the first person, it is said, to have taught Copernicus publicly in Bologna.[11] The university was persuaded to appoint Cassini to this prestigious post. He began teaching there in academic year 1651–52, at the age of twenty-five or twenty-six.[12]

The Mathematicians of Bologna

Bologna was then one of Europe's leading centers for mathematics. The bookstores were full of the subject, especially its applied arm, astrology. Every year one of the university's three professors of mathematics had to publish an astrological almanac, whether he believed in his forecasts or not.[13] Cassini's tenure lasted until he left for Paris in 1669. Both he and the university were sufficiently pleased with his service (he avoided the almanac duty) that it retained him in its course list until his death in 1712.[14]

Cassini's first colleagues were a Carmelite priest, Giovanni Ricci, followed in 1665 by Geminiano Montanari, who came from Malvasia's finishing school. A native of Modena, Montanari had earned degrees in civil and canon law at the University of Strasbourg in 1656. A most unlikely encounter in Vienna with one of Galileo's last direct disciples turned him toward mathematics. Returned to Italy, he began practicing law in Florence in 1659 and again fell in with a Galilean set. This was the famous Accademia del Cimento, the little band of experimental philosophers kept by Cardinal Leopoldo dei Medici. With this preparation, Montanari went back to Modena as philosopher and mathematician to its Duke, Alfonoso IV d'Este, who sent him to work with Malvasia. Montanari lived and worked in Malvasia's villa in 1662 and 1663, when he moved to the university.[15] He was a good technical astronomer (he was one of the inventors of the micrometer eyepiece), a good observer (he and Cassini were perhaps the first to see, though not to identify, variable stars), and a good organizer (he set up his own small group of experimenters, the Accademia della Traccia, which, after several transformations, became the nucleus of the Bologna Academy of Sciences).[16]

Montanari was an outspoken opponent of astrology. He had had to practice it to conform (as he put it) "to the custom that students of mathematics must learn it, however false or useless it may be, and unworthy of comparison with true mathematics, because it is the standard by which the public calculates the value of men in this profession." Against the fatalism of strict astrology, he pointed to his own free choice in leaving law for mathematics.[17] No doubt he enjoyed *Astrological jokes,* published in Bologna in 1667, a facetious concoction of advice and prediction that breaks off "for fear of terrifying the world." He later issued a similar hodgepodge to show that false forecasts were no less accurate than ones deduced from the rules of art. But whereas the earlier joke book aimed to wean doctors from their reliance on astrology, Montanari's had the higher objective of reinforcing a sense of personal responsibility as taught by the true religion. "Without free will [he wrote, with himself in mind], what remains to distinguish men not only from beasts . . . but even from stones?"[18]

The third professor of mathematics was Pietro Mengoli, a student of Cavalieri's, who began teaching at the university in the 1650s. His specialties were mechanics and a peculiar axiomatized theology, which he "built into an infallible science from concepts of the divine attributes."[19] He held unusual ideas, for example, that the moon is a perforated, sponge-like, hemispherical vase; that the sun slowed from its original speed, which took it around the earth in exactly 365 days, to its present pace after it stood still for Joshua; and, boldest of all, that Cassini had made errors at San Petronio. Mengoli's odd ideas did not recommend him to his colleagues. He explained his intellectual isolation as a consequence of "the lowness of my birth and the frankness of my conversation"; and also of the opinion of the local booksellers that "my books are unreadable."[20]

It was possible to learn something from these three men, especially from Montanari, who gave special attention to teaching.[21] They delivered the elements of their science from the good old books, like Euclid's geometry and Ptolemy's astronomy. But they also allowed themselves to bring their subjects up to date.[22] Several of their students were to work in or build church observatories.

The predominant authority among the strong-minded mathematicians in Bologna belonged not to the university, however, but to the Jesuit College of Santa Lucia. This was Giambattista Riccioli, born in 1598 in Ferrara, a Jesuit from the age of sixteen. He had the stature and also the mantle of Clavius, having learned his mathematics at the Jesuit college in Parma from Clavius' student Giuseppe Biancani, the first Jesuit to write an astronomy based on the Tychonic model. This text, *Sphaera mundi*, dates from 1620; Riccioli belonged to the first generation of Jesuits taught astronomy in a moderately modern manner. Biancani's modernity did not stop with Tycho. Like Bellarmine, he criticized Aristotle's ideas about the construction of the heavens and, unlike Bellarmine, favored some of Galileo's. In none of this, apparently, was Biancani opposed by his colleagues; "the attitude of the philosophers at Parma corresponded to the maximum possible openness (if not adherence) to non-Aristotelian positions."[23]

Riccioli left Parma in 1636 to teach scholastic philosophy in Bologna. He protested the transfer, which the Jesuit General ordered because he thought that the friendship between Riccioli and Grimaldi, then still a student, had grown too intimate. The transfer proved productive. The unusually high valuation of mixed mathematics, and the relatively tolerant attitude toward cosmology, that Riccioli had learned from Biancani helped him to profit from two resources in Bologna unavailable in Parma: Cavalieri, with whom Riccioli formed a firm friendship, and the university library, in which he found books he had only read about. Through Cavalieri he had an indirect contact with the work of the Galilean school, including the use of a telescope made by Galileo; through the library, the mind-bending experience of reading Kepler and other foreign moderns. He took it all in, but could give little of it out: just as his horizon was widening, the proscription against teaching heliocentrism accompanying Galileo's condemnation restricted his freedom of expression.[24]

Riccioli began to fulfill his ambition to provide his order with an encyclopedia of astronomy around 1640, while still professor of scholastic philosophy in Bologna. The huge work was ready for submission to the Society's censors by 1646. They gave it some trouble, not because they opposed the study of astronomy, but because they feared that Riccioli, who had been trained to teach theology, might not know enough to complete his project to the honor of the Society. "What new might be brought out after so many great practitioners — Tycho, Kepler, Lansberg [all Protestants!] — have devoted their lives to the subject with the backing of the Emperor?" The censors demanded that Riccioli send them, separately, whatever he had written about his own innovations and a full account of the instruments he had used. He replied that his innovations were scattered throughout the book, as the subject required, and could not be extracted without great expense and trouble; that whatever he had contributed "in no way would tarnish the glory of Tycho"; that he had used the best instruments, including Galileo's telescope and Danti's *meridiana;* and that, since his introduction to astronomy by Biancani, he had given it more attention than he had theology.[25]

The documents in the case also hint at a concern by the censors that Riccioli's need for large sums to make or acquire instruments, hire engravers, and engage assistants was diverting money that might otherwise have gone to support more worthy projects. Riccioli replied that he had received what he needed from his wealthier students, from the very rich relatives of Grimaldi, from the rector of his college, and from individuals devoted to astronomy like Malvasia. This list as well as the objection it answered suggest not only the opportunities open to a well-placed Jesuit with a gift for getting gifts but also the constraints in using the proceeds imposed by the corporate character of the Society.[26]

Riccioli's replies, and the advocacy of Athanasius Kircher, the influential polymath professor at the Jesuits' college in Rome, convinced the censors to approve the great project.[27] The Jesuit General switched Riccioli's teaching assignment from theology to mathematics and, beginning in 1649, to nothing at all, to give him time to make revisions and supervise the printing. Nonetheless, satisfying what he called his "addictions" — astronomy, chronology, and geography — claimed more time than he had to give.[28] The Society supplied him with a collaborator in the agreeable form of his great friend Grimaldi, who by this time was suffering from tuberculosis. When Grimaldi became seriously ill, his superiors transferred him from the arduous teaching of philosophy to the easy life of mathematics. "And so Divine Providence gave me, although most unworthy, a collaborator without whom I never could have completed my [technical] works." Thus Riccioli, in a eulogy of Grimaldi, who succumbed to his disease in 1663. "I could never express in words what I owe him." Cheerful, serious, affable, pious, accurate, reliable, Grimaldi earned from his fellow Jesuits what must be the rarest of their praises: "Vixit inter nos sine querela," "he lived with us without complaining."[29]

The encyclopedia of astronomy, issued as one volume in two parts in 1651, bore the title *Almagestum novum* as an indication of its up-to-date obsolescence. It is a deposit and memorial of energetic and devoted learning. It was also an introduction to the research frontier, which, in Riccioli's opinion, surrounded astronomy. "Not one part of [it], as it has come down to us, is complete." Among the most pressing problems he mentioned were several suitable for investigation at cathedral observatories: the sun's apparent diameter, the length of the year, the obliquity of the ecliptic. No serious astronomer could afford to ignore the *Almagestum novum.*[30] To pick one example of its circulation to stand for many, John Flamsteed, the Astronomer Royal of England, a Copernican and a Protestant, relied upon it for much of the information in the lectures on astronomy he gave in the 1680s. Grimaldi contributed substantially to the *Almagestum,* which contains many of his observations and some of his writing, and which he saw through the press "with tireless labor, inviolable faith, sharp judgment, complete candor, and obstinate vigilance."[31]

Among the enduring contributions of the partnership of Riccioli and Grimaldi was a map of the moon, the most detailed and accurate of its time, drawn from their own observations and their corrections of descriptions by others. Following a convention proposed by Gassendi and first realized by the Belgian cartographer Michael van Langren, they called the observable features after famous astronomers, including themselves, although, to be sure, they modestly gave their names to spots much smaller and less central than the great craters they assigned to Ptolemy and Copernicus. Danti has his little pit, as does Sacrobosco, Brahe, Clavius, Magini, Galileo, and Kepler.[32] Contrary to van Langren, Riccioli (for it was he who devised the scheme) chose his nomenclature without regard to religious confession and omitted reference to Catholic sovereigns that might have provoked Protestants into a war of names.[33] All the astronomers active before 1650 who played a part in the story of meridian lines had, and, happily, still have, a place on the moon.[34]

These few ingredients may suggest that the *Almagestum novum* was a landmark in more ways than one. Not only did it provide a compendium of astronomy as practiced in 1650, it also indicated how far by then the Jesuits had moved from the strict teaching of Aristotle. Riccioli and Grimaldi showed no reticence about treating the moon as if it were made of earth or about honoring cosmologists, even Protestants, who subverted the geocentric system the Jesuits were required to defend. The Order tended to tolerate widely accepted old innovations. As its censors wrote in 1649, concerning the teaching that the heavens are fluid: "This doctrine, as it is now common, can be allowed, although it was prohibited in the Society when it was not yet common."[35]

As Riccioli told the censors, the *Almagestum novum* contains a great many observations made by himself and his students and assistants. Most of the instruments they used had been given to their college. One of these instruments was the college church. Riccioli and Grimaldi took advantage of its unfinished vault to set up a temporary *meridiana* some 23.7 meters high with which to determine their latitude.[36]

The Grand Meridiana

The *fabbricieri* of San Petronio were in the middle of a major building campaign when Cassini took up his professorship in Bologna. Their plan removed the wall pierced for the *meridiana* that Danti had installed seventy-five years earlier. They wished to retain this unusual and perhaps useful ornament. Since, despite the Council of Trent's order that the Archbishop of Bologna should oversee the material as well as the spiritual improvement of San Petronio, the *fabbricieri* had retained control of their cathedral, nothing prevented them from asking Cassini to reset the instrument.[37] He replied that Danti's line was too far from true to be salvaged for anything other than decoration. Instead, he proposed to put another hole in the roof higher than the old one and run a line due north, avoiding the nave piers. He had carefully compared the architectural plans of the church with his own survey and insisted that he would miss every obstacle. Moreover, he planned to put his hole not in the new part of the nave, which he feared might settle, but in the old fourth vault, at a height that allowed the entire length of the line to come within the church.[38]

The *fabbricieri* worried about the cost and the practicability of Cassini's project. They consulted other members of the prolific tribe of Bolognese astronomers. Not having proposed the project themselves, the mathematicians doubted that it could be done. Cassini invoked Riccioli and Senator Malvasia, who persuaded the president of the *fabbricieri* to take the gamble. The Fabbrica paid sculptors, carpenters, and masons to take up the pavement, insert the line, and remove a corner of an overhanging buttress on the roof.[39] This cost them 2,000 lire, plus 500 lire for their consultant Cassini, which made in all about 15 percent of their annual income; a sum that imperiled their souls, since they faced excommunication if they did not repay soon the large debt they had incurred to complete the nave.[40] To reassure themselves that, if excommunicated, they would have spiritual advice, they

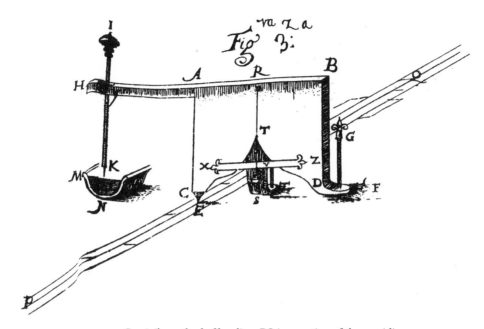

FIG. 3.2. Cassini's method of leveling. PQ is a section of the *meridiana.*
From Cassini, *Meridiana* (1695).

insisted that Grimaldi, Riccioli, and Ovidio Montalbano oversee the project. This last guarantor was the most conservative of all. He had taught Ptolemaic astronomy, and issued the astrological almanac for doctors required of Bolognese mathematicians, before becoming a moral philosopher and the Archbishop's censor of mathematical books.[41]

Cassini dug a ditch along the calculated course of the line, placed a wooden canal in the ditch, and filled the canal with water to provide an accurate level. The device used in releveling the line in 1695 under Cassini's direction will give some idea of the trouble he took. In Figure 3.2, the frame HABF, which rests on the feet E, F, and T, carries the adjustable hollow cylinder IK, through which runs a wire supporting a weight. The mason put E on the marble to be leveled and adjusted the various screws so that HAB was horizontal and the board AETD stood in the same vertical plane as IK. Then he turned the knob I until the weight just touched the middle of the surface of the water in the canal. The distance that the weight had to be raised or lowered from a reference mark in order to touch the water indicated how much the marble at E had to be lowered or raised. An auxiliary calculation showed that the earth's curvature could be neglected in using the canal.[42]

Cassini had the plate of metal containing the hole through which the sun's rays would enter the church cemented into the roof parallel to the pavement and gave it a diameter equal to one-thousandth of its height above the ground. He determined the height with extreme care and found the *meridiana's* vertex—the point directly under the hole's center—using a weighted string, whose swings he damped in a pot of water. He penciled in the places where, by his observations and calculations, the sun's image should fall at noon. The workmen then laid out the iron *meridiana* with the help of the water level and prepared marble plaques to mark the sun's entry into each of the zodiacal signs.[43]

All being thus prepared, Cassini invited everyone interested to attend the definitive setting of the rod on the day of the summer solstice in 1655. The text of the invitation, which Cassini had printed up in Latin, ran:

> This summer the first stone will be placed in the church of San Petronio for building celestial science from the ground up: the current solstice will be observed, and the sun's path across the meridian defined; on the pavement in the Church is a meridian line, on which the sun, admitted through the highest part of the eastern vault, will shine throughout the year precisely at noon; this line, which is suited to daily observations of the sun, moon, and principal stars, and to physical experiments, will be put down and exposed to public comment on the 21st and 22nd of June, at 15 hours civil time.[44]

"Thus," wrote Fontenelle, "there was established in the temple [of San Petronio] a new oracle of Apollo, where the sun can be consulted with confidence about all the difficulties of astronomy"; a bon mot with a bite, since the Jesuits then enjoyed the status of "oracles and prodigies of science."[45]

Many mathematicians came to see the inscription of the line some thought impossible: Grimaldi and Riccioli, of course; Mengoli and other doctors of the university; and a handful of priests. At 15 o'clock, or local noon according to the counterintuitive Italian civil time beginning half an hour after sunset, Cassini drew out a few circles with the *meridiana's* vertex as center. He marked the places where the sun's center crossed the circles. The sun met each circle twice, at equal times before and after noon. Cassini drew the chords between the places marked and bisected them (Figure 3.3). Then he drew the best straight line he could through the vertex and the points of bisection of the chords. This line, the noon line, was the *meridiana.* It went within a whisker of the pillars, but, as advertised, did not hit anything (Figures 3.4 and 3.5).[46] It appeared that San Petronio had been designed to serve God as a solar observatory.

Once the crowd had certified success, Cassini had the zodiacal marbles cemented in, the plaque for the summer solstice where the sun had been observed, the others where he calculated the image would fall at the appointed times. He also set out four scales, two on either side of the line. They gave the distance from the point under the center of the hole in hundredths of the height and in fractions

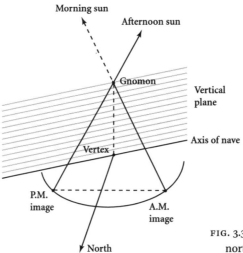

FIG. 3.3. Method of determining a
north-south line by the sun.

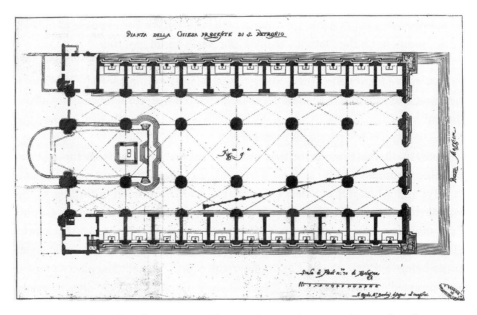

FIG. 3.4. Floor plan of San Petronio showing the *meridiana* just clearing the pillars.
From Cassini, *Meridiana* (1695).

of the earth's circumference (confirming the negligibility of its curvature for observations at San Petronio); a measure (the cotangent) of the altitude of the sun's
center, and the hour of sunrise corresponding to the place of the sun's image on
the meridian line (Plate 4). So great was Cassini's success, we learn from Riccioli,
that it overcame the jealousy aroused by his boasting and preferment, and earned

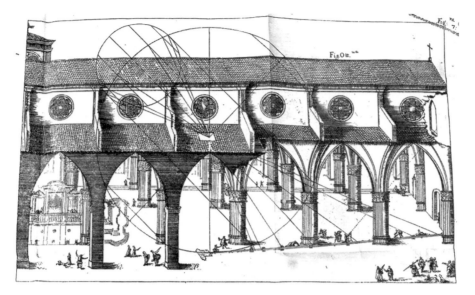

FIG. 3.5. People admiring Cassini's handiwork. The circles imposed on the diagram allow a geometrical location of the zodiacal plaques as demonstrated in Appendix B. From Cassini, *Meridiana* (1695).

him the goodwill of "all the nobles of Bologna, and the doctors [of the university], and all the other important men in the town."[47]

Cassini had the good news printed for those who had opened their temple to Apollo. "Most illustrious nobles of Bologna," he addressed the *fabbricieri,* "the kingdom of astronomy is now yours, will remain yours for ever, because you gave it a fixed and stable seat in a most august temple, with a more than royal munificence . . . ; nothing has been done before under the patronage of princes so worthy of the majesty of astronomy or so suitable to advancing its dominion." No one had thought an instrument so subtle, delicate, and exact could be built. "This glory was reserved . . . for your patronage, most illustrious senators." Cassini would have his glory too. He would establish Bologna as a center of a new astronomy. "Despite the Herculean labors of all astronomers from Ptolemy on," he told his patrons, astronomy had not yet submitted entirely to the yoke of geometry. He would put an end to "the very hard tyranny of 'Nongeometry.' "[48] We shall soon examine the meaning and basis of this boast, and Cassini's efforts to redeem it.

More Geometrico

The *fabbricieri* of San Petronio, the Senate and University of Bologna, and the Marchese Malvasia shared a deficiency as patrons. They were not princes. Cassini's first conquest in this line was Christina, the philosophical Queen of Sweden. With the help of a Jesuit sent for the purpose, she had abdicated her throne but not her regal ways, agreed to join the Church of Rome, and set off for the Vatican with six thousand books and manuscripts shipped on twelve men-of-war. She stopped in several Italian cities while the Pope prepared for her arrival in Rome. As she approached Bologna with her retinue of two hundred late in November 1655, she was met by the Papal Legate Giangirolamo Lomellino and the flower of Bolognese society, in forty carriages; with whom, and a regiment of infantry, she entered the city to a blast of artillery and a douse of holy water. The Legate gave a banquet, and fireworks, and an entertainment for which a special theater was built, all at the cost of the Pope and the Senate. The Queen made a tremendous impression, according to an eyewitness, with her majestic bearing, big eyes, strong nose, northern dress, and military bearing, a confusing combination of Mars and Venus.[49]

The learned lady proposed to visit the university. Since she had a taste for astronomy, or at least astrology, several of the mathematicians were invited to meet her. They included Mengoli, who gave her a primer, a *Royal road to mathematics*, from which she could have learned the laws of trigonometry in Latin verselets. "Moderation is necessary in the sciences as in everything else," she used to say, "but those who practice them get carried away, and instead of making themselves more clever, they become more stupid and ridiculous."[50] Cassini knew how to moderate his mathematics in the presence of princesses. Introduced by Legate Lomellino, Cassini offered Christina a picture and description of the great *meridiana*, drawn on a large piece of satin.[51] Apparently they had time for some conversation about astronomy during which Cassini would have had the grace to hide his disapproval of astrology.[52]

Christina then went off to Rome, where she was received by the entire college of cardinals sitting on mules. Cassini's satinized astronomy helped her to understand her surroundings. She lodged in the Tower of the Winds, which contained the *meridiana* that Danti had drawn for Pope Gregory. It is said that the Pope chose the Tower of the Winds for her accommodation not because of her love of science but because of its distance from his own apartments; and that he had Danti's assessment that all evil came from the North (Danti was thinking of winds) covered over lest it offend her.[53]

Cassini's line in San Petronio had an appeal not only for educated people like Christina but also for unfortunates "without a tincture of astronomy" (as Cassini characterized the intellectual underclass). It became a landmark. The author of a life of Lomellino, which appeared in 1659, enriched his account of the Cardinal's legacy in Bologna with an exact but irrelevant description of the *meridiana*. By 1667, a compatriot of Cassini's could write in a literary history of Liguria of "the marvelous heliometer most celebrated throughout Europe . . . erected in Bologna in the Basilica of San Petronio."[54]

Let us not overdo. In 1663, Jan Blaeu published his *Grand atlas* in twelve volumes. That for Italy contains a notice of the sights of Bologna. In San Petronio Blaeu had eyes only for the blessed Caterina de' Vigri, of the Order of Santa Clara, who had her nails trimmed there once a month, in public, though she had been dead for two hundred years. The following year the great traveler Balthasar de Monconys passed through Bologna in quest of culture. "We went into the great Church [of San Petronio]," he recorded, "where there is nothing to see."[55] But these were foreigners. The Bolognese thought that the heliometer and its maker belonged among the stars. Domenico Guglielmini, a student of Montanari's who became a professor at the university and observed at the *meridiana* after Cassini left for Paris, wrote in explanation of the willingness of the *fabbricieri* to repair the instrument in 1695. "[It is] one of the principal ornaments of this country [Bologna], both for its usefulness in astronomy and because it is the work of Sig. Gio. Domenico Cassini, one of the greatest mathematicians and men of letters of our century."[56]

· IN PARTIBUS ·

Cassini's combination of exact science, practical ability, diplomacy, and influential acquaintance brought him into the same sort of engineering projects that had engaged Danti. In 1657 he served as the technical expert for Bologna in the perennial dispute between Bologna and Ferrara over the river Reno. Cassini made a close study of the river, which he issued, with customary flamboyance, as a "new hydraulics" *(Idrodinamica nuova)*. The case came before the Pope, Alexander VII, who liked Cassini's performance. Cassini took the opportunity to unpin Riccioli from a hook in the Holy Office. Riccioli had written a discourse on the obscure but dangerous subject of the conception of the Virgin. The censor to whom he submitted it had not approved publication and refused to return the manuscript. When Alexander, who had an interest in astronomy and amused himself making sundials, mentioned Riccioli's *Almagest* favorably, Cassini told his story and the Pope ordered the manuscript returned.[57]

Alexander let the Bologna Senate know his high opinion of Cassini through the Legate, Cardinal Giulio Rospigliosi, "a man of outstanding merit and sublime mind," who will reappear presently as Pope Clement IX. The Senate, thus re-alerted to Cassini's capabilities and connections, appointed him superintendent of the waterworks of Bologna.[58] The Pope's brother then called on him to advise about the fortifications of Urbano, a papal stronghold in the shape of a pentagon on the road from Bologna to Modena, and the Pope himself asked him to represent the Holy See's claims to water rights in the Chiana against the Grand Duke of Tuscany. In 1665 Cassini became superintendent of the waters of all the Papal States.[59] It may be recalled that he was also a professor at the University of Bologna.

The professor did not neglect all his students. He viewed the comet of 1664 in Rome with Christina, who sent her carriage to bring him to her palace early enough to dispute with her favorite cardinal before the comet occupied their attention. A rival of Christina's, Signora Colonna, also used to pick him up to go star gazing and to recite his descriptions of the constellations in Italian verse, "which she amused herself learning by heart."[60] The household of Pope Alexander VII also had his attention. He would search the skies from the terrace of the headquarters of the congregation De Propaganda Fidei (a nice image, that) together with the papal siblings and nephews. His Beatitude himself once kept him an entire day, "talking about astronomy and diverse other sciences." As Cassini observed, with the Jesuit missions to China in mind, astronomy was of the greatest value for the propagation of the faith. "For it is under the confession and protection of this science that those who devote themselves to carrying the gospel to the infidels penetrate to the most distant countries . . . , that they acquire the admiration of the people, that they insinuate themselves among the great, and that they win even the favor of sovereigns."[61]

The Grand Duke of Tuscany now wanted Cassini's services and showered him with hospitality whenever he came to Florence, where, also, Cardinal Leopold's Accademia del Cimento would always hold a special session for him.[62] Cassini preferred to retain his professorship, his control of the papal waters, and his easy access to the Vatican. But then there came a competitor equal to and even greater than the Pope. Between and on his various assignments, Cassini had managed to do some astronomy that interested Louis XIV's astronomers. It came about in this way. During his visits to Rome he had come to know Giuseppe Campani, a clock maker turned lens grinder. That was in 1662 or 1663, when Campani was waging an underhanded campaign to snatch the primacy in the making of telescope lenses from his fellow Roman Eustachio Divini. Campani gained the edge in 1664 by an account of new observations, particularly of Saturn's rings. These observations confirmed decisively the theory of the appearances of Saturn put forward in 1659 by Christiaan Huygens, then at the beginning of his brilliant career. The confirmation put an end to a wrangle initiated by Divini's anxiety to defend his lenses from the implication of inferiority to Huygens'. With the help of Honoré Fabri, S.J., an overly clever and influential controversialist, of whom much more will be said, Divini had sought to demolish the ring hypothesis. Fabri devised a complicated alternative explanation of Saturn's appearances and published it under Divini's name; Huygens countered; Fabri-Divini replied; the Accademia del Cimento found for Huygens after building models of the opposing hypotheses; and Campani settled the matter by observing the ring directly.[63]

The publication of Campani's observations and the favorable reviews of them in the first numbers of the *Journal des sçavans* and *Philosophical transactions* would have undone Divini were it not that direct comparisons between his and Campani's best lenses proved inconclusive. The situation changed dramatically when Campani produced a telescope 50 palms (11.25 m) in length. When tested in Florence in July 1665 it outdid everything previously made for clarity and magnification.[64] With his customary luck, Cassini had been present in Rome when the first inconclusive comparisons were made. With his customary good judgment, he did not doubt that Campani would prevail. For his early allegiance, Campani allowed him use of the telescopes and, in 1664 or 1665, gave him one of 23 palms (5.2 m, 17 ft.).[65]

Cassini confirmed Campani's observations of Saturn's ring and of Jupiter's stripes.[66] But that was nothing. In June 1665 Cassini saw spots move across the surface of Jupiter, which he interpreted, rightly, as shadows cast by its moons; "the discovery of which [wrote Campani] redounds no less to his glory than to my lenses." Cassini's sharp eyes also spied a spot that did not move relative to Jupiter's surface, which he interpreted, again rightly, as a permanent feature. From observations of this blemish, he argued that Jupiter spins and gave the rate of rotation. Campani confirmed both sorts of spots.[67] Cassini looked again through a Campani glass and detected the rotations of Mars and Venus. The Paris astronomers followed these feats closely. Huygens, who had been brought to Paris as the leading luminary of the then new Academy of Sciences, confirmed the shadows of Jupiter's moons and the revolution of its body, once in ten hours. He wanted Cassini as a colleague, "because he is a very great astronomer and also for his telescope, which has made beautiful discoveries."[68]

Cassini could offer two additional items to the Parisian academicians. One, an *Ephemeris* of Jupiter's moons dedicated to Rospigliosi, gave their orbits and periods much more precisely than other astronomers had managed to do. These tables, published in 1668, culminated fifteen years of work begun in 1652, with a telescope provided by Malvasia. Cassini's breakthrough was to deduce the angles between the planes of the moons' orbits and the ecliptic. An accurate account of the orbits made possible accurate predictions of the eclipses of the moons as they moved across and behind the body of the planet; and an accurate table of these eclipses made possible the determination of the difference in longitude between the place for which the table was calculated and any other place with a competent observer. Cassini's tables raised the hope that Europe was about to gain that great desideratum, a reliable way to find the longitude at sea.[69]

Although the method failed on the ocean, it succeeded well on land, where astronomers had a steady base for observation and enough time for calculation. Using Cassini's tables, Parisian geodicists found that the channel coasts of France had been placed too far west by many leagues. Cassini was the instrument that pared metropolitan France of more territory than it would lose in all its subsequent wars.[70] Thus he fulfilled the advertisement in his *Ephemerides* of Jupiter's moons, to make astronomy additionally useful, and aligned himself with the Paris Academy of Sciences, whose principal goal in undertaking astronomical observations, he supposed, was their "application to the advancement of geography and navigation."[71]

The second item in Cassini's portfolio interesting to his colleagues in Paris concerned corrections that had to be applied to precise observations to make them suitable for testing hypotheses about planetary motions. There is no way to avoid the effects that necessitate the corrections. One of them is the familiar refraction of light when it passes from an optically rare medium, like air, to a denser one like water. Refraction causes objects beyond the earth's atmosphere to appear higher in the sky than they are. The second irksome effect arises from the fact that the earth, though small in relation to distances between the sun and the planets, is not infinitesimal. Since the center of observation (a point on the earth's surface) does not coincide with the center of the theoretical system (the earth's center), an angular error, or parallax, results that causes bodies to appear lower in the sky than they are. The effects of refraction and parallax require corrections in opposite directions and of different magnitudes. Since the effects were combined in the observations, untangling and correcting them made a difficult and frustrating business.

To the armchair geometrician, however, they give no trouble. Because the density of the atmosphere changes continuously with height, a light ray originating outside it describes a curved path to the observer rather than the zigzag that occurs at an air-water boundary. Just as by following the line of sight one would misplace the coin in the cup from B to A in Figure 3.6, so the astronomer who did not correct for refraction would accept the apparent altitude of a celestial object, ∠S'OH, for the true altitude, ∠SOH (Figure 3.7). The necessary correction changes with altitude. An object in the observer's zenith requires none; one on the horizon claims a maximum; between refraction varies in a complicated way with altitude.

Tycho was the first to measure refraction, though very roughly, by comparing the apparent positions of circumpolar stars at upper culmination (where refraction is negligible) and lower culmination (where it can be significant). "So many and so large are the effects of refraction, especially for the sun and moon," wrote Riccioli, "that by ignoring them astronomy was, so to say, missing one eye." Before Cassini's time, astronomers corrected for refraction only for altitudes under 45°. At first he followed the common practice. But he discovered at San Petronio that ∠SOS' (the correction) exceeded one minute of arc at 45°.[72] That meant something to astronomers who affected to give latitudes accurate to seconds.

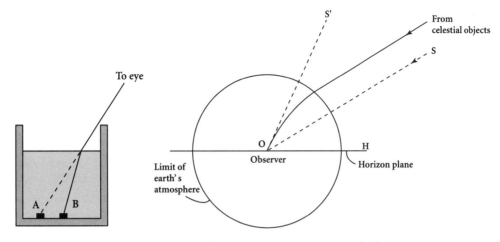

FIG. 3.6. A familiar example of refraction in water; the coin appears to the left of its true position.

FIG. 3.7. The effect, much exaggerated, of refraction on astronomical observation; the star appears at S' rather than at S to the observer O.

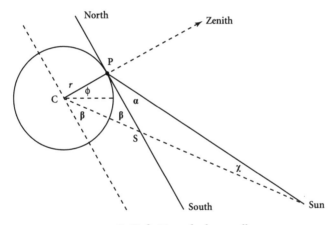

FIG. 3.8. Definition of solar parallax χ.

The Lord giveth and taketh away. Cassini found the correction for parallax (angle χ in Figure 3.8) had been much exaggerated. An observer P at the earth's surface measures the altitude of the sun's center as α; were he stationed at the earth's center C, he would have found an altitude $\beta = \alpha + \chi$. The parallax vanishes for objects in the zenith and reaches its maximum value for objects on the horizon. In the case of the sun, this "horizontal parallax" amounts to only nine seconds of arc. The correction for parallax is a simple trigonometric function multiplied by r/s, the ratio of the earth's radius to its distance from the sun.[73] It did not take a Cassini to do the arithmetic. But determining r/s required skill, judgment, and luck. Available techniques employed as an intermediate step the ratio m/s of the lunar to the solar distance and calculated r/s as $(m/s) \cdot (r/m)$. To obtain m/s, one measured the angle between the sun and moon precisely at half-moon, an instant difficult to determine. Astronomers consistently made the angle, and consequently s, too small.

Tycho set the horizontal parallax at three entire minutes, that is, about twenty times too big. Copernicus gave the same error. Kepler came down to one minute and called for better methods. By far the best value before Cassini was Riccioli and Grimaldi's, some 28 seconds, off by only a factor of three. These large numbers, applied as corrections, vitiated measurements otherwise accurate to under a minute. Cassini picked his way through this swamp of bad numbers and once again surprised his colleagues, this time by setting the maximum value of the solar parallax at the unprecedentedly low value of 10 seconds of arc. Cassini's correction of the correction resolved many apparent discrepancies among the observations of the best observers.[74]

Impressed by his heliometer and his tables, and by his skill and judgment as an observer, the Parisian astronomers were willing to overlook Cassini's crazier ideas, like placing the center of the path of a comet in the star Sirius, "something bizarre and incredible."[75] They alerted Louis' all-seeing minister, Colbert, who recognized Cassini's value as an ornament and also as the likely inventor of a practical means of finding the longitude. The negotiation for Cassini's removal to Paris involved two sovereign states and the Senate of Bologna. He would accept only with the Pope's blessing—and only with a salary half again as large as Huygens'. The Pope, Clement IX, who as a cardinal had sung Alexander VII's praise of Cassini to the Senate, did not wish to release him from papal service. The Senate wanted to keep him at their university. It took a direct appeal from the Sun King to the Pope and the Senate to procure the release of the solar astronomer, and then only on the understanding that the relocation would not be permanent.[76]

For Cassini, a sojourn in Paris held the opportunity of dominating astronomy from the center of European science. His means would be the Royal Observatory, which had risen to its first story when he arrived. Unfortunately its architect, not anticipating his coming, had not provided space for a meridian line the equal of San Petronio's. Cassini argued that no respectable observatory would be complete without a great *meridiana* and demanded that the architect terminate the towers planned for the corners so as to leave room for a big camera obscura on the second floor. The architect declined to do so. Both parties appealed to the King. The *nouveau arrivé*, unable yet to express himself easily in French, lost. Cassini was most disappointed that the Paris Observatory would be only a house for equipment and observers and not, like San Petronio, "a great instrument itself."[77] Still, as his son Jacques later expressed the family's satisfaction with its headquarters, "everyone knows that of all the edifices so far built for astronomical observations, none equals, or even approaches, the magnificence of the Royal Observatory of Paris."[78]

Under the influence of his new colleagues, Cassini's interest shifted from gnomonics to geodesy and regular observations of stars; the Observatory was not outfitted with a *meridiana* of any consequence until 1729, when Jacques Cassini installed one as a monument to his father. But from the beginning the Observatory was fully outfitted with Cassinis. Gian Domenico lived there almost from his first days in France; his family, including his nephews the Maraldis, also astronomers, lodged there in all for over a century; whence, no doubt, has arisen the common assumption that the Cassinis directed the Observatory from its completion. In fact, it had no official head, budget, or organization before Cassini's grandson was named director in 1771, and no paid assistants before his great-grandson took over in 1784.[79]

The founder of this dynasty, Cassini I as he is known to historians, quickly ingratiated himself with his new employer. The heavens helped him as they had Galileo. In a famous gesture reported earlier, Galileo had offered the first fruits of his telescope, the four main satellites of Jupiter, to the Grand Duke of Tuscany, whose patronage he sought, as "the Medici stars."[80] Others had tried the same trick but failed, notably poor père Rheita, who saw five spurious satellites of Jupiter, bigger than the Medici stars, which, in a literally empty gesture, he baptized "urbanoctarians" after Urban VIII. Lucky père Cassini, "who did not lose a single clear night in observing the heavens," spied two real moons around Saturn soon after moving to Paris. He presented them to Louis not by name but by number. The ancients knew seven planets, he said, and astronomers furnished with telescopes had added another five (Galileo's four moons of Jupiter and Huygens' moon of Saturn, discovered in 1655). With his gift the total reached "the number of XIV, which now has the honor of being united to the august name of Louis."[81]

Astronomia Reformata

An essential parameter in laying out a meridian line is its latitude. The standard way to determine latitude took the elevation of the polestar at its lower and upper culminations and averaged the two measurements. The method must be practiced in the winter, because only then are nights long enough that both crossings, which take place twelve hours apart, are clearly visible.[82] In this way, at the turn of the year 1655–56, Grimaldi and Riccioli, observing at the *meridiana* in their church of Santa Lucia, obtained $\phi = 44°30'9''$, and Cassini, prevented by illness from fixing up San Petronio for such measurements, did the same at Malvasia's, for a whole week in January 1655, with the result of $\phi = 44°30'24''$. He then betook himself to the Torre Asinelli, the remnant of a medieval palace built when height was proportional to prestige, and the more prestigious enjoyed the option of pouring molten pitch on their lesser neighbors. From angles measured there, he worked out that San Petronio lay 2'' south of Malvasia's observatory and that its latitude was $44°30'22''$. Subtracting this value of ϕ from the result of the famous public observation of the summer solstice of 1655, Cassini found the obliquity of the ecliptic, ε, to be $23°30'30''$.[83]

Contrary to Toscanelli and Danti, Cassini's main substantive purpose in installing a meridian line was neither to correct the calendar nor to determine changes in the obliquity of the ecliptic. Indeed, by comparing his observation of the equinoxes with those of Tycho Brahe from the end of the sixteenth century he made the year to be $365^d5^h49^m0^s$ (he would have been closer had he adopted the value obtained by Riccioli and Grimaldi), with which the Gregorian year of $365^d5^h49^m12^s$ agreed well enough to require no further adjustment. If, however, a more punctilious or pettifogging age should think one desirable, Cassini advised that all the necessary measurements could be made at San Petronio, without reference to earlier observations like Tycho's. Cassini estimated the accuracy of observation of the equinoxes at San Petronio at about one minute of time/year, or 15 seconds per leap-year cycle.[84] After only sixty years, he supposed, observers at the new Apollonian oracle could establish the length of the year to a second of time. But their concern was not his.

· THE ORBIT OF THE SUN ·

Cassini proposed to redeem his promise to free astronomers from "ageometria" by giving them an accurate and simple procedure for establishing the "orbit" of the luminaries and the planets directly from observation. "Orbit" carries quotation marks to indicate two differences from its modern meaning: to Cassini it signified a representation of the apparent motions reduced to the simplest plausible geometry and not necessarily a real path in space; and (to specialize to the two most important objects) it was meant indifferently of the sun and the earth. Copernican astronomers could transfer Cassini's solar "orbit" directly to the earth with no expenditure of thought. With this understanding, "orbit" will appear henceforth without quotation marks. This convention, combined with the designation of all astronomical systems as hypothetical, gave Catholic writers scope to develop mathematical and observational astronomy much as they pleased despite the tough wording of the condemnation of Galileo.[85]

Greek astronomy rested on the rules, which amounted to a definition, that all orbits treated in astronomy should be circles or components of circles, and that the planets and luminaries should move on their circles with uniform motion around the circles' centers. In practice, however, the rule could not be observed fully without undue complications. In a brilliant technical move, Ptolemy altered the rule to allow circular motions uniform around a point that did not have to be the circle's center. A capital point about this point, which Ptolemy called the equant, is that he did not use it—because he did not think he needed it—to regulate the orbit of the sun. A few illustrations will make all this clear.

The simplest geometrical representation of the sun's apparent annual course would be a circle in the plane of the ecliptic centered on the earth. The simplest representation of its annual motion would be a uniform increase in its angular distance from some reference point in the ecliptic, say the vernal equinox. Naturally this uniform increase would be understood as seen *from the earth,* since the earth's center is the center of the ecliptic. We would have the situation depicted in Figure 3.9. The angular distance of the sun from the vernal equinox increases regularly at the rate of $\omega = 360°/y$ per day, where y is the number of days in a year. With the Julian value for y, $\omega = 0.9856°/$day. Since the solstices and equinoxes are situated 90° apart in the ecliptic, the orbit depicted in Figure 3.9 makes the interval between them exactly $y/4$ days; that is, it represents the lengths of all the seasons as the same.

Now, if there is anything plain in astronomy, it is that all the seasons are not equal. In the Northern Hemisphere, summer is several days longer than winter, and spring a few days shorter than fall. The scheme of Figure 3.9 gives a poor representation of the sun's motions; it needs to be altered so as to make the sun appear to move more slowly around the summer than around the winter solstice. One possibility would be to keep the diagram but to change the rule regulating the motion: the sun would speed up and slow down in a manner determined by regular observation. Such a scheme would have little to offer beyond a reproduction of the observations; and it would have the drawback of supposing the sun to violate

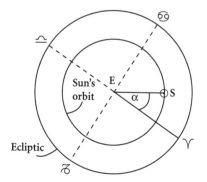

FIG. 3.9. Sun's annual orbit depicted as concentric with the ecliptic.

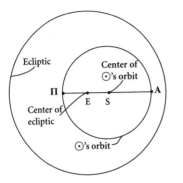

FIG. 3.10. A more realistic version of Figure 3.9 with eccentric orbit, eccentricity *e*, apogee A, and perigee Π.

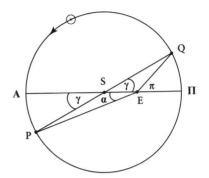

FIG. 3.11. Apparent speed of the sun around perigee and apogee (the apsides).

capriciously the wholesome rule that it move at constant speed around the center of its orbit. A real speed-up and slow-down of the sun would have been unintelligible; that is, it would not have been Greek.

Therefore the ancient astronomers regarded the apparent inequalities in the sun's annual motion as an optical illusion. That was an admirable demonstration of the power of mathematics and method over experience. The Lord who made everything by number, weight, and measure had so arranged matters that a simple scheme, obedient to the rules of astronomy, exists for representing the seasonal inequalities as an optical illusion. One needed only to suppose that the center of the sun's orbit S (figure 3.10) does not coincide with the center of the ecliptic E. Let their separation be ae, where a is the radius of the orbit and e is a number much smaller than one. The diameter of the sun's orbit that runs through the center of the ecliptic (or of the earth, for they are the same) is called the "line of apsides." The point Π on this diameter closest to the earth is the "perigee"; that furthest distant, A the "apogee" (Figure 3.10); or, to speak with Copernicus, the perihelion and aphelion.

If the sun moves uniformly about its orbit, it will appear from the earth to slow down around the apogee and speed up around perigee. Figure 3.11, in which the labels have the same significance as in Figure 3.10, illustrates the effect. Let the sun go from A to P through an angle $\gamma = \omega t$ in the time t as seen from the center of the sun's orbit S; it will appear to move through the smaller angle $\alpha = \angle AEP$ as seen from the earth. (That $\alpha < \gamma$ follows from Euclid I.32: an external angle of a triangle, like γ, equals the sum of the opposite internal angles, in this case $\alpha + \angle SPE$.) In the time t, therefore, the sun will have moved in the ecliptic over a smaller angle than ωt. That is precisely the behavior it displays around the summer solstice. Hence EA must point somewhere around the beginning of Cancer. Similarly, if at

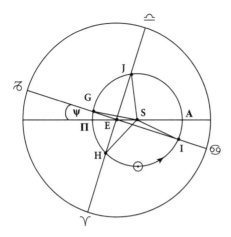

FIG. 3.12. Ptolemy's method for obtaining the solar eccentricity; ψ is the offset of the line of absides from the line of the tropics.

perigee the sun ran through the same angle γ = ωt as seen from the center of its orbit, it would move through an angle π = ∠QEΠ in the ecliptic, where, evidently, π > γ. Around Π, therefore, the sun appears to move faster than the mean speed ω. That is how it behaves in midwinter.

From the observed lengths of the seasons the ancient astronomers worked out values for the "eccentricity" e and for the angle ψ between the line of apsides and the line joining the solstices. The method gave very good results, as may be illustrated by comparing e and ψ, computed by it using the lengths of the seasons given in current almanacs, with their values as delivered by modern astronomy. Figure 3.12 shows the ancient method applied to modern seasonal lengths. Since fall is longer than spring (in antiquity it was the other way around), EA must point somewhere between the summer solstice and the autumnal equinox, and, consequently, EΠ between the winter solstice and the spring equinox. We know the differences in time between the sun's arrival at G, H, I, and J on its orbit, when, viewed from the earth E, it appears at the first points of Capricorn, Aries, Cancer, and Libra, respectively: these are, to the rough approximation of a standard almanac, 92^d18^h, 93^d15^h, 89^d20^h, and 89^d0^h.[86] Hence the angles GSH, HSI, JSI, and GSJ are easily computable: they equal the known time differences multiplied by ω. That follows from the fundamental assumption of the model, that the sun goes through equal angles in equal times on its orbit around S.

Trigonometers need nothing more to work out e and ψ. Since the calculation will be obvious to them and tedious to everyone else, it will be enough to give the result: e = 0.0334, ψ = 12°58', that is, A lies toward 13° Cancer and Π toward 13° Capricorn.[87] Modern astronomy uses the values e' = 0.0167, ψ' = 13°34'. The agreement for ψ is good, but that for e very bad, out by a factor of two and—what is of first importance—exactly a factor of two. This doubling had already reared its

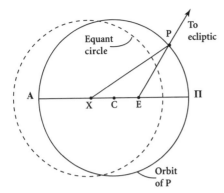

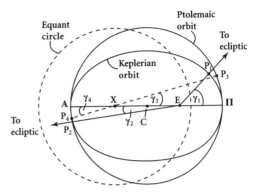

FIG. 3.13. Eccentric and equant X; the eccentricity is bisected, XC = CE = $ae/2$.

FIG. 3.14. A Ptolemaic circular orbit with eccentric and equant compared with a Keplerian ellipse.

head when Cassini built his *meridiana* in San Petronio. It informed the *Ephemerides* of Malvasia and lay behind Cassini's boast that he and his *meridiana* would reform astronomy.[88]

The inspirational factor of two related to the equant point. In Figure 3.13, E as usual is the center of the earth and the ecliptic. C is the center of the circle on which P moves while running through equal angles in equal times around the equant point X. To help visualize the motion, the figure has an "equant circle" on which the point would revolve with constant velocity if its orbit were centered on X instead of C. From E, P appears in the zodiac in the direction EP. With what in retrospect appears to have been uncanny intuition, Ptolemy set X as far on one side of C as he put E on the other. This specification seems uncanny because it was the very best approximation to Keplerian motion available to him.

According to Kepler's first two "laws," planets move in elliptical orbits around the sun at one focus in such a way that the line joining the planet to the sun sweeps out equal areas in equal times. This complicated rule of motion is illustrated in Figure 3.14, where the curvilinear areas $P_1E\Pi$ and P_2EA are supposed to be equal. Evidently it makes the planet move more quickly near perihelion (P_1) than near aphelion (P_2), since $\gamma_1 > \gamma_2$, and thus recovers a fundamental feature of Ptolemy's eccentric solar orbit. The equant model comes very close to reproducing motion in a Keplerian ellipse because a planet moving around the focus occupied by the sun in accordance with Kepler's laws moves almost uniformly around the unoccupied focus X. (In the figure, the Ptolemic prescription would put the planet at P_3 and P_4, with $\gamma_3 = \gamma_4$: the apsidal distances and timing agree exactly on the two theories.) The two prescriptions differ in where they appear to put the planet only by small quantities proportional to the square of the eccentricity. Except for Mars, the eccentricities of the planets are so small that the difference between Keplerian

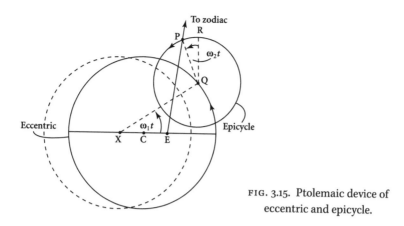

FIG. 3.15. Ptolemaic device of
eccentric and epicycle.

motion and uniform circular motion with equant is undetectable by naked-eye observation with the usual means of angular measurement. With the great meridian line, however, detection became possible, even in the case of the almost circular orbit of the sun.

Ptolemy did not mount planets on circles eccentric to the sun. His equant regulated the motion of a point Q that served as the center of a second circle, or "epicycle," centered on Q, around which the planet P revolved uniformly with respect to the fixed line QR (Figure 3.15). In effect, the epicycle represented the earth's revolution, and the eccentric the planet's revolution, around the sun. Copernicus disliked the equant and did away with it, at least in principle, to the disadvantage of his system. Kepler restored it, and used it effectively in analyzing planetary motions before he discovered his laws. In this analysis Kepler introduced the equant for the first time for the regulation of the motion of the sun, which gave predictions indistinguishable from those of the final Keplerian theory with elliptical orbit and areal law.[89] Many astronomers objected to a solar equant. Thus began what Flamsteed, writing a generation after its resolution, called a "controversy ... of no small moment." It was this great and obscure controversy between adherents of Ptolemy's traditional solar theory and proponents of Kepler's "bisection of the eccentricity," with its Copernican associations, that Cassini proposed to settle at San Petronio.[90]

The resolution has already been revealed. The modern value of e is half the ancient: the distance between the two foci of Kepler's ellipse just equals Ptolemy's offset of the earth's center from the center of the sun's orbit. But the eccentricity of an ellipse is the distance of either focus from the center, not their separation. Cassini's *meridiana* — or, as it was aptly named for this purpose, "heliometer" — confirmed the bisection. That in itself did not confirm Copernicus. But it suggested the following syllogism: all the planets have equants; the sun also has one; therefore the sun might be a planet. However, the sun does not look like a planet, and its orbit can be ascribed to the earth without any change in the appearances; the earth would go around the sun in precisely the same sort of orbit that the planets describe according to Kepler; therefore, from the point of view of mathematical astronomy, the earth can be treated as a planet. Or, to put the point in a few words (they are those of Astronomer Royal Flamsteed), "the Suns Excentricity is bisected as the Copernicans affirme." This connection of ideas — the bisection of the eccentricity implies that the earth is a planet — became commonplace among Copernicans.[91]

The bearing of observation on the great question of the factor of two can be visualized easily. On Ptolemy's theory, where the entire eccentricity lies between the earth and the center of the sun's orbit, the distance between the two bodies at perigee is less, and that at apogee greater, than on Kepler's theory, where only half the eccentricity comes between the observer and the orbit's center. These distances cannot be observed directly. Fortunately, a convenient substitute exists in the sun's apparent diameter, which is inversely proportional to its separation from the earth. By 1650, when debate raged between the whole- and the half-eccentrics, astronomers disposed of many observations of the sun's size taken at different times of the year.

Riccioli reviewed all the data and all the arguments and concluded that no observations were good enough to require acceptance of Kepler's bisected eccentricity for the orbit of the sun (or earth). Astronomers from Ptolemy on had given as the sun's apparent diameter at mean distance nothing larger than 32'44" (Copernicus) and nothing smaller than 30'30" (Kepler). Riccioli and Grimaldi had found 31'56", within a minute of all the competing numbers. Moreover, some of the same astronomers gave 2', others around 1', as the difference between the sun's apparent diameters at the absides. Riccioli: "I have not approved [of the bisection of the eccentricity] because the difference of a minute in the apparent diameter of the sun at apogee and perigee, which is one of the main foundations of Kepler's theory, does not agree with my observations and with most of the evidence."[92] In a matter of such importance, it would be rash to prefer one set of data, even one's own, over another. Thus Gassendi taught Ptolemy's solar theory although he had found the difference of the solar diameters to be just under a minute, unequivocally favoring Kepler.[93] It is hard to be an empiricist when observations do not determine the question. A closer look at the measurement Cassini made at San Petronio will indicate the empirical difficulties.

The difference between the absidal separations of the earth and the sun is $a(1 + e/2) - a(1 - e/2) = ae$ on the equant theory and $a(1 + e) - a(1 - e) = 2ae$ on the pure eccentric or perspectival theory (Figure 3.16). The relevant observation of the

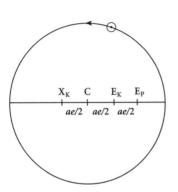

FIG. 3.16. Comparison of apsidal distances in Ptolemy's and Kepler's representations of the sun's orbit.

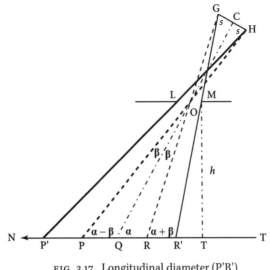

FIG. 3.17. Longitudinal diameter (P'R') of the sun's image in San Petronio.

sun's apparent diameter is illustrated in Figure 3.17. The difference between the apsidal diameters $2(\beta_\Pi - \beta_A)$ is $(2s/a)[(1 - e)^{-1} - (1 + e)^{-1}] \approx 4se/a$ on the perspectival theory and half that on the equant theory. According to Kepler, $e = 0.036$.[94] Taking $2s/a = \sigma = 30'$, the mean angle subtended by the sun at the earth, Cassini would have expected to find $2(\beta_\Pi - \beta_A) = e\sigma = 1'$ if the sun agreed with Kepler or $2e\sigma = 2'$ if the sun preferred Ptolemy. These were just the values around which the measurements that Riccioli deemed inconclusive happened to cluster. The instruments that had made the measurements worked on too small a scale to settle the matter. The new observatory of San Petronio was big enough to be decisive. The quantity of interest, the angle $e\sigma$, translated into a length $e\sigma h$ in the diameter of the sun's image along the *meridiana*, where h is the height of the gnomon. Owing to the great magnitude of h, this length amounts to a little under a centimeter (see Appendix C), an amount readily detected in principle.

· THE SPOT ON THE FLOOR ·

A tolerance of a centimeter would have made the observations easy, were the object of measurement sharp and stationary. But the sun's image stuttered and fluttered upon the cathedral floor. Did its appearance menace its measurement? Earlier determinations by Tycho, Gassendi, and others using tubes or plane sights without lenses gave inconsistent results owing, probably, to diffuseness in the boundary of the image.[95] "I would have men only consider, how much the sight is deceived, while the same thing is measured, according to the several degrees of light and darkness, where with the sight of the eye is affected."[96] The eye faced the situation depicted in Figure 3.17, where light from the solar disk GH reaches the floor from P' to R' beyond the central image PR.[97] The figure relates to San Petronio, where the hole is mounted on a horizontal plate set in the roof rather than, as at Santa Maria Novella, in a window. This placing simplifies the geometry a little.

We wish to know from Figure 3.17 how much the real image with fuzzy contours, P'R', exceeds the sharp ideal image PR. Since the rays from a given point on the sun reach the earth almost parallel, we can consider P'POL a parallelogram (that is an advantage of making the hole parallel to the floor). Consequently P'P = LO = p, the hole's radius. Similarly, RR' = p. The bright central image of radius PR/2 appears surrounded by a dimmer band with external radius PR/2 + p. In the most unfavorable case, at the summer solstice, width of the penumbra/radius of the central image = (diameter of the gnomon/height of gnomon) · 100. Since Cassini made the diameter of the hole one-thousandth part of its height, the width of the penumbra at the summer solstice is a tenth of the radius of the central image.[98]

Well, not exactly. The geometry in Figure 3.17 refers to the diameter of the sun's image along the meridian line. Perpendicular to the line the diameter is shorter since the limbs of the sun have the same elevation α rather than, as in the longitudinal case, elevations differing by σ. Figure 3.18 presents the geometry.[99] Figure 3.19 shows to scale the resultant idealized central images at the solstices (in the church each would be surrounded by a penumbra of annular width p). The dimensions given derive from the rough values $h = 27.1$ m, $\sigma = 30'$, $\phi = 44°30'$, $\varepsilon = 23°30'$. Since $2p = h/1000$, $p = 1.35$ cm. In his full description of the *meridiana*, which he did not publish until he readjusted it in 1695, Cassini included a folding plate almost two meters long on which the sun's images at the solstices are drawn life size.[100]

If the atmosphere is not perfectly calm, the image flickers even on a clear day. And, since its penumbra always shades off to an indistinct boundary, fixing its diameter was often as much art as science. The standard practice required two observers, one to mark each limb of the sun's image as its center crossed the line; for best results, one would tell the other where to place his mark so that points equally illuminated to the eye of the senior observer could be taken. Then, at leisure, the observers obtained the distances of the marks from the vertex in hundredths of the height from the scale inscribed in the marbles encasing the line, and in hundred-thousandths by taking the distance of each mark from the nearest scale division by a pair of dividers and measuring the openings against a graduated plate.[101]

Cassini said that he could spy differences of a minute of arc in the sun's altitude easily, and of seconds occasionally. Now a minute of arc amounts to about 9 millimeters on the ground at the summer solstice and 6 centimeters at the winter sol-

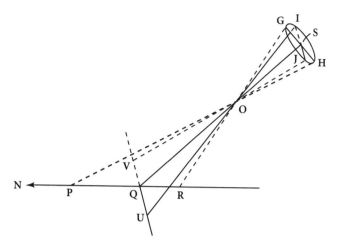

FIG. 3.18. Transverse diameter (UV) of the sun's image in San Petronio. The penumbra is not shown. IJ and UV are perpendicular to the plane of the paper; the other symbols have the same significance as in Figure 3.17.

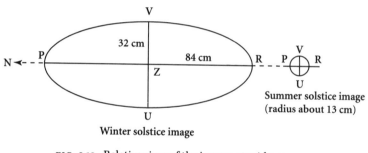

Winter solstice image

Summer solstice image
(radius about 13 cm)

FIG. 3.19. Relative sizes of the images at midsummer and midwinter at San Petronio.

stice. Even the smaller could be distinguished easily were it not for the trembling of the image and the uncertainty in identifying corresponding spots. Hence Cassini's directive: "If you always take the same limit of light and the edge of the fluctuation furthest from the center, at least you will have the same proportion in comparing the apparent diameters at different times, and you will not err in the apparent altitude of the sun's center." Still, you must record the state of the atmosphere and the vagueness of the boundaries, "in order that these factors can be taken into account in using the observations."[102]

And these were not the only factors. The constant ring of the penumbra had to be subtracted from the calculated image in order to obtain apparent solar diameters comparable with one another.[103] (The penumbra constitutes a smaller fraction of the total image the larger the image is.) Even this nicety did not produce the desired result, since the eye does not pick out the rim of the penumbra, but a point within it, as the apparent boundary. Hence, the astronomer who subtracted the entire diameter of the hole from the diameter of the whole image underestimated the size of the central image. Cassini corrected the correction by increasing the diameter of his central images by one part in sixty; but his recipe did not work well in general and the observers at San Petronio never found a reliable rule to compensate for their underestimate of the size of the whole image.[104] One obvious way to minimize the error was to make the hole negligibly small. But Grimaldi had ruled out that contrivance by his discovery of diffraction, which spreads, rather than restricts, the boundaries of the image. He expressly warned astronomers against making their pinholes too small and blamed past failures of his fellow Jesuits to obtain reliable and reproducible results with the camera obscura on ignorance of this rule.[105]

Nevertheless, experienced observers knew how to get exact and consistent results. Cassini found the sun's apparent diameter at apogee to be 31'8"; Grimaldi and Riccioli measured 31'0". The corresponding values for perigree were 32'10" and 32'4". Thus the Jesuits confirmed the bisection of the eccentricity, the cornerstone of Kepler's version of the Copernican theory, and "destroyed Aristotelian physics in the heavens," by observations made in the Church of San Petronio in the heart of the Papal States. Others confirmed the reliability of the instrument. Mengoli: "I know that it deserves every kind of confidence." Antonio Masini (another contemporary Bolognese observer): "It is certain that the celestial observations [taken at the *meridiana*] are much more exact than any made earlier."[106]

The great controversy over whether the sun obeys an equant, "a question of such importance that, without [the answer to] it, we cannot proceed profitably any further in astronomy," had been submitted to "Apollo himself at the heliometer of San Petronio."[107] As soon as Apollo answered, probably just after the winter equinox of 1655, Cassini informed the Senators of the fabric, in a flowery feuilleton:

> We have won, illustrious Maecenuses. I bring you in triumph the horses of the sun, now tamed for the first time . . . by the force of your arms . . . We are now in alliance with the sun. You will be admitted more freely to its secrets than any other mortals have ever been, and also to the secrets of the other planets it regulates. In return, you transmit to it forever a new light from the great heliometer you caused to be built in the most august temple of San Petronio.

Cassini transmitted the same news to his colleagues in the sober language of mathematicians: "The theory of the sun that results from observations at our heliometer is uniform with the theories of the other planets."[108]

Although it sounds preposterous even in a century that has known elementary particle physicists, Cassini's claim to have refounded astronomy would not have been beyond belief in his time. Newton did refound the subject thirty years later. Ever since Kepler had knocked astronomy free from the ancient prescriptions about circular motion, mathematical astronomers had enjoyed the license, and suffered the uncertainty, of lawlessness. On the grand cosmological level, there were the competing systems of Ptolemy, Copernicus, and Tycho, and as many more, ancient and modern, all reported in the pages of Riccioli.[109] He himself taught a modified Tychonic plan, in which Mercury, Venus, and Mars went around the sun, and all the rest around the earth. And, in a playful mood for once, he mentioned a new system devised by Giovan Battista Baliani, an influential citizen of Genoa, an engineer, and a philosopher with close ties to the Jesuits. Baliani had proposed putting the moon, yes, the moon, at the center of the solar system. The rationale was to save the theory of the tides that Galileo had put forward as a conclusive proof of the double motion of the earth. Baliani discussed his system at length with his Jesuit friends but did not publish it. That disappointed Riccioli, who had called for its delivery under the auspices of Lucina, the goddess of childbirth and bad dreams, so that he could add it to his collection of odd astronomical systems.[110]

In principle, each of these systems could use either circular or elliptical orbits, and, to regulate the motion, eccentrics, equants, or the area law.[111] Again, nothing required those who favored ellipses to stop there: perhaps some other curve would be better, an ovoid maybe, or an egg, which Kepler himself had tried. Cassini was willing to entertain any sort of curve other than a circle and, after his removal to France, "carried away by the blind desire to give his name to a discovery that would go down to the most distant posterity," he proposed to replace Kepler's ellipses with a new class of ovals.[112] Although later astronomers have scoffed at these curves, which are defined as the locus of points the product of whose distance from two fixed points is a constant, they earned Cassini high marks from Newton and were preferred to Kepler's, for a time, in Paris. The sculptor who inscribed these ovals on the monument to Cassini in the Paris Observatory chose a fair symbol for his subject although, no doubt, the meridian line would have been better: the curves express Cassini's boldness of thought and mastery of geometry, as well as his ambition for fame, indebtedness to Kepler, and indifference to the opinions of others.[113]

Everyone hoped that nature made use of a principle of motion less unpleasant than the area law and that diligent astronomers could discover it. Meanwhile, diverse new observational techniques and calculations further perplexed theorists already overwhelmed with choices. Riccioli wrote in the introduction to his *Almagest* that not even the basic parameters of the sun's orbit had been established and that astronomers had to rely on ephemerides constructed on opposing principles since some proved better for one set of predictions and others for another. "I say this ... not to cause posterity to despair, but to excite their industry toward this heap of difficulties." In this opinion he had the concurrence of Francesco Levera, a man often and aptly critical of the Bolognese school of astronomy. He wrote in 1663, in a huge *Prodromus* to astronomy, that "the truth of celestial motions hides itself more deeply every day, and is more and more palpably obscured."[114] A philosopher, astrologer, theologian, doctor of both laws, and formerly a hanger-on in the entourage of Urban VIII, Levera had replaced Cassini as director of Christina's celestial studies. The *Prodromus* was part of a major project "beyond human expectation," funded by Christina, which would have established all the sciences of Urania, from astrology through geography, on firm principles. Unfortunately, Levera did not complete his design. His epitaph, by Christina: "to him we owe our knowledge of the true motions of the sun; [only] death prevented him from doing the rest at my expense."[115]

Levera, Cassini, and a few others undaunted by uncertainties and hard work, fancied that a royal way to the stars existed and that they would know it when they found it by its simplicity and directness. They therefore rushed in where Kepler himself had failed, and (as was characteristic of him) had admitted failure. Kepler could not find a simple geometrical method of deducing the elements of his elliptical orbits (their eccentricities e and the direction of their lines of absides ψ) from observations of the planets or, for that matter, determine geometrically from his area law where to find them once he had obtained the elements by trial and error. He doubted, rightly, that an exact geometrical method existed and he knew that anyone who found one would be a great mathematician. "Whoever shows the way to my erring self will be a great Apollonius to me."[116]

Kepler's problem may be understood most directly by considering the case of the earth (Riccioli and Levera would have said the sun) and assuming the e and ψ of its orbit known. Then the problem amounts to finding the "true anomaly" θ (Figure 3.20), that is, the earth's angular position a time t after passing perihelion, from the curvilinear area ΓΠΡ, which, according to Keplerian astronomy, is proportional to the time since passage through Π.[117] Since, again according to Keplerian astronomy, the orbit is an ellipse, the calculation involves an equation that cannot be solved exactly. Indeed, a direct attack on the true anomaly θ proved to be too difficult, so Kepler had recourse to the "eccentric anomaly" η, which is defined implicitly in Figure 3.20.

To obtain "Kepler's equation" for η requires what was advanced mathematics in

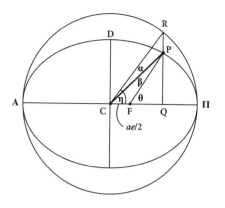

FIG. 3.20. Diagram for the deduction
of Kepler's equation; θ is the true, η
the eccentric anomaly.

the seventeenth century. Let b be the semi-minor axis CD of the ellipse and a the semi-major axis CΠ, which is also the radius of the circle centered on C and passing through A and Π. Drop the perpendicular PQ from the earth's position P at the time t onto the line of absides at Q and continue QP to intersect the circumscribing circle at R. Draw CR; ∠RCQ = η, the eccentric anomaly. The elliptical calculator was given FΠP, the fraction of the total area swept out in time t. The total area, πab, is described in a year; hence, if t is measured in days, FΠP = $(t/y)(\pi ab)$, where y as usual designates the number of days in a year. But also, FΠP = ΔFPQ + PQΠ. According to a well-known proposition in Apollonius' *Conics*, PQ = (b/a)RQ; hence the area PQΠ of the ellipse is (b/a) times the area RQΠ of the circle, and all calculations can be made using η instead of θ. With this substitution, the equality of areas, ΔFPQ + PQΠ = FΠP = $\pi abt/y$, yields Kepler's equation.[118] Although simple to write down, at least for a trigonometer, it can be solved only by guesswork and successive approximations, "by a method that is not natural, but alien [to astronomy]." This was the opinion of Seth Ward, professor of geometry at Oxford, who later became an Anglican bishop. It appears in his version of geometric astronomy, published around the summer solstice of 1656, that is, at about the same time that Cassini issued the first results from his heliometer.[119]

By the late 1650s many, perhaps most, mathematical astronomers had adopted ellipses in place of the old circles. The generalization applies not only to believers in the truth of the "Copernicano-elliptical" system, like Ward, but also to anti-Copernicans like Riccioli (whose account of the area rule was then the fullest outside Kepler's own) and conservative agnostics like Cassini.[120] The first elliptical astronomer to claim to have solved Kepler's problem geometrically was Ismael Boulliau, who offered a method that he did not recognize as the equivalent of installing an equant at the empty focus. That was in 1645. Eight years later, Ward demonstrated the equivalence and corrected several serious errors in Boulliau's geometry in the gentle manner then used by mathematicians. "[Bouillau] wrote

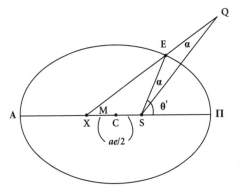

FIG. 3.21. Motion in an ellipse
regulated by an equant X.

rashly and prematurely about things of which he perceived and understood very little (which is human) and published them precipitately to the great injury of students." That left vacant the position of Kepler's Apollonius. Ward again: "Although Boulliau never solved the calculation of the first inequality [that is, the elliptical path] legitimately and geometrically, and Kepler judged it impossible to find the true and apparent anomaly from the mean motion accurately and almost deductively; yet methods exist, indeed several of them, for doing so."[121]

Ward's elegant methods require prior determination of the line of apsides and the eccentricity, say by observations around the solstices, and the placement of an equant in the unoccupied focus. Here is its principle. In Figure 3.21, $A\Pi$ is the line of apsides, E and S the earth and sun or vice versa and X the equant in the unoccupied focus. Let the mean anomaly be M, the true anomaly (the quantity sought), θ'. Ward's trick: extend XE to Q making EQ = SE; then, by a basic property of the ellipse, XQ = $A\Pi$ = $2a$. The rest is trigonometry. It delivers the true anomaly in terms of the absidal distances and the mean anomaly. "And thus is supplied what has been lacking in elliptical astronomy ... in the calculation of the first inequality."[122] (See Appendix D.)

Three years after correcting Boulliau, Ward published an *Astronomia geometrica* dedicated to him, and also to Gassendi and Riccioli. "Should this book come to you," so Ward addressed Riccioli, "you might want to favor it, with your characteristic kindness, since you have expressed the hope that your and Grimaldi's labors will not be infinite and that astronomical calculations will no longer have to be done over and over again. I flatter myself that now that I have shown the way by a geometrical method, which makes immediate use of the easiest and most accessible observations, astronomy will quickly reach perfection." And if this did not suffice to put the authoritative compiler of the big new *Almagest* in his place, Ward added: "Do not despise my book because it is small, since the seeds of the greatest things are little, and little things also have a certain charm; and this work stretches out the thread by which you can escape from the labyrinths in which, in your candor, you confess that you are sometimes caught."[123]

Was this enough to claim the mantle of Apollonius? Ward preferred to use his triumph like a modern academic, to justify the time that he (and Riccioli) had for research. "If I've delivered astronomy from a scandal, if I've shown the method by which everything can be done accurately and geometrically . . . then I hope that candid readers will wish the best for this flourishing academy [Oxford] and that they will not be mislead by a frothy enthusiast or an unsound circle squarer to envy us our learned leisure; and that they will accept more indulgently the lapses of the author (who hardly ever reread anything he wrote) and the printer, who took long enough about his business."[124]

Ward probably thought that he had an exact solution to Kepler's problem.[125] If he did, he was mistaken. But not by much. The difference between the true anomaly (θ in figure 3.20) and the pseudo-true anomaly of Ward's theory (θ' in figure 3.21) has a maximum value of $e^2/16$, a mote well-nigh undetectable at the time (see Appendix E). In the case of the sun, it amounts to 14 seconds of arc and occurs a month and a half before and after passage through the absides. Still, it is not zero. The all-seeing Kepler had known that his sort of elliptical motion could not be duplicated by an equant. Ward's followers in England had reached the same conclusion by 1658, but continued to use equants to ease their calculations.[126]

Ward published his criticism of Boulliau twice in 1653 — alone and together with a treatise on comets — and again in 1654, in the company of the comets and a little piece on trigonometry. Neither it, nor Ward's *Astronomia geometrica* of 1656, could long have eluded the Jesuit network that kept Riccioli current. Very likely Cassini knew something about Boulliau and Ward by 1655 and, perhaps, made some use of their approaches in his crusade against "ageometria". He said that he had worked out his ideas by 1653, when he wrote to Gassendi requesting observations of oppositions of the superior planets to the sun. Gassendi complied, with raw data. "I give no calculation, lest I introduce a bias in whatever you have in hand."[127] The data would have been useful for establishing the elements of orbits. If Cassini then had a geometrical method for the reduction, as his first biographers aver, he did not disclose it.[128]

Cassini's colleagues in the Paris Academy, who had heard about his boast to have ended ageometria, may have forced him to redeem it. That would explain why, shortly after setting up in France, he lectured about geometrical astronomy in the King's library. An abstract of his boasted method, with no geometrical justification, appeared in the *Journal des sçavans* for September 1669, with the usual promise that a full treatise, which never materialized, would follow. Here is the method.[129]

In Figure 3.22, A, B, and C are three zodiacal positions of the sun as seen from

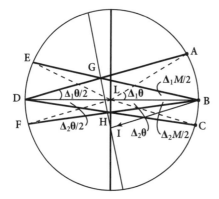

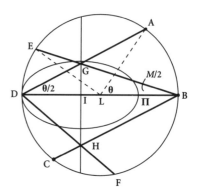

FIG. 3.22. Cassini's method of determining elliptical orbits geometrically via three observed positions A, B, C.

FIG. 3.23. Cassini's method simplified when the line of apsides is known.

the earth at L. Choose one of these positions, B say, as a reference point and draw the reference diameter BD. Connect D with A and C. If $\Delta_1\theta$ and $\Delta_2\theta$ are the angular differences between A and B and B and C, respectively, as seen from L, then, by Euclid III.20, $\angle ADB = \Delta_1\theta/2$ and $\angle CDB = \Delta_2\theta/2$. Next, being very clever, strike off the arcs DE and DF equal to the differences Δ_1M and Δ_2M in mean anomaly as seen from L (that is, make $\angle ELB = 180° - \Delta_1M$ and $\angle BLF = 180° - \Delta_2M$). Join B to E and F; then $\angle EBD = \Delta M_1/2$ and $\angle DBF = \Delta_2M/2$. Let G be the intersection of BE and AD, and H that of BF and DC. Draw the chord through GH. Drop the perpendicular BI on this chord. Then I is the center of the orbit, LI its eccentricity, and LI extended the line of apsides. The method implies that any number of observations of the sun can be employed, all of them producing a point on the same chord GH.[130]

Cassini was not allowed to redeem his boast so easily. What might have passed for a refoundation of astronomical technique in 1653 or even 1655 could not do so after Ward's treatise of 1656. Nicholas Mercator, a Continental geometer living in England, who later gained fame by designing and building the fountains at Versailles, immediately supplied a demonstration based on the teachings of the "most illustrious professor of astronomy . . . of the most celebrated university of Oxford." More important by far, however, was Mercator's insistence that neither Ward's nor Cassini's nor any other theory using an equant point could duplicate Kepler's laws or save the phenomena. On this showing, which English astronomers accepted, Cassini was neither first nor right in claiming to have devised a method that made astronomy geometrical again.[131]

A paraphrase of the first steps in this demonstration will suggest the rationale of the method and its connection with elliptical astronomy. Assume first that BD contains the line of apsides (Figure 3.23). Locate the point G according to Cassini's prescription and drop the perpendicular GI. The point I is the center of an ellipse of major axis 2*a* and eccentricity IL = *ae* to the accuracy of Ward's approximation (see Appendix F). The general case considered by Cassini brings nothing new except some clever and troublesome geometry. Declared without proof, it was a mystery. "At first you do not see the reason for the construction, and the author does not give a hint of any." Thus Delambre, who, digging as usual at Cassini, apologized in his *Histoire de l'astronomie moderne* for mentioning the matter. "But the reputation of the author, the importance he attached to it, the way he boasted about it, demand a closer study." Delambre's way had been cleared by Mercator, who gave an unnecessarily complex demonstration of what he called disdainfully Cassini's *lucubratiunculae,* his little midnight labors; by other exponents of Ward's approach; by Jacques Cassini, who finally published, in 1723, what may have been his father's demonstration of 1653; and by Eustachio Manfredi, one of Cassini's successors at Bologna, in 1736. Delambre gave the most direct proof.[132]

After working through the geometry, Delambre permitted himself a moment's appreciation of the cleverness of the method, especially the bit about the perpendicular. But for Delambre, the manner of its delivery outweighed the merits of its technique. "Are we not tempted to see a sort of charlatanism in the emphasis in which the method was announced, in the incomplete idea of it given at different times, in the unnatural way it was definitively explained thirty years [in fact ten] after its author's death?" Still there are other opinions. A few months after the first sketch of the method appeared in the *Journal des sçavans,* a geometer better even than Delambre declared himself "much pleased at Monsieur Cassini's invention for finding y^e. Apogaea & eccentricitys of y^e. Planets." The pleasure was Isaac Newton's, who, in the 1660s and 1670s, was trying his hand at equant theories.[133] Astronomers laboring ageometrically to find lines of apsides and eccentricities from Kepler's equation could not help but be impressed by Cassini's method of obtaining them to a very close approximation merely by drawing a circle and a few straight lines. Newton's worthy rival Huygens also developed schemes similar to Cassini's and Manfredi's. More must be said before taking the measure of the grand Cassini.

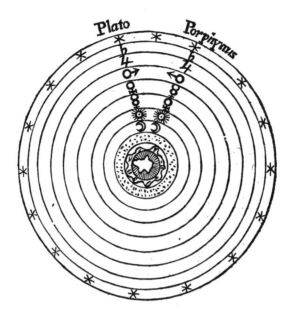

4: Normal Science

The first assiduous observers at San Petronio were Cassini, Grimaldi, and Riccioli. Their combined observations fill about twenty-five pages, at fifteen observations per page, in a register published in 1736. Each entry includes a description of the weather, the distances of the sun's limbs from the vertex corrected for the penumbra, and the apparent diameter of the sun, all given to seconds of arc. In total, the register, which runs from 1655 to 1736, records some 4,500 observations. So large a cornucopia could not have been collected without a community of collectors and a regular means of recruiting them. They included the original observers Cassini, Grimaldi, Mengoli, Montanari, and Riccioli; Cassini's students Augustino Fabri, who complied astrological ephemerides for physicians and, like Malvasia, cast horoscopes for the equinoxes and solstices, and Flaminio Mezzavaca, likewise a maker of almanacs, but also a lawyer and, eventually, governor of several papal dominions; and those who perpetuated the observational routine, Montanari's student Domenico Guglielmini and Guglielmini's student Eustachio Manfredi, who drew up the register of 1736.[1] These last two deserve an introduction.

Guglielmini earned an M.D. from the University of Bologna in 1678 at the age of twenty-two. He won an enduring reputation as an authority on hydraulics, which he pursued both theoretically, as a professor, and practically, as head of Bologna's waterworks and advisor to the Pope on the meandering of the Po and the flooding of the Reno. Leibniz may be listed among his admirers. In 1698, nine years after succeeding to Montanari's professorship, Guglielmini left Bologna to become Galileo's successor several times removed at the University of Padua.[2] In all this his career resembled Cassini's. But he lacked his predecessor's flair and grace. The biographer of Bolognese, Fantuzzi: "He had by nature the temperament of a scholar, healthy, robust, and melancholy . . . also rough and difficult." The biographer of savants, Fontenelle: "He had the sort of conversation characteristic of the study, a little crude and savage, at least for those unused to it." Guglielmini had entered the republic of letters with a fierce and prolix defense of Montanari against a rival mathematician. "Two or three pages would have been enough for the truth," sniffed Fontenelle, "[but] passions make books."[3] Guglielmini also was faithful to Cassini, as will appear.

Eustachio Manfredi, son of a notary of Bologna, pupil of the Jesuits, earned a degree in law at the university in 1695, at the age of eighteen. Seeking greater enlightenment, he formed a little academy, the Inquieti, which followed Montanari's Traccia and preceded Bologna's Accademia delle Scienze; and he took up the study of geography, history, and astrology. That brought him to mathematics, to Guglielmini, and to a break with astrology. His new teacher, who had had a hand in Montanari's famous lampoon of astrology, soon convinced him of the shortcomings of the art. He was left with astronomy. He helped Guglielmini observe at San Petronio and, after Guglielmini's transfer to Padua, made the instrument his own. In 1699 he succeeded his teacher as public lecturer in mathematics at the university. That pleased Cassini, as it ensured the continuation of observations at the old heliometer. In due course, Manfredi added to his teaching the charges of superintendent of waters and combatant in the unceasing battle of the Reno and the Po.[4]

Manfredi made astronomy a family affair. His younger brother Gabriele, perhaps the best Bolognese mathematician of his generation, often observed at San Petronio. An early practitioner of Leibniz' calculus, which he had learned through Guglielmini, Gabriele became professor of mathematics at the university in 1720 and, after Eustachio's death in 1739, his successor at the waterworks. Their brother Eraclito, who also became a professor of mathematics at Bologna, likewise followed the sun on the floor of the cathedral. Their sisters Maddelena and Teresa acted as their computing bureau. The sisters did most of the calculations for Eustachio's *Ephemerides bononienses* (1715), which was so excellent an illustration of its genre that the Jesuits showed it in China as an exemplar of European strength in astronomy.[5]

Manfredi was a perfect group leader: generous, affable, compassionate, literate (he was a fine poet), a good trencherman, he differed from Mengoli and Guglielmini, and from Cassini too, in possessing humility, "which is not very common among the learned."[6] The most notable of the many collaborators drawn to San Petronio by Manfredi's good nature and wide reputation was Andreas Celsius, the Swedish savant now universally known for his temperature scale, who worked at San Petronio for about seven months in 1733–34.[7] Celsius' measurements flowed into Manfredi's big register. The learned world, through its mouthpiece the *Acta eruditorum,* praised this compendium on its publication in 1736 for its current value: "There is scarcely anything else among solar observations more suited for the basis and foundation of the rest of astronomy."[8]

One of the most useful results of these systematic observations was Riccioli's conversion to Kepler's approach to planetary orbits. Riccioli could not resist Cassini's demonstrations when confirmed by his own laborious measurements. He accordingly bisected the eccentricity and calculated elliptically in his updated Almagest of 1665, to which he gave the Cassini-like title *Astronomy restored;* and no doubt he would have put all the planets around the sun had holy writ and papal edict not abundantly proved that God was not a Copernican.[9] The supplement to Riccioli's astronomical encyclopedia therefore spread the news not only of the confirmation of Kepler, but also of the excellence of the *meridiana* as an instrument of exact measurement. In both points he had changed his mind since writing the *Almagestum novum,* which had pointed out the faults of Danti's line and rejected the observations made there.

When the group around Francesco Levera in Rome criticized the theories that Cassini drew from his observations of the flickering images in San Petronio, they cited Riccioli's condemnation of Danti's instrument in support of their views.[10] Riccioli undertook the defense of the observatory on which he and the mathematicians of Bologna had lavished their labors. The heliometer of San Petronio, retorted the Jesuit father, was constructed with a care "more angelic than human." It gave the opportunity for measurements far more refined than those of Tycho Brahe, in which Levera put his faith. Yes, the big image flickered. "But if any little error should slip in, it will be just about insensible, and quite negligible." With Riccioli's strong endorsement, the reputation of San Petronio and its observers for extreme accuracy in solar measurements spread beyond Italy. "The very exact [observations] made in their great church . . . are something extraordinary and to say the truth the most beautiful ever made in this field." Thus one of Huygens' Parisian correspondents. Even the skeptical and exacting Flamsteed was impressed. The height of the *meridiana,* its precise orientation, its perfect leveling, "abundantly convince us of the curiosity [care] and diligence of [Cassini] & may induce us to believe his observations very accurate."[11]

Despite this acclaim, Levera's criticism had a sound basis. He complained, rightly, that Cassini relied upon observations contaminated not only by the unsteadiness of the images but also by the unknown effects of parallax and refraction. As we know, the error in determining the position of the sun's center owing to the flickering could be rendered negligible by appropriate observational protocols. The tangled effects of parallax and refraction, however, could swamp the measurements needed to decide the delicate question of the bisection of the eccentricity. As Levera observed, quoting again from Riccioli, an error of 16 seconds in declination amounted to a whole day around the solstices. Since astronomers still differed by a minute or more in their assignments of parallax, Cassini might have been off by four days in his solstitial measurements, or even more, since he assumed a very small parallax (this in 1656) while the sacred Tycho had plumped for 2 minutes or more. Also, comparisons of the sun's apparent diameter at different times suffered from seasonal changes in the atmosphere that were supposed to modify the effects of refraction. In this way, Cassini had obtained an apparent solar diameter of 31'40" at the vernal equinox of 1656 whereas Tycho had never recorded one larger than 30'50". Cassini's results could not be trusted to within a degree let alone the minute or two he claimed.[12]

These criticisms, which told against the measurements of 1655–56 from which Cassini demonstrated the bisection of the eccentricity, had lost their force by the time Levera's *Prodromus* was published in 1663. Ongoing observations at San Petronio, analyzed brilliantly by Cassini, had helped to untangle the effects of parallax and refraction and establish a solar theory far better than Tycho's or even Kepler's. Cassini had published his improved tables of parallax and refraction in 1662, in Malvasia's *Ephemerides*. Levera's ignorance of these results and his inexperience with practical astronomy (as Riccioli was informed by people who knew Levera well) exposed him in turn to the criticism and ridicule of Riccioli and Cassini.[13] There follows an account of the work that gave them confidence, work that (to quote Manfredi) "raised up and set aside most of the difficulties of the motions of the sun, which had stymied the old astronomers."[14]

Perfecting the Parameters

A good value for the obliquity of the ecliptic (ε) is prerequisite to any exact astronomy. Together with the local latitude, it fixes the place of the celestial equator and the tropics, the directions of sunrise and sunset, and the sun's declination at every point of the ecliptic. Its approximate measure may be obtained without much trouble either by taking half of the difference between the sun's noon altitudes at succeeding solstices, or the whole difference between the altitudes at the summer solstice and an equinox. The two results, which should be the same, may be written $\varepsilon_1 = (\alpha_{ss} - \alpha_{ws})/2$ and $\varepsilon_2 = \alpha_{ss} - (90°-\phi)$. In the second determination, the noon altitude of the equinoctial sun is inferred from observations of the height of the pole.

The values obtained for ε_1 and ε_2 in this way by observers in Europe would disagree irrespective of the quality of their instruments. To improve the values, the raw measurements of altitudes had to be corrected for parallax, refraction, and the noon deficit, to mention only the annoyances known in the seventeenth century. Parallax and refraction, which, as we know, plagued astronomers before Cassini, were, as we shall see, brought under control largely through his work at San Petronio. The noon deficit, the difference in declination of the sun from a solstice at noon on the day of that solstice, might seem to require important corrections to α_{ss} and α_{ws}. But a miss of even twelve hours amounts to a mere 3.5 seconds of arc (see Appendix G). Let us turn to more serious matters.

· SOLAR PARALLAX ·

The maximum correction for parallax, $\chi(0)$, turned out to be little larger than the maximum size of the noon deficit. That was a great discovery. A much larger value for $\chi(0)$, a little under 3', had been accepted by most prominent astronomers from Ptolemy up to, and including, Copernicus and Tycho. What gave it such endurance was an unfortunate coincidence between the distance from the earth to the mean sun, calculated on the assumption that the circles and epicycles that carried the planets Mercury and Venus fit as tightly as possible between the orbits of the moon and the sun, and the same distance calculated from direct observation. The second method, one of the most ingenious contrivances of Greek geometrical astronomy, is indicated in Figure 4.1. Here ZD represents the sun's radius, YC $= r$ the earth's, and XB the approximate length of half the journey of the moon through the earth's shadow during a total lunar eclipse. BC $= m$ and CD $= s$ are the average

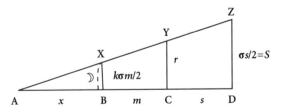

FIG. 4.1. Diagram for finding the solar distance from a lunar eclipse. C and D are the centers, and CY $= r$ and DZ $= S$ the radii, of the earth and sun, respectively.

distances from the earth of the moon and the sun, respectively. To measure XB, the Greeks used the unknown radius of the moon and a trick. They timed the moon's transit from complete disappearance to first reappearance and also the interval between first touch and full immersion; the ratio k of the transit to the interval gives the number of lunar diameters in XB.

The analysis of Figure 4.1 requires the information indicated in Figure 4.2. The apparent diameter of the moon is almost exactly equal to that of the sun. Hence the three similar triangles in Figure 4.1 yield the relationship

$$m = rs/[(\sigma s/2)(1 + k) - r]. \tag{4.1}$$

It may encourage the geometrically challenged to know that Copernicus and Tycho both slipped up in this calculation. "How careful the astronomer must be," sighed Riccioli, "to bind the apparent radii of the luminaries and the shadow, the parallaxes, distances, and angle of the shadow together by a geometrical chain on [this] hypothesis."[15] To complete the binding, the astronomer measured the angle γ between the sun and the moon precisely at half moon, when $\angle$EMS in Figure 4.3 comes to 90°. Exploiting the labor of mathematicians who had busied themselves with calculating the ratios of the sides of right triangles containing any given acute angle, the astronomer could convert his observation of γ into a value for s/m. Trigonometers call this particular ratio the "secant of γ," written "sec γ."

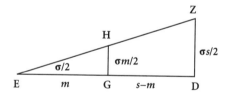

FIG. 4.2. Equal apparent sizes of the sun and the moon.

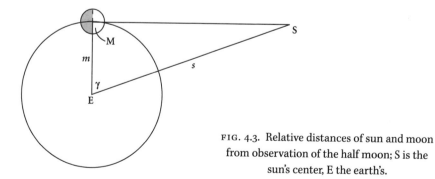

FIG. 4.3. Relative distances of sun and moon from observation of the half moon; S is the sun's center, E the earth's.

With this substitution, Equation 4.2 provides the desired result,

$$s = [r(1 + \sec \gamma)]/[(1 + k)(\sigma/2)]. \tag{4.2}$$

The value of s is very sensitive to small changes in γ since γ is not far from 90°, where the secant goes to infinity. The value of m does not suffer a similar fitfulness. The larger γ, the larger s and the smaller the parallax $\chi(0) = r/s$; if r/s is dropped in comparison with σ, equation 4.1 gives

$$m = [229/(1 + k)]r. \tag{4.3}$$

The value of k varies depending on m at the time of eclipse and, of course, on the accuracy of the timing. A convenient average is 2.8, since it gives the neat result $m = 60r$.[16]

With $m = 60r$, $s = m\sec \gamma = 60r\sec \gamma$. Since $\sec 87° = 19$, a measurement of γ of just over 87° produced another neat answer, $s = 1200r$. Of course, if γ came out 86° or 88°, the result would have been far different, 860r in the first case, 1720r, exactly twice as much, in the second. This sensitivity cloaked the most important cosmological constant in uncertainty. Astronomers understood that the solar distance obtained by the method of the half moon depended on the exact measurement of a quantity they could not measure exactly. Taking a pessimistic approach to the problem, they assigned to $\chi(0)$ the largest angle they could not measure at all, that is, the angle just under the least their instruments could resolve. Ptolemy set this limit at 3'. He then had $s = r/\chi(0) = 1146r$.[17] All roads converged on 1200.

The first to whittle down the parallax was Kepler, who, as usual, gave a crazy reason for going in the right direction. He invented the law that the ratio of the volumes of the earth and sun should be in the ratio of their distances from the earth's center. That made $s = 3438r$.[18] The solar system had begun to expand. Kepler hoped to check his law—which, incidentally, gave $m = 60r$, and so not only confirmed the received value but made the lunar distance the mean proportion between the earth's radius and the solar distance—by observation of the half-moon through the newly invented telescope.

New tools are not an unalloyed good. The telescopes of Kepler's time confused the judgment of the moment of dichotomy by showing fuzzy shadows cast by the lunar mountains.[19] Eventually, the improvement of lenses, the enlargement of instruments, and familiarity with the lunar landscape permitted the observation of the half-moon that Kepler recommended. In this way the moon-mountain men Grimaldi and Riccioli obtained the bold new result, $\chi(0) = 28''$, $s \approx 7400r$.[20] Riccioli gathered these and other values together in one of his invaluable tables (Table 4.1). Meanwhile, the practice of estimating $\chi(0)$ as the limit of observation also pro-

Table 4.1 Solar distances in terrestrial radii

Observer	Apogee	Mean	Perigee
Ptolemy	1210	1168	1126
Clavius	1210	1168	1126
Copernicus	1179	1142	1105
Tycho	1182	1150	1118
Kepler (1610)	1800	1768	1736
Kepler (1629)	3438	3381	3327
Kircher	1940	1906	1872
Riccioli/Grimaldi	7600	7300	7000

Source: Riccioli, *Alm. nov.* (1651), *1:1*, 110.

gressed with the improvement of lenses and the first use of micrometer eyepieces. The limit could be reduced not only by direct nonobservation of χ but also by failing to find a parallax in Mars at opposition (when the earth stands between it and the sun). Since in this position Mars is closer to the earth than to the sun, $\chi(0)$ can be no bigger than the horizontal parallax of Mars diminished by the ratio of their apparent diameters. In this way, an English astronomer, Jeremiah Horrocks, set an upper bound to $\chi(0)$ of 15″, a value adopted by Thomas Streete, who published in 1661 a "new theory of the coelestial motions" for the newly restored Stuart regime, and then by Flamsteed, who arrived at the fundamentally correct idea that no sensible error would be committed by setting the solar parallax equal to zero.[21]

By the 1660s the problem of the parallax had become acute and embarrassing. Something better than guessing at γ or setting upper bounds was needed. Streete: "[χ] is of so great concernment in Astronomy, that without it we can never make any such Theory and Tables of the Coelestial motions, as shall be proved near enough concentaneous unto truth." The Parisian astronomers broke their heads over the subject, some staying with Kepler, others, like Huygens, supposing a value near zero. Cassini could show them that Huygens was right. "It is very difficult to say anything precise in this matter," he wrote, and may also have said to his new colleagues, "it is one of the most troublesome in astronomy."[22] "However," he continued, "I have a way to find out how to speak truly as well as precisely about it."

· REFRACTION ·

An irksome feature of parallax was its entanglement with refraction (ρ). The quantity that must be applied to the sun's apparent position to allow comparison

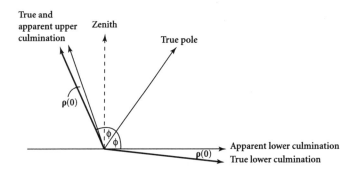

FIG. 4.4. Effect of atmospheric refraction on the apparent heights of stars.

with its predicted position according to solar theory is the difference between the corrections for parallax and refraction. The adjustment, which involves only the net, provides no information about either correction separately. To advance, the astronomer had to make an arbitrary assumption.[23] A plausible one was that the stars have no discoverable parallax. Circumpolar stars therefore made good indicators of refraction. The first astronomer to provide quantitative values for refraction was Tycho.

Tycho assumed that he could neglect refraction for the polestar and obtained in the usual way, which he deemed unproblematic, the height of the pole ϕ. A star that appeared to graze his horizon at lower culmination would cross the meridian at upper culmination at an altitude of $180° - 2\phi$ (Figure 4.4) *if there were no refraction.* But in fact, upper culmination occurs a little bit further from the pole than $180° - 2\phi$. This bit, $\rho(0)$, is the amount by which refraction lifted the star at lower culmination to bring its light into Tycho's instruments. Since the star was under the horizon, its true polar distance was not ϕ, but $\phi + \rho(0)$, and, hence, its observable distance at upper culmination was $\alpha = 180° - 2\phi + \rho(0)$. Since Tycho knew, or thought he knew, ϕ, his measurement of α gave him $\rho(0)$. He made it out to be 30'. Examination of circumpolar stars closer to the pole showed the refraction to be 10' at 5°, 5'30" at 10°, 3' at 15°, and 0' at 20° and above.[24]

Tycho could not use these numbers for the refraction of the sun because he assumed a horizontal parallax twenty times too large. Since parallax depresses the sun while refraction raises it, Tycho had to increase $\rho(0)$ by 3' (in fact he made it 4') to kill his excessive parallactic "correction." Since he had the same problem at higher altitudes, he allowed sensible solar refractions up to 45° (whereas stellar refractions ceased at 20°). Higher up, the still large parallactic correction reigned unopposed.[25] The final test of the refraction tables was a comparison of the corrected observed altitudes of the sun with the predictions of solar theory. Since the parameters of the theory — e, ψ, and ε — in turn depended upon the corrections applied to the observed altitudes, progress could come only through an intricate circular play with numbers. San Petronio was to supply the raw data for the most telling confrontation between the solar theories and the competing proposals for correction of the observations that underlay and tested them. An indication of the complexity of the play is that Riccioli could make things balance only by using three different tables of solar refraction, one valid for the summer, a second for the winter, and a third for the times between.[26]

In 1656 Cassini broke through the difficulties by declaring that the solar parallax did not exceed 12". On this brave assumption, he obtained a solar theory in good agreement with the measurements at San Petronio. The oracle seemed again to have vouchsafed to him information hidden from other mortals. How great therefore must have been his shock to discover that his beautiful solar theory, drawn up with such toil, conflicted with the height of the pole as he and his Jesuit mentors had fixed it in 1655. The difference, over 2', was a disgrace for an astronomer who claimed to be working to an accuracy of 15". Naturally, he sought the cause of the discrepancy not in his hard-won theory, but in the determination of the pole height. What had failed was the equation

$$\alpha_c(ss) = 90° - \phi_c + (\alpha_c(ss) - \alpha_c(ws))/2, \tag{4.4}$$

where the subscript indicates compensated values.

Assume that Cassini's corrections for refraction corrected for his miscorrection for parallax. Then the preceding equation should have held if ϕ_c had been corrected properly. But, as we know, Cassini, following the standard practice, ignored refraction at 45° and so supposed the pole in Bologna to be higher than it is by more than a minute. This neglect threw out $\rho(0)$ by two minutes.[27] (In Figure 4.4, replacement of ϕ by $\phi - 1'$ will make the polar distance at upper culmination $\phi + \rho(0) + 1'$; hence at lower culmination, when it appears on the horizon, the star's true position is $\rho(0) + 2'$ below the horizon.) The proper value of the stellar refraction at $\alpha = 0$ was then not $\rho(0)$, but $\rho(0) + 2'$. By introducing a minute of refraction at 45° and recalculating the other refractions, Cassini obtained a solar theory that agreed not only with the observations at San Petronio but also with the independent determinations of its latitude.[28]

The agreement made him uncomfortable. "The resulting distance of the sun was so incredible." Always wishing to be a believer, Cassini fell back on Kepler's hypothesis, on which he calculated the tables of refraction that Malvasia extracted from him for publication in the *Ephemerides* of 1662.[29] Following Riccioli, Cassini presented Malvasia with a table for each season of the year, the summer one being identical with the single table composed on the basis of a solar parallax under 12".[30] The equinoctial table implies $\chi(0) = 30"$, the winter, $\chi(0) = 1'$. With this hedged bet Cassini gave ρ(0) = 32'20", 32'40", and 33'0", and ρ(45°) = 59", 1'6", and 1'13", for the summer, spring-fall, and winter, respectively.[31] The difference between the extremes at 45° would have been at or beyond the limit of dependable measurement in 1660. Malvasia, who was very pleased with Cassini's tables, dismissed the worry that the ascription of refraction to solar altitudes above 45° conflicted with the received wisdom of astronomers. After all, he wrote, the supposed refraction fell to under a minute over 51°, "which I would not think immediately observable."[32]

Cassini's triple table of refractions resembled Riccioli's only superficially. Whereas his teacher allowed himself to assign values to the refractions at each angle and season as best fit the phenomena, Cassini restricted himself to only two parameters. He calculated the *stellar* refractions physically, on the basis of Snel's rule for the bending of light at the interface of different optical media. This rule had been the common property of mathematicians since Descartes had demonstrated it, and its uses, in 1637.[33] In applying it to the atmosphere, Cassini made one of those bold and even arbitrary assumptions he delighted to make. He decided that for his purposes the earth's atmosphere could be assigned a constant density and, therefore, a constant index of refraction μ for the application of Snel's law, and that the atmosphere extended only as far as needed to harmonize theory and observation. Although crude, the method can give excellent results with only two parameters—μ and the thickness t of the atmosphere—in place of adjustable refraction coefficients for each degree of altitude.[34]

Cassini's approach is indicated in Figure 4.5. The star that appears to an observer at O at altitude α stands at an altitude β = α − ρ(α) to the true horizon. Its rays that reach O enter the atmosphere at P, where they suffer their entire refraction according to Snel's rule. Things fit best if t ~ 2.6 miles (see Appendix H). Then, if ρ(0) = 32' (the result obtained by correcting for refraction at 45°) and t/r = 2.6/4000, ρ(45°) is 1'9", the amount Cassini had fixed on to bring his solar theory into agreement with the latitude of San Petronio. As we know he gave three tables of refraction in Malvasia's *Ephemerides*. That required different heights of the atmosphere for different seasons, which, as Flamsteed later remarked, "seemed absurd."[35] Putting the best face on things, Cassini observed, in the style of Kepler, that he had found very beautiful numerical coincidences. For example, recalculating the solar theory after correcting observations for an assumed horizontal parallax of 59.5" and the consequent refractions, he found the best fit by decreasing the eccentricity to 0.0170.[36] That put the center of the sun's orbit a distance $se = re/\chi(0)$ = 59r from the center of the earth. That was near enough to the traditional value of the mean separation of earth and moon. Hence, with Kepler's wrong value for the parallax, $\chi(0)$, se, and m all had the same measure, "a most elegant symmetry of the orbits."[37]

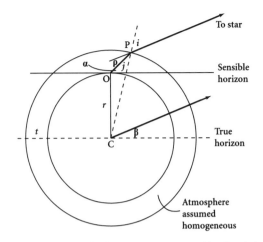

FIG. 4.5. The effect of refraction as computed by Cassini.

Riccioli approved neither this numerology nor the arbitrariness of the assumptions underlying the tables from which it resulted. He objected particularly to assigning finite stellar refractions at 45° and above, for which he and others had found no evidence. Also, Cassini's choice of the thickness of the atmosphere was purely opportunistic and everything suffered from the uncertainty in solar parallax, which still eluded specification "to the exactness of the few seconds of arc required by modern astronomy."[38] The publication of this rejection in the influential *Astronomia reformata* caused Cassini to publish his method in open letters to Montanari and other friends in 1666.[39] He had rechecked everything, he said, and found no discrepancies that could not be attributed to "either the trembling of the solar image or uncertainty in its boundary." Together he and Montanari had marked where the sun's image should fall on the heliometer of San Petronio according to the theories of Riccioli and of Cassini with and without correction for refraction at the pole. "And it was evident to the eye that the sun passed ordinarily very close to the marks made according to my latest hypothesis [with refraction], and very far from those derived from my first hypothesis, or from Father Riccioli's." With this confirmation, Cassini proposed the fundamental parameters $\phi = 44°29'5''$, $\varepsilon = 23°29'5''$.[40] They were used at Bologna until the astronomers there gave up Kepler's oversized parallax—once Cassini and Montanari had defeated Mengoli, who had arrived at different numbers based on much the same data and a misapplication of Snel's law.[41]

To go further with refraction, "without which," as Manfredi wrote in 1736, "nothing true can be set down exactly [*subtiliter*] about the position of the stars and their motions," Cassini needed greater facilities than the Senators of Bologna had provided.[42] He needed to check his solar theory where observations would not be confused by refraction. That meant a trip to the equator, where the sun never strays more than ε from the zenith. When Cassini arrived in Paris, the academicians may already have been discussing the desirability of such an excursion; after his arrival, it became one of their priorities. They proposed an expedition to the island of Cayenne at a latitude of 4°40'N. They deputed one of their junior members, Jean Richer, to lead it, and they asked the King to support it. The rhetoric of their appeal has been preserved by Cassini:

> It was necessary to undertake a painful trip, and to remain for a long time in an unbearable climate. But of what is the French nation not capable in the service of so great a king! Is any undertaking impossible for a prince like him, who spares nothing for his glory either in arms or in the arts?[43]

They divided up the chores. Richer would do the suffering, Louis the paying, the Academy the directing, and Cassini the calculating. The first three assignments on Richer's list of scientific activities were to find the true obliquity of the ecliptic, the true times of the equinoxes, and the parallaxes of the sun, Venus, and Mars.[44]

By the time of the expedition, in 1672, Cassini had again come to favor a small solar parallax. His test involved a comparison of extremes: predictions of ε derived from his refractions and new low parallax ($\chi(0) = 10''$) and from Tycho's refractions (null) and high parallax ($\chi(0) = 3'$). Tycho had made the true distance between the tropics, or 2ε, 47°3'; the corresponding figure from the Malvasian tables was 46°58'. Hence the observed values, $2\varepsilon_{obs}$, on the two hypotheses would be 47°5'23'' and 46°57'15'', respectively. In the first case the true value has been uncorrected by adding Tycho's large values of $\chi(ws)$ and $\chi(ss)$ computed for Cayenne; in the second case, by subtracting the net of refractions over the very small χ's of 3 or 4 seconds:

$$2\varepsilon = 180° - [\alpha(ws) - \rho(ws) + \alpha(ss) - \rho(ss)]$$
$$= 2e_{obs} + \rho(ws) + \rho(ss). \qquad (4.5)$$

That χ must be added and ρ subtracted to go from true to apparent values in this case, whereas the reverse holds in northern latitudes, follows from Figure 4.6.

With an octant of six-foot radius furnished with a limb divided to minutes and readable to 10'', Richer found $2\varepsilon_{obs} = 46°57'4''$, differing from Cassini's prediction by just over 10'' and from Tycho's by more than 8'.[45]

Turning the calculation around, Cassini corrected Richer's apparent obliquity by the refraction coefficients in the Malvasian tables to produce the overprecise ε = 23°29'54.5'' and a definitive, single table of refractions beginning with $\rho(0) = 32'20''$. A little further fiddling corrected the eccentricity of the solar orbit by 1 part

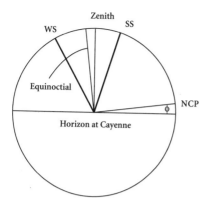

FIG. 4.6. Intersections of the tropics and equinoctial with the meridian at Cayenne.

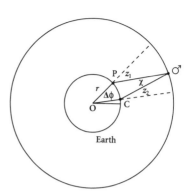

FIG. 4.7. The parallax χ of Mars between two stations on earth.

in 18. The resultant solar theory predicted the sun's declination to within a minute throughout the year, "which suffices for the use of geography, and navigation, and most astronomical work."[46] Copernicus' theory of the sun gave a maximum error in declination of 18', about twenty times Cassini's maximum.[47] This time all astronomers stipulated the superiority of Cassini's approach to refraction and parallax. No doubt, Cassini's solar tables broke through a major barrier to the advance of exact astronomy even though, as Riccioli and many others observed, it incorporated a caricature of the physics of atmospheric refraction.[48] A more faithful calculation, in which the density of the atmosphere decreases continually with height and a ray of starlight bends into a smooth curve, took decades to achieve, "the *Curve* which a Beam of *Light* describes, as it approaches the Earth, being one of the most perplexed and intricate that can well be proposed."[49]

While roasting in Cayenne, Richer made another set of observations that supported, or were made to support, the large solar system implied by the small value of $\chi(0)$ he had confirmed. In 1672 Mars stood unusually close to the earth. The circumstance gave the Paris astronomers a fine opportunity to measure its parallax by comparing Richer's observations made near the equator with Cassini's made some 45° to the north. The principle is illustrated in Figure 4.7, in which Paris (P) and Cayenne (C) are supposed, for simplicity, to lie on the same meridian. At the same instant, the observers take Mars' zenith distances z_1 and z_2. Since $\angle POC = \Delta\phi$, the difference in latitude between the stations, they would have, on comparing observations,

$$\chi = z_1 + z_2 - \Delta\phi. \tag{4.6}$$

From χ the distance to Mars, $r_{mars} = \text{O}\,\text{♂}$, easily follows on the supposition, which is very nearly correct, that Mars is equally distant from Paris and Cayenne.[50] From r_{mars} the distance to the sun s easily follows on the Copernican hypothesis, which, contrary to his usual practice, Cassini here employed freely. Figure 6.4 shows the general case. Beginning at opposition, when the sun (S), earth (E), and Mars (P in the figure) lie in the straight line SE_1P_1, and ignoring the eccentricities of the orbits, the almost exact astronomer computes the angles E_2SE_1 and P_2SP_1 at any later time from the known periods of revolution of the earth and Mars around the sun, and measures $\angle SE_2P_2$ between the sun and Mars. That provides all the angles in ΔSE_2P_2. Since one of its sides, r_{mars}, is also known, so, by trigonometry, is the side $E_2S = s$, the desired solar distance, and the maximum average solar parallax $\chi(0) = r/s$.

In practice the calculation of Mars' parallax left much room for fiddling, especially in the choice of pairs of zenith distances to combine to form χ. Naturally Cassini picked the numbers that came out to agree with the solar parallax that he had already adopted for his refraction tables. It appears therefore that it was not the Academy's expedition to Cayenne that fixed the solar parallax by observation of Mars, as is sometimes said, but Cassini's struggles to find the key to the solar motions documented at San Petronio that fixed the parallax of Mars.[51] Even Delambre, who understood the maneuver, applauded the end, the refraction table, "a beautiful piece of work, the fruit of much trial and error," though not the means. "Let us admire or, rather, congratulate Cassini for having known how to make his gnomon pay off so handsomely, but in future [Delambre was writing over a century after Cassini's death] we must refrain from recourse to such methods."[52]

Accepting that the determination of refraction was the critical problem of astronomy in the 1660s,[53] Cassini's refraction tables must be considered the most important single result of all the work at San Petronio. Updated in 1672, they were used for almost a century in France as well as in Italy. Elsewhere they were modified by various amounts. For example, for $\alpha = 30°$ Cassini gave 1'42", Flamsteed 1'23", Newton 1'32", Halley the same, and the abbé Lacaille, an excellent observer of the mid-eighteenth century, 1'55". These discrepancies hurt; as Lacaille remarked, astronomers who boasted accuracy to within 3" "corrected" their results by a number known only to within 20".[54] The slow and painful progress by which the accuracy of the corrections came to exceed the accuracy of the observations would be slow and painful to review. It was not completed before the nineteenth century.

We return to that capital quantity, ε, the obliquity of the ecliptic, and to that "celebrated debate, and leading problem in the work of almost all astronomers," whether it is, or was, constant in time, and whether, if not, it changed in one direction only or oscillated to and fro.[55] As usual, Riccioli collected all the relevant data and arranged it in a table, of which Table 4.2 is an excerpt.

What to conclude? The empiricist might decide that ε declined from the time of Ptolemy to that of Copernicus, when it leveled off at 23°30', or, perhaps, began slowly to increase, as further indicated by Riccioli's later measurement, in agreement with Cassini's, of 23°30'20''.[56] That is what Copernicus did, owing, in Flamsteed's opinion, to "the extraordinary & unreasonable veneration [the restorer of astronomy] had for ye assertions of the ancients."[57]

The conservative Riccioli had no such piety. Observing that the ancients did not have the modern passion for precision, he decided that all the values in his table were the same. He thus came again to agree with Kepler that ε does not change. "Indeed [he said] I think it more likely that God would have wanted the royal way, that is, the ecliptic to be one and the same forever."[58] In this opinion he had the happiness to be joined by Levera, who thought that a drifting ecliptic would imply a sloppy creator; and by Montanari, Mengoli, and Flamsteed, who reasoned nontheologically that the numbers favored constancy. Drawing on Riccioli's data corrected by his version of Cassini's refraction tables, Flamsteed found

Table 4.2   Riccioli's values of ε from Ptolemy to Riccioli

Date	Observer(s)	Amount over 23°
140	Ptolemy	51'20''
880	Albategnius	35'
1460	Regiomontanus	28'
1525	Copernicus	28'24''(30'47''[a])
1570	Danti	29'0'' (30'30''[a])
1586	Tycho	29'30''
1630	Gassendi	30'0''
1643	Riccioli/Grimaldi	30'0''
1646	Riccioli/Grimaldi	30'0''

Source: Riccioli, *Alm. nov.* (1651), *1:1*, 161–62.
a. As corrected for parallax and refraction by Riccioli.

Table 4.3 Cassini's values of ε from Regiomontanus to Cassini

Date	Observer	Value over 23°
1460	Regiomontanus	29'20.5"
1492	Walther	29'48"
1525	Copernicus	30'40"
1570	Danti	29'58.5"
1572	Danti	29'18.5"
1589	Tycho	29'45"
1589	Tycho	29'32.5"
ca. 1590	Tycho	29'20"
1632	Gassendi	30'45"
1656	Cassini	29'5"
1658	Cassini	28'55"
1660	Cassini	28'42"

Source: Excerpted from Cassini, in Malvasia, *Ephem.* (1662), 184.

that all previous determinations could be made to agree with ε = 23°29'. He needed only to assume that the ancients erred by as much as a degree, Copernicus and his contemporaries by 5', and Tycho by 2', in observing or reporting the noon altitudes of the sun.[59]

So dull a result and harmonious a choir could not recommend itself to Cassini. He made his own table of obliquities, corrected by his refraction tables of 1662, and inserted it in Malvasia's *Ephemerides*. His numbers differed considerably from Riccioli's (Table 4.3).

It would be rash to conclude much from these numbers beyond placing ε around 23°29'. So Cassini said just after presenting his table. But then, true to form, he ignored all previous determinations and deduced from the last three observations, all his own, that ε was decreasing at the rate of 6" a year. That was worth knowing. "This one thing I should not fail to warn you," Malvasia announced to users of his ephemeris, "the obliquity of the ecliptic is now manifestly decreasing." According to him, Cassini did not think that the ecliptic was rushing toward the equator, with which it would coincide in only 150 centuries, but rather that it librated with a frequency yet to be determined. "This [magnitude] is to be investigated further by the Instrument of San Petronio."[60] Further investigation, or the conservatism of age, caused Cassini to backpeddle. In answer to an inquiry from Flamsteed, he replied, in 1673, that he took ε to be 23°29' and that he did not know whether it changed. "An annual variation of a few seconds appeared at the great gnomon of Bologna, as appears in Malvasia's *Ephemerides,* but whether this arose from differences in refractions in the different years or from a real change I do not dare to determine." Twenty years later he was bolder and less truthful. "To these *Ephemerides* were added some observations with various examples of the use of refractions in solstitial observations made over two centuries, which show that in all that time the obliquity of the ecliptic has not altered sensibly."[61]

The final pronouncement of the oracle of Bologna went against its maker. According to its most faithful amanuensis, Eustachio Manfredi, anyone could see plainly, without calculation, that each succeeding year the solstitial images fell closer to that of the equinoxes. Consulting his big register of the labors of the observers at the heliometer, he made out that between 1656 and 1733 ε had progressively shrunk by 69", or just under 1"/year in 77 years, "which is so long a period that unless we deny the force of observations or judge that the observers acted altogether negligently, some cause of the decline must be sought in the heavens themselves."[62] The oracle spoke truly, but, as is often the case with oracles, inexactly. There is much more to say about the rise of cathedral observatories and the decline of the obliquity.

Repairs and Improvements

Cassini's method of obtaining the change in the obliquity (if any) from comparison of measurements made only at San Petronio depended for success on the faithfulness of the heliometer as well as on the diligence of its observers. Hence they watched for signs that the settling of the church was impairing the instrument. Just after Cassini left for France, Mengoli found that the plate containing the hole had fallen over 4 percent of its original height. By comparing observations of the sun's noon altitude made by Grimaldi and Riccioli at their *meridiana* in Santa Lucia with those made by Cassini at San Petronio, Mengoli worked out that the slippage had occurred not long after the original installation in 1655. Montanari measured a decline much less significant, of just over 1 percent, in the presence of many witnesses. A battle ensued among the mathematicians of Bologna over the magnitude of the correction for displacement that they had to apply to earlier and current observations to make them intercomparable.[63]

Mengoli and Montanari were satisfied with correcting their observations and did not bother to restore the hole to its full height. That was done by Guglielmini in 1689. Six years later, when Cassini and his son Jacques visited Bologna, it again needed raising. Guglielmini joined the Cassinis and a draughtsman-mechanic, Egidio Bordoni, in a complete overhaul of the instrument. The *fabbricieri,* persuaded that the misalignment of the instrument was a stain on their church as well as an injury to astronomy, paid the bills.[64] The restorers made a flexible wooden rod on which they marked precisely 100,000 Parisian inches against a standard Cassini had brought from France. They set the hole at exactly 100,000 inches above the pavement; releveled the meridian line in the manner described earlier; lifted and reset the marbles bearing the scales; and in all, as Guglielmini put it, made "a secure and royal road to celestial observations."[65] And they diversified the measurements. The *fabbricieri* gave "the first astronomer of his age" permission to knock a hole in the great window over the main entrance to San Petronio in order to observe the polestar.[66] On the sill Cassini fixed a foresight; on the ground he rigged up a telescope in such a way as to be able to read the altitude of the pole from the scale of the *meridiana*. Because the window stood west of the line, Cassini designed a rigid mounting that allowed the telescope to be moved parallel to the *meridiana;* a pointer on the foot of the mounting indicated the position along the scale. In this way he and his collaborators found $\phi = 44°30'15''$ when corrected for refraction, about $1'15''$ higher than Cassini had measured it in 1655. That became the accepted value in Bologna for 75 years.[67]

To maintenance there is no end. In 1722, Manfredi, alerted by measurements made at a meridian line set up in Rome, found it again necessary to raise the hole. According to his analysis, rainwater, dust, and mud flowing over the gnomon plate gradually loosened and lowered it. Many curious among the learned came to watch and help him adjust the great heliometer using the equipment left by Cassini and Guglielmini. The detection of the drop made it necessary, yet again, to correct measurements made since 1695 for the estimated slippage per year.[68]

Fifty-four years later Manfredi's former assistant and successor, Eustachio Zanotti, undertook another restoration, commissioned by the *fabbricieri,* in order to realign the instrument and to adapt it to modern time telling. With two assistants, Zanotti took up the marbles and the line, excavated the pavement, replaced the iron rod with one of brass, and raised the hole, which had slipped 3 percent of its height since Manfredi's time. The new *meridiana* opened for public inspection on 4 October 1776. Curiosity and civic pride brought people out in quantity. "It is certainly no exaggeration to say that rarely has such a crowd of citizens been seen; and it is equally rare in public works that such approbation prevails, everyone commending the wise resolution of the *fabbricieri* to preserve a *meridiana* celebrated above all others, and one of the main attractions of this city."[69] Reviewing the record of observations since 1655, Zanotti made the length of the tropical year $48^m47.0^s$ over 365^d5^h. The modern, retrospective calculation gives $48^m47.03^s$ for the excess. The heliometer of San Petronio eventually produced the result for which Danti had built the first *meridiana* there, and to an accuracy beyond his wildest imaginings. But by the end of the eighteenth century improving knowledge of the length of the year to the next decimal place no longer served a useful public purpose.

In 1779 an earthquake shook San Petronio. Zanotti immediately repaired the "famous instrument to whose installation the science of astronomy owes so much." He thought that he could increase the debt by detecting the change in obliquity in a single year.[70] He reported an *increase* of 1'25" from March 1777 to September 1778, or about 0.83"/year. How good was that? When Frederigo Guarducci examined the *meridiana* in 1905, he found that its height had not changed sensibly since Zanotti had refixed the plate in 1780; that it had remained level; that it ran off true north by only 1'36.6"; and that the solstitial images had moved toward one another from their positions during Cassini's time. The displacement was by 2 centimeters for the SS and 12.6 centimeters for the WS. Attributing the change entirely to an alteration in the obliquity of the ecliptic, one has $\Delta\varepsilon = 133"$ in two centuries, or $-0.65"$/year.[71] How good was that? Let us put off a little longer satisfying that overeagerness to know that Saint Paul diagnosed as one of the flaws of humankind.

Meanwhile—while Cassini's *meridiana* underwent its renewal repairs—the Bolognese built other, lesser gnomons all over town. In 1674 Montanari set one up in the Palazzo Pietramellara at the expense of a senator who wished thus to honor an ancestor who had taught mathematics at the university when Columbus was sailing the ocean sea. It had (and has, for it still exists) a height of 6 meters, a subsidiary hole for viewing the northern sky, and a clear run down an upper corridor in the palace. Mengoli, Guglielmini, and Montanari observed there.[72] In 1741 Ercole Lelli, a master artisan working under Manfredi, made a *meridiana* 2.5 meters high for the Palazzo Poggi, the home of the Bologna Academy of Sciences and its observatory. The opportunity was presented by the refurbishing of the observatory to accommodate instruments that Manfredi had ordered from London. The line had the unusual feature of a wire stretched above it to give better definition to the encounter between the sun's image and the rod.[73]

To end with the last public *meridiana* built in Bologna in the eighteenth century, in 1788 Ferdinando Messia, an Olivetan monk who professed mathematics at the University of Naples, installed a line 20 meters long (height of 8 meters) in a corner of the dormitory of the hospital of San Michele in Bosco, probably to regulate clocks for regulating prayers.[74] The lines in Palazzo Poggi and San Michele in Bosco may still be seen but, unlike the great heliometer in San Petronio, they are not in working order.

One of Cassini's conceits was to give the length of his gnomon as a fraction of the circumference of the earth. It was an obvious association of ideas, this comparison of Bologna's *meridiana* with the terrestrial meridian running through it. He estimated the ratio of their lengths as 1/600,000 when he set out the line in

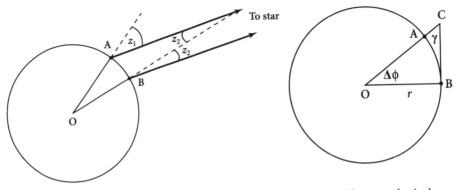

FIG. 4.8. The earth's radius the Greek way, from the stars.

FIG. 4.9. The same the Arab way, from a mountain AC.

1655, but, uncharacteristically, did not have his guess inscribed in stone. He had reason to hedge since his calculation of the earth's radius differed from Riccioli's, who, characteristically, was writing a textbook on the subject. When the book appeared in 1661, it gave a dozen different methods of measuring the earth, three of which Riccioli judged superior to the others. One of these, the Greek approach, takes the zenith distances z_1 and z_2 of the sun or a star at the same instant from two places A and B on the same meridian (Figure 4.8). After measurement of the arc AB along the ground, the earth's circumference C follows from the rule of three: $C = AB[360°/(z_1 - z_2)]$. From his determinations of the latitudes of Malvasia's observatory and San Petronio, and of other places around Bologna, Cassini worked out the earth's circumference at a little under 23,000 Bolognese miles, just under 8 percent too large.[75] Grimaldi and Riccioli obtained $C = 26,000$ Bolognese miles in the same manner, by taking the height of the pole between two churches and surveying the distance between them. When they performed this operation in 1645 or 1646, the distance, some 2 kilometers, was probably the longest ever measured accurately.[76] For twelve years they perfected the method, using, among other stations, the Torre degli Asinelli, campanili, and city gates. In the process they shrank the earth by an eighth.[77]

The second of Riccioli's preferred methods came from the Arabs: one climbs a mountain at the seashore and measures the angle γ, the complement of which averaged 35'28" (Figure 4.9). The height of the mountain, found by triangulation, was 0.1955 Bolognese miles. That made $C = 2\pi r = 23,000$ Bolognese miles, a little larger than the updated average by zenith angles.[78] To settle the business, he and Grimaldi tried a third method, which gave what they deemed their best results. They used the Jesuit summer house atop a mountain near the sea and the tower of the cathedral in Modena. Let a represent the height of the gazebo on the mountain E and b that of the cathedral tower B, both taken from sea level; and let α and β indicate the angles between the verticals at the two stations and the lines of sight to the bases of the tower and mountain (Figure 4.10). By measuring a, b, and the difference in latitude between the stations, Riccioli and Grimaldi could obtain two values of r for every pair of angles α, β. They gave as their very best value 23,170 Bolognese miles.[79]

In 1654, just before beginning work on the *meridiana* and while the Jesuits were climbing mountains and multiplying measurements, Cassini tried his hand without moving from the Torre degli Asinelli. He took the dip angles γ_1, γ_2, from two points on the tower, B, E, to the horizon at F, G and determined the vertical distance $d = $ BE between his two stations (Figure 4.11). The measurements: $\gamma_1 = 89°40'55''$, $\gamma_2 = 89°46'50''$, $d = 0.0284$ Bolognese miles. The calculation: $\sin \gamma_2 - \sin \gamma_1 \approx d/r$. The result: $C = 22,000$ Bolognese miles. "A most ingenious method," wrote Riccioli, "and worthy of such a man; it is most remarkable that it comes so close to the truth."[80] And what is the truth? Table 4.4 will assist comparison.

Since the earth's circumference along the meridian through San Petronio is around 40,000 kilometers, Cassini's value was by far the best. The length of the *meridiana* of San Petronio is 67.7 meters. On Cassini's determination of C, the *meridiana* made up one part in 611,000 of the earth's circumference; on Riccioli and Grimaldi's best value, one part in 643,000; on the modern value, one part in 590,000.

If the Jesuits were right, Cassini's *jeu d'esprit*—that the *meridiana* made 1/600,000 of a circuit of the globe—would have been too far out to do him much credit. Were they right? Malvasia had no doubt that Cassini was more reliable than his teachers, owing to his "experience, skill, and utmost diligence devoid of impatience."[81] While agreeing with this estimate, Cassini awaited more and better mea-

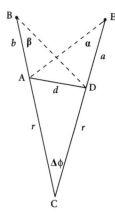

FIG. 4.10. The same Riccioli's way, from two towers AB, DE.

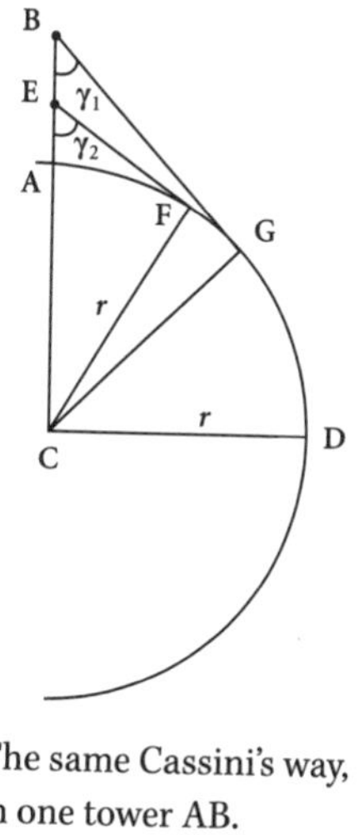

FIG. 4.11. The same Cassini's way, from one tower AB.

Table 4.4 The earth's circumference in Bologna, 1654–1656

Observer	Method	Miles	Kms[a]
Riccioli/Grimaldi	Greek	22,840[b]	42,939
Riccioli/Grimaldi	Arab	23,009	43,240
Riccioli/Grimaldi	Tower/mtn.	22,945[c]	43,136
Riccioli/Grimaldi	Tower/mtn.	23,170[d]	43,560
Cassini	Tower	22,000	41,360
Moderns	Various		40,130[e]

Source: Riccioli, Geogr. ref. (1661), 163, 176.

a. At 1 Bologna mile = 1.88 km.
b. Average of three.
c. Average of two.
d. Their best result.
e. At equator.

surements before pronouncing. He found them almost immediately after his arrival in France, where Jean Picard and others, following the Paris Academy's mission of advancing geography and navigation, had in hand a trigonometrical survey of more than a degree of longitude through the new Paris Observatory.

The result of this painstaking and pioneering effort, which included measurement of a baseline five times as long as Riccioli and Grimaldi's, was 57,060 toise (the standard French geodetic unit, about two meters) per degree, very close to the final metric measurement. That made $C = 4 \cdot 10^7$ meters and the ratio of Cassini's *meridiana* to the earth's circumference 1/592,000, as close to the desired result as Cassini could reasonably have hoped to come. He sent the good news to the *fabbricieri*. "I [!] have found by calculation a most marvelous thing, which adds an extraordinary charm to the properties of the *meridiana*." Its length truly is $C/600,000$. "Thus this line is fixed in such proportion by the solstices as if . . . the church had been designed in accordance with the very exact [solar] measurements made [there]." Cassini requested that the fraction 1/600,000 be incised in marble. That was not done, perhaps because Cassini's correspondents did not think the business as important as he did.[82]

He set things right during his working visit to Bologna in 1695. He had inscribed along the line a scale giving the distance from the vertex in seconds and thirds of the earth's circumference (that is, in (1/360)(1/3600) parts and their sixtieths) and, on a plaque, an announcement that the entire line occupied 1/600,000, or about 2″, of a meridian of longitude.[83] Here is most of the self-serving, nonstop text:

With the authority of the illustrious senators, president, and *fabbricieri*, this horizontal meridian line, which throughout the year catches the noon image of the sun thrown from the vault onto the inscribed places of the celestial signs; which, forty years earlier, was run obliquely between the columns, through a very narrow path, and prepared for ecclesiastical, astronomical, and geographical uses by Gian Domenico Cassini, the primary astronomer of the University of Bologna and papal mathematician; which, when again laid out by him most diligently during a trip to Italy from the Royal Parisian Academy of Sciences, where, with the approval of Pope Clement IX, he had gone to the most Christian King Louis the Great, is found to agree marvelously with the celestial meridian; and which, when accurately reset to the horizontal whence it had receded owing to a small movement of the church and uneven wear of the pavement, includes, from its vertex to the furthest point of the image of the midwinter sun, one six hundred thousandth of the circumference of the earth.[84]

Zanotti removed the scale in seconds and thirds of C during his restoration in 1776, but left the plaque.[85]

Cassini continued in the meridianal line after his return to France. He designed a gnomon for the second floor of the Paris Observatory (he had already placed a small one on the first floor). When completed by his son Jacques in 1729, the upper meridian occupied $1''20'''$ of the earth's circumference. Also, Cassini extended the arc of the meridian of longitude through Paris laid out by Picard. When completed by Jacques and his cousin Giacomo Maraldi in 1718, it occupied about $8°$ of the earth's circumference. And thus, wrote Fontenelle, to within the accuracy allowed obituarists, "M. Cassini had the glory of ending as the only creator of the *meridiana* of Bologna, and creator of most of that of France, the two most beautiful monuments that practical astronomy has ever raised on this earth, and the most glorious to the skillful inquisitiveness of mankind."[86]

5: The Pope's Gnomon

Calendrical and Other Politics

· EASTER AGAIN ·

Among the inconveniences of travel during the seventeenth century was a paper loss of ten days on crossing from a Protestant to a Catholic country, and a like gain in going the opposite way. Correspondence between Europeans living under different calendars usually bore two dates, for example, "10/20 January," or an indication whether a single date should be understood as Julian or Gregorian. The English added to the merriment by beginning their year on 25 March, so that during the first quarter of a Continental year they appeared to be 355 days behind Catholic Europe. Yes, 355 days: 20 January 1660, English style, corresponded to 10 January 1661 Gregorian. It is therefore untrue what some say, that the soul of Galileo, who died on 8 January 1642, passed that year to Newton, who was born on 25 December 1642; for Newton, had he been alive when Galileo died, would have recorded the date as 18 January 1641, whereas Galileo, had he not been dead at

Newton's birth, would have reckoned the arrival of the "Light of the World" not at Christmas 1642 but on the preceding 15 December.[1]

This merely civil confusion was aggravated by a sharp contradiction in the religious calendar. During the seventeenth century the Protestants as well as the Catholics placed the vernal equinox on 21 March, whether it occurred then or not, and notwithstanding that the date signified a cosmos ten days older on the Julian than on the Gregorian system. For this reason alone, the same rules had picked out different Sundays for Easter in the Catholic and the Reformed churches. An English mathematician, writing in 1664, reviewed the observances of the two parties during the eighty years that the Gregorian calendar then had been in effect. Catholic Easter had coincided with Anglican Easter less than half the time (36 years out of 80) and had anticipated it by one week 26 times, by four weeks 5 times, and by five weeks 13 times.[2]

All mathematicians knew that the discrepancies would only grow worse in the eighteenth century. The civil calendars would diverge by eleven days rather than ten owing to the Gregorian suppression of the Julian leap day of 1700. The religious calendars would move further out of step owing to the error committed by Clavius and signaled earlier, which, if uncorrected, would on its own cause fifteen false Catholic Easters during the Age of Enlightenment.[3] This powerful error provided an opportunity. Its discovery by Cassini put pressure on the Catholic Church to alter the Gregorian calendar in ways that might entice the Reformed princes into calendrical communion with Rome. That was the hope of Leibniz, who imagined that his countrymen would not reject out of hand "anything intelligent and reasonable." "There is no want of material for a new Papal bull," he wrote to the Danish Astronomer Royal, Olaus Rømer, formerly a colleague of Cassini's in Paris, "to get us out of this labyrinth respectably."[4]

Even the English teetered on the edge of change. The Archbishop of Canterbury considered whether it might not be better to bury bygones in the jubilee year 1700 and go Gregorian. That was a good idea. He then made the grave mistake of asking the experts in the person of John Wallis, Savilian professor of mathematics at Oxford, formerly a founder and then an ornament of the Royal Society of London. Wallis advised that the slip of the vernal equinox in the calendar was "very inconsiderable," certainly no reason to accept the Gregorian civil year. And there was a very strong ground for rejecting it: "I see not why we should admit it, after having so long renounced it." To this already sufficient argument, Wallis added a consideration that had not weighed on the Archbishop of Canterbury. "I cannot but think there is, at bottom, a latent Popish interest, which (under other specious pretenses) sets it on foot; in order to obtain (in practice) a kind of tacit submission to the Pope's Supremacy." To be sure that the Archbishop could not ignore his advice, Wallis wrote also to the Bishop of Worcester and published both letters in the

Royal Society's *Philosophical transactions*.[5] The English followed their expert rather than their archbishop and stuck with their peculiar system for another fifty years.

The German states proved more flexible. After a sustained discussion, they agreed to what Leibniz called a middle and provisional way. They would adopt the Gregorian reckoning from 1700, with one exception. They planned to calculate the Paschal moon from the Rudolphine tables composed by their co-religionist Kepler rather than from the cycles adjusted by Giglio.[6] This compromise, which brought Protestant Germany into civil agreement with Catholic countries while preserving religious independence and the opportunity for further liturgical disputes, was the brainchild of a professor of mathematics at the University of Jena, Erhard Weigel. By separating the civil from the religious, Weigel achieved a civility that contrasted favorably with the hell-fire rejection of Gregory's "loathsome and abominable errors, his sacrilegious and idol-worshipping practices, his vicious, perverse, and impious dogmas," that is, his calendar, by Protestant professors of the previous century.[7]

The plan to use Kepler's tables, to employ true moons and exact equinoxes rather than tabulated moons and the fixed equinox on 21 March, had its own inconveniences. They had received a full airing at Rome in connection with the Easter of 1665, which some computers, including Levera, placed over a month before the canonical date. That year the vernal equinox came on 19 March, which was a Friday. A full moon fell a few hours after the equinox. Hence if you believed, as did Levera, that the Nicene fathers had set the true equinox as the marker for Easter, you would have placed Easter on 21 March. But if you held to the rules, 14 Luna would have to come after the 21st. You would have waited for the full moon of 18 April, which was a Sunday, and, still faithful to the rules, you would have celebrated Easter on 25 April.[8] The discrepancy amounted to thirty-five days.

A similar situation occurred in 1696, the last leap year before the omitted bissextile of 1700. That year, 1696, the equinox fell as far as possible from the canonical date, namely the afternoon of 19 March, and, until its restoration to the 21st in 1704, true astronomical calculations, such as the Protestants proposed to make, would give different Easters from the Gregorian ones in 1700 (off a week), 1701 (a month), 1703 (a week), and 1704 (a month).[9] Anyone who watched the lunar eclipse of 4 March 1700 could see that the tables, which gave it over eighteen hours later, were seriously awry.[10]

The great discrepancy of thirty-five days between the astronomical and tabulated Easters of 1665 had caused a scandal in proportion. Riccioli came forward to clear it up. He wrote, as his confrère Honoré Fabri put it, "in his usual way, wonderfully learnedly," and so required translation for comprehension by ordinary intellects. Here is Fabri's translation: the church has had to fix the vernal equinox out of ignorance. "Even the most celebrated astronomers do not know for certain [when it falls]; for, believe me, astronomical tables, although revised a hundred times, still contain errors, and the true length of the tropical year is not yet fully known."[11] By the 1680s, Kepler's tables, "supposedly the most exact," were anticipating the spring equinox by three hours. Leibniz agreed. The famous Rudolphine tables, though a marvel in their time, no longer gave a good account of the motion of the sun.[12]

And so it came about that, half a century after Cassini had confirmed the length of the Gregorian year at his heliometer in San Petronio, a call went out to improve the parameters of the Easter canon. That was the chief rationale that Cassini offered for refurbishing the meridian line in San Petronio in 1695.[13] Leibniz tried to promote similar instruments, for the same purpose, in the cathedral of Regensburg or Halberstadt. But he could think of no one willing and able to observe in either of them.[14] The Catholics proved more resourceful. They energized the Pope, Clement XI.[15] Cassini's observation about the consequences of Clavius' error were referred to Rome by the King of France, who offered the Pope the advice to consult the Parisian academicians.[16]

The Pope asked for the Oracle. Cassini declined to go to Rome but promised advice from a distance and the loan of his *primo nipote*, Giacomo Maraldi, who had been working with him in Paris since 1687.[17] The Pope invited Maraldi to join a new commission on the calendar under the presidency of Cardinal Enrico Noris, an old protégé of Queen Christina's, the founder of an academy of sciences in his bishopric of Rimini and an expert on the primitive church and its chronology. "How many memorials of ecclesiastical history he could deduce from a single [old calendar] stone!"[18] Noris' commission consulted Cassini, who replied that there was no major error in the Gregorian reckoning, either of the year or of the month, and that the best reform possible was the elimination of Clavius' error about the tabulated moons.[19] Nonetheless, the Pope persevered in his project. Rome was perplexed again by numbers and tables. "After many centuries the names of Sosigenes and Meton, who, respectively, invented the solar and lunar cycle, are again heard in the world."[20]

The Pope chose S. M. degli Angeli, a church designed by Michelangelo within the Baths of Diocletian near what is now the central railroad station in Rome, as the site of his *meridiana*. Two reasons recommended Clement's choice besides the fundamental requirement of an unobstructed view to the south. For one, the walls of the church, being those of the ancient baths, had long since stopped settling. For another, the church had great symbolic value. Diocletian had been a maker of martyrs. The memory of his persecutions already had inspired one calendrical improvement, namely, Dionysius' recalibration of the Easter cycle from years Diocletian to years Anno Domini. Now the Pope would produce another, by stabilizing the celebration of the holiest day of the victorious Church, "a new triumph of our faith," as a contemporary chronicler of Roman life put it. "The reigning Pope Clement XI desir[ed] that the holy city of Rome no longer be deprived of so necessary and appropriate an ornament for one of the most solemn obligations of the Roman Pontiff and supreme Priest, which is to regulate the principal fixed and moving feasts throughout the year."[21] "Clement XI wanted this affair, always esteemed most serious in the church, and full of mysteries, to be treated most diligently, lest God's business be performed negligently or carelessness slip into divine things; he thought that everything should be tried, academies consulted, and the heavens themselves, and the luminaries, established by God, 'ut essent in signa, & tempora, & dies, & annos,' skillfully observed."[22] The author of these last words, the builder of Clement's *meridiana*, was Francesco Bianchini, "the greatest man that Italy has produced in our [eighteenth] century."[23]

· BIANCHINI AND HIS BOSSES ·

Francesco Bianchini was a gentleman of Verona. A tireless antiquary, superintendent of all the antiquities of Rome, he liked to spend his time looking at old pictures visible from high ladders, "up where foreigners will not climb."[24] An enthusiastic philosopher, a leading member of the Accademia Fisicomatematica founded in Rome by Giovanni Ciampini, another cleric who rose to prominence and wealth by ability and papal preferment, he liked to travel with a barometer and often needed a separate carriage to haul around his instruments.[25] A compulsive historian, he wrote folios on the Romans and an incomplete universal history. This last won reputation and influence for its use of artifacts and emblems, for its illustrations of objects Bianchini considered representative of each era (an abacus for the Chinese empire, for example), and for its exploitation of myth and legend in the manner of, and prior to, Giambattista Vico.[26] A born teacher, he recycled the Zeitgeist illustrations of his *Istoria universale,* many of which he had drawn himself, into a card game, one card for each of the forty centuries from creation to Christ, five for each of the sixteen subsequent centuries, divided into suits and played, in order of events, according to the rules of the Italian game of *stuppa.*[27] An assiduous astronomer and cartographer, he mapped the appearances of Venus with the curious results described earlier, improved the telescope, began a trigonometrical survey of the papal states, and, of course, laid the meridian line in the church of Santa Maria degli Angeli. "A most learned astronomer," was the judgment of Bianchini's friend Leibniz, "and a man distinguished in other things."[28]

And in mathematics? "After all this," wrote Fontenelle, slipping from necrology into self-analysis, "you would hardly expect that Bianchini was a great mathematician." Mathematics and scholarship do not mix. "They exclude one another, they despise one another, it is rare to have them together, and even then it is almost impossible to have the time to satisfy both."[29] Still, he was a great man, this "never sufficiently praised Monsignore Francesco Bianchini."[30]

Bianchini's preparation in learning began early. When very young he received a curious annuity, to continue until he was thirty, which could be used only for buying books. In 1673, stuffed with erudition beyond his years, which then numbered ten, he was sent to the Jesuit college in Bologna, the very one at which Riccioli had taught until his death in 1671. There he learned to do mathematics, speak Latin, draw pictures, be religious, and admire the Society of Jesus. He aspired to join it, but his father, thinking him too young and overly influenced, sent him to the University of Padua to study anything he liked. He liked everything, particularly mathematics and physics, which he learned from Montanari, who had migrated to Padua from Bologna in 1678 to take up a special chair in astronomy and meteorology.[31]

Bianchini and Montanari got on perfectly together. The pupil was open, free of guile, exceptional in his combination of "profound learning and remarkable modesty and sweetness of disposition"; the professor, "[eager] to make the sciences useful to the public welfare rather than evidence of private industriousness," preferred teaching to writing. As we know, the preference did not prevent him from being publicly useful by writing a book against astrology. Bianchini contributed a prefatory poem, "To astrology accused of falsehood [the book's title], rightly damned by a very wise teacher."[32]

Montanari's unassertive openness to Galileo's instrumentalism and experimentalism, and to Descartes' alternative to Aristotle, suited Bianchini's capacities and character. Under Montanari's influence he wrote an unpublished essay on gravity based on Cartesian vortices, learned to do experiments and to observe the heavens, and, perhaps, found his historical method, which, in a manner similar to Galileo's separation of natural science from theology, divided history into an analytical, secular and an inspired, sacred part.[33] Bianchini attributed Montanari's achievements to a true religiosity. "He gave the first and proper place in his mind to the study of religion and to divine things, which is the true and highest wisdom of Christian philosophy."[34]

With Montanari a Christian life of public service did not imply penury or deference. He left Bologna because of deterioration in the university's funds or because he believed that his efforts on behalf of fisicomatematica were insufficiently appreciated. He pointed to his Accademia della Traccia, which did experiments in the style of the Accademia del Cimento and, like them, avoided declaring the causes of things; he had taught all branches of mathematics; and he had tried to convince his students of the futility of astrology. He was not able to foresee that his crusade would make him some influential enemies.[35] At Padua he did not have to cast horoscopes and earned enough to afford public service. He had (as Bianchini expressed it) "the decent and secure income that scholars measure more by the outlay required by the mind, to procure books and other helps to study, than by the expenses demanded by the body, to minister to the necessities of life."[36]

In fact, he did much better than that. He had an apartment in a palace on the Royal Canal whose owner, Girolamo Correr, put at his disposal whatever was needed to outfit an observatory to rival Cassini's. Among its instruments was a *meridiana* with sights to the pole as well as to the sun. Noris hunted Montanari out. "He had his own kingdom, with servants, books, and so on." Montanari had been most encouragingly successful in securing the necessities of a scholarly life. Authorities both clerical and lay supported his astronomy: in Bologna and Padua, the universities provided professorships and instruments; in Venice, a layman equipped an observatory and library; in Padua, Cardinal Gregorio Barbarigo, a frequent near-miss candidate for Pope and eventual saint, promoted Montanari's plans for an observatory and meridian line in the local seminary, "whence it can be said," said Bianchini, "that [Montanari] had forced the realm [*università*] of science into the immediate service of religion." Bianchini profited immediately from the same network. He enjoyed the patronage of Correr and inherited Montanari's books and instruments.[37]

Bianchini's teacher of theology recommended that he go to Rome to begin an ecclesiastical career. Bianchini went. He applied to Cardinal Pietro Ottoboni, "one of the greatest minds of the century," who knew Bianchini's family and recognized his calling. Ottoboni put Bianchini in charge of his library, a choice collection begun by Cardinal Sirleto (who had headed the committee that recommended Giglio's principles of calendar reform) and acquired by the Vatican a few years after Bianchini's death. As Ottoboni's librarian Bianchini had little to do besides adding to his mountain of erudition and doing experiments at Ciampini's academy, which, from 1689 on, met in Ottoboni's palace.[38]

This academy, established in 1677, brought together laymen and ecclesiastical scholars to discuss the usual range of contemporary polite subjects: anatomy (including the structure of inanimate objects, that is, chemistry); philosophy (including speculation about the nature of bodies); mathematics (including geography, navigation, and hydrology); mechanics (including optics, perspective, painting, and architecture); and history (including numismatics and inscriptions). Prominent among its productive members were Francesco Eschinardi, S.J., faithful to the Aristotelianism of his order; the master lens-grinder Giuseppe Campani, who became a close friend of Bianchini's; Bianchini, who would inherit Ciampini's mantle; Cassini and Montanari, as correspondents; their former student Agostino Fabri, the medical astrologer; Leibniz, as a visitor; and Giovanni Ciampini himself, whose main interest was sacred rites and architecture.[39]

Ciampini came to be a patron of arts and sciences with the help of his brother, a senior official in the curia, who introduced him to Ottoboni. He acquired important offices under Clements IX and X, who encouraged his study of antiquities. He pursued them so vigorously that, like Thales of old, he fell into a well while loftier thoughts filled his mind. Fished out and mended, he turned his attention to building up his library and museum and to founding his academy. Queen Christina, to whom he may have had access through her good friend Clement IX Rospigliosi, gave the fledgling academy moral if not financial support.[40]

The academy met the first Sunday of each month for discussion and, sometimes, to witness experiments on the spring of the air, the rise of liquids in small tubes, and life in a vacuum, fare also on the menus of Leopold's Cimento and Montanari's Traccia. Several prelates helped defray expenses for instruments and materials. Discussions ranged far; no respectable subject in natural philosophy was beyond bounds. Reports of the discussions also ranged far; Eschinardi published a book of them addressed to a friend abroad, no doubt Cassini. Together Eschinardi and Ciampini edited a short-lived revival of the *Giornale de' letterati di Roma*, an Italian version of the newsy French *Journal des sçavans*, which kept the Roman province of the republic of letters in touch with northern Europe. In short, Ciampini was a dominant figure in the high culture of Rome during the last quarter of the seventeenth century. Bianchini would take on a similar role after Ciampini's death in 1699.[41]

It appears that the generosity and curiosity that informed Ciampini's life may have hastened his end. He is said to have died of mercury poisoning after trying out a new Hermetic medicine on poor people hit by an epidemic. His will directed that his estate go to establish a college for twelve needy students of any subjects but theology, law, and medicine. The students were to cultivate Christianity and the sciences, do pious works, and advise the Congregation of the Index of Prohibited Books. This curious mixture of religion, science, freedom, and censorship never came into being. Relatives unwilling to try Ciampini's experiment at their expense successfully contested his will.[42]

Ottoboni proved as good a patron to Bianchini as to Ciampini. In 1689 he was chosen to lead the Roman Catholic Church. He called in his librarian. "Bianchini, siamo Papa noi! che volete, che vi diamo?" ("Bianchini, I am the Pope! What do you want me to give you?") The answer: "Your blessing." The new Alexander (Ottoboni took the name Alexander VIII) gave his blessing and also, as was his lavish wont with his favorites, two pensions, a canonry, and the post of librarian at the court of his twenty-two-year-old grand-nephew Pietro, whom he appointed cardinal and superintendent of the Papal States; the spendthrift younger Ottoboni remained Bianchini's patron for decades. These were but drops in the ocean of blessings that Alexander promised Bianchini if he would take holy orders. But Bianchini had enough for books and instruments, plenty of leisure, and no desire for power. He never advanced beyond the deaconate, which he took in 1699.[43]

Alexander died in 1691. His successor, Innocent XII, a much thriftier man, awarded Bianchini only an additional canonry, in the important church of San Lorenzo in Damasco, as an apology for not appointing him to the job he sought, the overseer (*Custode*) of the Vatican Library. After Innocence came Clemency, number eleven of that ilk, in 1700. Clement was far from innocent, however, in most other respects: he had come to the cardinalate through the curial diplomatic service and was ordained a priest about the time of the conclave that elected him Pope. Young and vigorous (he was fifty-one at election), he had the misfortune to preside over the Church when its loss of political power north of the Alps was laid bare by the War of the Spanish Succession. His attempt to maintain a strict neutrality while asserting papal prerogatives at a time when excommunications and anathemas had no effect resulted in one humiliation after another for the Holy See. He had to make do with cultural politics, encouraging learning and talent at home and emphasizing associations between modern and ancient Rome abroad. In pursuing this policy, the Pope depended on men like Bianchini, "the importance of [whose] contributions to papal scholarship and art patronage cannot be overemphasized."[44]

Bianchini had easy access to the new Pope through Clement's best friend and Bianchini's enduring patron, Cardinal Pietro Ottoboni.[45] The Pope gave Bianchini the title of *cameriere d'onore* and free lodging in an apostolic palace. This time the presents had a price. On the advice of Cardinal Noris, Clement made Bianchini secretary to the new commission on the calendar and architect of the *meridiana* in Michelangelo's church.[46] There was already a *meridiana* in Rome, in the cabinet of curiosities set up by Kircher in the main college of the Jesuits. But Kircher's line had not been built with the refinement of Cassini's.[47]

Bianchini had done some astronomy with Montanari and was familiar with Cassini's methods. He had tried to work Cassini's scheme for getting the parallax of Mars and learned if nothing else that numbers should not be taken too seriously. "I do not worry about very small fractions [he wrote] so that I am not thought to be uselessly captious in things so uncertain." He had tried to exploit Cassini's geometrical astronomy but, like many others, could not fathom it and applied to its author for help.[48]

Bianchini had admired the heliometer of San Petronio as "the greatest and the most exact [instrument] in astronomy to be seen in Europe." He had installed at least one *meridiana* himself, in 1692, in Ottoboni's palace. That brought him into direct contact with Cassini. During his trip to Italy in 1695, when he refurbished

FIG. 5.1. Pope Clement XI, his coin, and his gnomon. From Bianchini, *De nummo* (1703).

the heliometer in San Petronio, Cassini had visited Rome. He amused himself by constructing a *meridiana* in San Marco, perhaps with Bianchini's help. Together they compared the performances and orientations of the new line and that in Ottoboni's palace.[49] On receiving his commission from Clement, Bianchini consulted Cassini about the special problems presented by Santa Maria degli Angeli. Cassini supplied a copy of his critique of Clavius. Bianchini also sought the advice of Cassini's alter ego in Italy, Manfredi. He then went to work, day and night, for six months, constructing a line that contemporary connoisseurs rated the most beautiful, ornate, and versatile of all *meridiane*.[50]

The Pope came to see the sun cross the almost completed instrument on 6 October 1702. A coin was minted to commemorate the event (Figure 5.1). Bianchini described it all—the instrument, the coin, and the visit—in 1703, in an expensive booklet with the clever title *De nummo et gnomone Clementino*, which might be rendered, preserving the alliteration, as *Clement's medallion and meridian*. The Pope had taken a close interest in the book, reading it for heresies and infelicities himself, and directly authorizing its printing, "to satisfy the desire of so many great benefactors [*padroni*] and friends who want a description of the *meridiana*."[51]

The making of the meridian gave Bianchini a taste for big science. He proposed to Clement, who had made him president of the antiquities of Rome in 1703, another major undertaking, an ecclesiastical museum to house all inscriptions, medals, statues, amulets, and so on, having to do with the early church. The Pope agreed, until he learned the cost. The abrupt end of the project in 1710 distressed Bianchini.[52] In compensation he received another canonry, a lucrative one in Santa Maria Maggiore, and the right to draw on the income from certain church holdings in Sicily. He grew rich enough from his disappointments to set up as a patron and philanthropist in his own right. He spent generously, intelligently, and, occasionally, excessively, on books, instruments, antiquities, charity, a carriage, horses, and travel.[53]

In 1712 he made a grand tour out of a commission to deliver the hat of office to the newly appointed French Cardinal Armand de Rohan. Clement chose him especially for this task. The Pope was eager to improve relations between France and the Holy See as the War of the Spanish Succession drew to a close. He needed as an ambassador a man closely associated with himself and Rome, universally admired for his culture, probity, morals, and affability, and persona grata in Paris. Bianchini met the description: he had learning in abundance, an unusual combination of piety and savoir-faire ("he is a little saint," Noris reckoned, "but every inch a courtier"), and the high status, accorded in 1705, of corresponding member of the Paris Academy of Sciences.[54] Delivering the cardinal's hat and patching relations with France may not have been the only assignments the Pope gave Bianchini. He probably had the additional and trickier job of contacting and encouraging Catholic groups in Protestant countries.[55]

Thus programmed, Bianchini packed up his barometers and telescopes, his prints and books and other gifts, and set out for Paris, where he was met by Cassini's son Jacques. He visited old Cassini, who, with but a few days to live, talked briskly of the reform of the calendar; and also ministers, cardinals, savants, the Dauphin, with whom he became intimate enough to enter into correspondence, and the King himself, who, he says, treated him with great deference. He presented the Academy with his invention for orienting long telescopes, "which was praised to the skies by everyone so that every day it had to be shown to the princes and princesses who came to see it."[56]

Petted in Paris, Bianchini was jailed in Germany. By mistake. On his release the local police chief, now thinking him a great dignitary of the Church, knelt before him and asked his blessing, which he bestowed with his usual good nature. "It did not bother me at all to sleep in the straw," he said, in answer to the chief's apologies, "for being an astronomer I'm used to camping like a soldier." Throughout the Germanies he amused himself by counting and measuring everything denumerable: the number of pearls on a cope, the width of the Rhine, the depth of a well in Koblenz. Proceeding thus, he reached England, paced off the façade of Saint Paul's, which (his piety falsifying his quantity) he erroneously made half the size of Saint Peter's, stayed precisely forty days, met his old correspondent Flamsteed, and visited the great Newton. He was much surprised by the warmth of his reception. Made unusually gracious, perhaps, by Bianchini's report of the successful repetition in Rome of some of his optical experiments, Newton welcomed his Papist visitor to the Royal Society and gave him copies of the *Principia* for mathematicians in Italy. Returned home, his missions accomplished, Bianchini traveled up and down the peninsula, taking latitudes and longitudes wherever he went, with a view to tracing through Italy a meridian similar to the one Jacques Cassini was then completing from Dunkirk through Paris to Perpignan.[57]

Bianchini's compulsion to count almost killed him. A tomb was discovered on the Via Appia. He rushed to measure the site. While he bent over his rule, the pavement opened beneath him; he hung on to the edge of the hole, but could not sustain his weight, and fell to the gravel below. He was lame for the rest of his life. That did not prevent him from limping around Italy with a horse-load of instruments to pursue his trigonometrical survey; or from laying out a meridian line in a palace of the Duke of Parma; or from seeing through the press his sumptuous volume on the figure of (the planet) Venus. Bianchini died on 2 March 1729, dressed in clerical garb, after buying some antiquities and reading himself the last rites. He left a few mementos to his cardinal friends, notably the Newtonian reflector given him by his patron John of Portugal to a fellow former student of Montanari's, the enlightened inquisitor Gianantonio Davia; his instruments, books, statues, carriage, and horses to his nephew; and his manuscripts to a convent in Verona. The grateful citizens of Verona put up a monument to him in their principal church.[58]

The Meridian in Michelangelo's Church

The following notice, from a bull of Pope Pius IV, dated 10 March 1560, is inscribed on a wall of Santa Maria degli Angeli: "We have decided to convert to the purposes of religion the Baths of Diocletian, which were built with the blood and sweat of the faithful for the convenience and pleasure of idolaters by an impious tyrant and a most cruel enemy of the church." The work was entrusted to Michelangelo, then also a monument, over eighty-six years of age. It was a most unusual commission. The Roman Senate contested the Pope's right to dispose of ancient buildings. It withdrew its opposition to his giving the Baths to the Carthusians in exchange for the commitment to preserve as much of the ruin as possible in building the new church. This instinct for preservation ran counter to then current practice, which was to dissolve ancient structures into the fabric of the buildings that superseded them. Hence the constraint, which was to be important for Bianchini's work, of leaving the walls of the Baths intact.

The grand ground-breaking took place in 1561 in the presence of nineteen cardinals and all the magistrates of Rome. To cleanse the place of its pagan residue, Pius had put up a plaque that bore these powerful words: "What was an idol is now a temple of the Virgin / Its creator is the Pious Father himself / Demons begone!"[59] Michelangelo built his church within the Roman *frigidarium*, a huge space 58.8 m (length) × 24.15 m (breadth) × 30.15 m (height). The plan left four voluminous corners for chapels, which Paul IV wanted his cardinals to pay for. No one came forward. Paul's successor, Saint Pius V, did not push the cardinals for contributions and it fell to the unstoppable Gregory XIII to bring the project to completion. He wanted it finished quickly, in time for the jubilee year 1575.

Neither money nor time sufficed to finish the building according to Michelangelo's plan, however. The chapels were reduced in size, some surfaces stuccoed rather than marbled, and supernumerary fake pillars introduced to break up the space. The main altar was dedicated at the jubilee of 1700; the future Clement XI said mass there just before entering the conclave that elected him Pope. In the middle of the eighteenth century further liberties were taken with the design, which produced the church we see today (Plate 5).[60] Unfortunately, reworking of the cornices and pediments blocked off the sun's rays around the summer solstice, and resurfacing of the floor removed some inscriptions relating to the *meridiana*. Nonetheless, it remains what it was to the author of the standard eighteenth-century guide to Rome, the most notable object in the church of S. M. degli Angeli.[61]

The hole admitting the sun's rays stands 20.5 meters above the ground in the south wall of an arch across the southeast arm of the cross (Figure 5.2). As appears from the figure, another hole, in the main arch across the northeast arm in the plane of the meridian, passes light from culminating northern stars, a design Bianchini probably learned from Montanari. Bianchini chose the setting that would give him the longest possible complete *meridiana*, from which the height of the southern hole followed by trigonometry. The height of the northern hole is fixed by the length of the *meridiana* and the latitude of the church.[62] The pavement contains little brass stars that embellish the instrument. We have much here beyond the layout of San Petronio.

· THE SOUTHERN GNOMON ·

Bianchini's heliometer owed many technical details to the example of the "never sufficiently praised builder of the Bologna *meridiana*," for example, the method of leveling and the ratio of the diameter of the hole to its height.[63] And he used the master's *meridiana* in San Marco to check the layout in S. M. degli Angeli: he stood on top of the Baths of Diocletian and whistled to Maraldi down below when he saw a signal from San Marco indicating that the sun had arrived at the meridian there. Measurements of the altitude of the sun's limbs at the two places differed by only a few seconds of arc.[64]

It was in the embellishments and accoutrements that Bianchini showed his genius. Begin with the hole: at San Petronio the sun's rays enter through an opening

FIG. 5.2. The meridiana in S. M. degli Angeli, Rome. The ray from the right comes from the sun; that from the left, from a star near Polaris. From Bianchini, *De nummo* (1703).

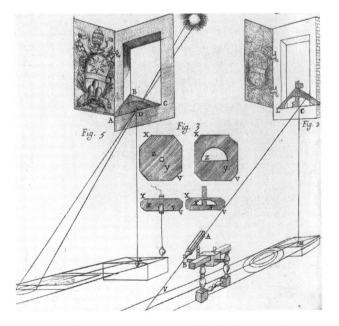

FIG. 5.3. The southern gnomon at S. M. degli Angeli, Rome. From Bianchini, *De nummo* (1703).

decorated by a simple solar motif; at S. M. degli Angeli, they pass through a hole in Clement's coat of arms, presented in high relief on a movable panel so hinged that, when opened, it allowed observation of the sun and moon for some distance on either side of the prime meridian (Figure 5.3). The panel carried a painted version of Clement's arms on its backside so that no user of the *meridiana* could be ignorant of its patron.[65]

The main feature of interest of the southern gnomon is the brass stars indicating the diurnal paths of prominent stars. Their layout offers a pleasing prospect to aficionados of conic sections. Figure 5.4 shows the diurnal circle CD of the sun when its declination δ is northerly; EF is the intersection of the plane of the equator with the plane of the paper. O is a small hole in a roof somewhere on the earth. The rays from the circle CD that fall through the hole form a cone COD with vertex O, and, beyond the hole, spread into a similar cone XOY. The axis of the double cone is the axis of the world. Figure 5.5 specializes to a roof at latitude 45° N; SN is the meridian line corresponding to O.

All the points of Figure 5.4 reappear in Figure 5.5 except that the horizon cuts out the rays from around D, which, if the earth were transparent, would shoot into the sky around A. The horizon plane, whose intersection with the vertical plane through O and NCP is NS, cuts the spreading cone XOY in the curve UV, which has an exactly similar piece on the other side of NS. The theory of conic sections gives only five possibilities for the curve UV: a straight line (which would be made by a section through the axis NCP·SCP); a circle (made by a section perpendicular to NCP·SCP); and (as shown in Figure 5.6) an ellipse, a parabola, or a branch of a hyperbola, depending on whether the section cuts NCP·SCP at an angle less than 90°

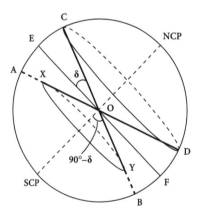

FIG. 5.4. Cone of rays created by the sun in its diurnal motion around the parallel CD.

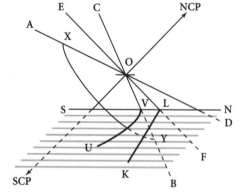

FIG. 5.5. Paths of the sun's image on the pavement in the afternoon at an equinox (LK) and at the summer solstice (VU).

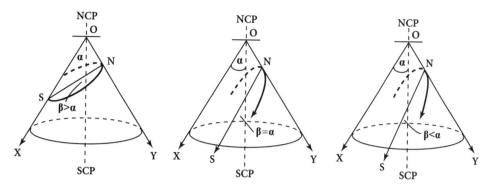

FIG. 5.6. The conic sections made by the pavement with the cone of rays OXY from the gnomon. The nature of the section depends on latitude and declination: a hyperbola (left), holds throughout the year at all latitudes between the Arctic and Antarctic circles; a parabola (center), can occur only at and above the circles; and an ellipse (right), only above the circles.

but larger than α; at α; or at an angle less than α. (Here 2α is the angular width of the cone, which, in figures 5.4 and 5.5, is $\angle XOY$.)

All five possibilities are realized in the intersection of the horizon plane with the spreading cone of rays made by the sun in its diurnal motion. It is a question of latitude. Let us set aside for the moment the singular case represented by the line KL in Figure 5.5, which occurs when the sun is at an equinox. With this exception, whatever the sun's declination, the locus of the image made by it through a small hole during its diurnal motion is a straight line at the equator (the horizon contains the poles); a hyperbola between the equator and latitude $\phi < 90° - \delta$; a parabola at $\phi = 90° - \delta$; an ellipse at $\phi > 90° - \delta$; and a circle at the poles. The truth of these assertions can be collected from Figures 5.4 and 5.6. Take the plane of the paper in Figure 5.6 to be that of the meridian. Then β, the angle between the NS axis of symmetry of the various curves, is just ϕ, since it measures the altitude of the pole. From Figure 5.4, the half-angle of the cone, α, is $90° - \delta$.

When the sun is on the equator, the hole O lies in the diurnal circle and the cone of rays widens into a plane. The intersection with the horizon therefore makes a straight line, independent of latitude. This intersection, the line LK in Figure 5.5, runs out to the east and west points of the horizon and so stands perpendicular to NS. Bianchini marked out this line, HI in Figure 5.2; H indicates the center of the sun's image at noon on the day of an equinox and I its place sometime in the morning. (H is marked in the *meridiana* just right of the center of the figure; I is between 7 and 8.) The curve NO lies on the diurnal locus of the image of the star Sirius, the Big Dog, which evidently has a declination a little larger than that of the first point in Leo (or the last point of Taurus); NO (picked out at the southern end of the *meridiana*) is a hyperbola very close to what the sun would trace about a month after (or before) the summer solstice. A more northerly star, Arcturus, has the diurnal locus outlined by the markers along the hyperbola QR (Q lies under the point of entry of the ray descending from the left), which, like NO, turns convex toward the equinoctial line HI. The drawing shows that Arcturus' declination is about that of the midpoint of Scorpio (or Aquarius). That leaves only the curve GT. If it indicates the course of the solar image, it signifies a day shortly before the sun's entry into Virgo or shortly after its entry into Taurus. The right guess is just before Virgo. It commemorates the visit of the Pope to S. M. degli Angeli on 10 August 1702 to see the work he had commissioned.[66]

He saw many things no longer visible today. These included the original zodiacal signs, drawn after Baeyer's star catalogue by two local professors, one aptly named Paradiso; the present ones are worn and some have been reworked. There were once scales of tangents, of the lengths of the days, and of the true time of equinox, at the equivalent of a minute of declination to an hour of time, all placed exactly against the meridianal point to which they referred. There were brass stars on either side of the line, with their names and right ascensions, indicating where the corresponding star's moving image crossed the line. These too are gone. And the remaining brass stars, designating the diurnal paths of Arcturus and Sirius, shine not as they once did, against a brick firmament, but on a marble floor, put in in 1772, which diminishes their lustre.[67]

Bianchini obtained the diurnal loci of the stars by telescopic observation through the hole in the south wall of the church, from which a part of an architrave had to be removed to give entry to the rays. Arcturus and Sirius and other bright stars could be seen through the telescope on clear days even in bright sunshine. The telescope had lenses commissioned from Campani, still going strong with the help of his daughters, whom he had taught his secrets.[68] The apparent paradox of stars visible in daylight much pleased the Queen of Poland when she visited S. M. degli Angeli in September 1703, and, probably, the Pope himself when he officially inaugurated his heliometer on 6 October. This play with the stars, Bianchini wrote, was most proper in a church. "[It] seems in a way to add to the feeling of veneration of the faithful in the church: while they perceive an image of the heavens serving as a floor in the house of God, they also see the stars He formed, still lit by day in obedience to His commands, as if everlasting lights for fixing the times of singing His praises."[69]

· THE NORTHERN GNOMON ·

"In the center of the [northern] gnomon, through which we comprehend the manner of dividing time into centuries, years, months, hours, and days, is placed the

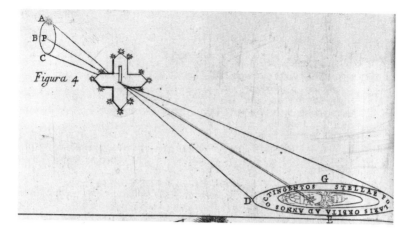

FIG. 5.7. The northern gnomon at S. M. degli Angeli, Rome. ABC indicates Polaris'
diurnal orbit, GDE its image on the pavement. The Latin says "the polestar's orbits
for 800 years." From Bianchini, *De nummo* (1703).

wholesome sign of the cross, by which we bear witness that everything goes back
to Him by whom all things were made and that all times serve the Lord of Time."[70]
This holy hole (Figure 5.7) allowed telescopic observation of Polaris and other circumpolar stars against the scale divisions of the *meridiana.* Bianchini would sight
the star through the telescope and find the spot on the scale so defined by looking
down at the instrument from its objective lens. The telescope carried backsights
on two external surfaces; by following where the visual rays directed by them in-
tersected with one another at the *meridiana,* he could tell an assistant where to
read the altitude of the star from the *meridiana*'s scale.[71]

These observations, corrected as needed by Cassini's table of refraction, established the latitude of S. M. degli Angeli as 41°54'27"N, thus 2°36' south of San Petronio. (By the method of the eclipses of Jupiter's moons, Bianchini found the
difference in longitude of the two stations to be 4ᵐ15ˢ of time, or 1°15'45" of arc.) A
masterly grantsman *avant la lettre,* Bianchini further justified opening the church
to the north as a means of fixing midnight with great accuracy (when Polaris
crossed the meridian) and thus the start of the ecclesiastical day. The times of divine office would be known to exquisite exactness; the faithful who regulated their
observances by the Pope's gnomon would neither feast nor fast a second longer
than required.[72]

The system of ellipses around the "image" of the pole (the point where the light
from a star at the pole would fall, if there were such a star) bound still closer the
astronomical and the liturgical uses of the Roman *meridiana.* Astronomically, the
ellipses represent the locus of the images made by the polestar during its diurnal

motion on days taken at twenty-five-year intervals (Figure 5.8). That the locus is an ellipse follows from the discussion of the light cone in connection with Figure 5.6; since the polar distance of Polaris was taken to be 2°18' in Bianchini's time,[73] $\beta \geq \alpha$ at Rome then. The dimensions of the axis of this ellipse may be obtained from the geometry of Figure 3.18, which concerns the size of the solar image at San Petronio. One need only imagine that the diurnal circuit of Polaris is a sun centered on the polar axis and subtending an angle $\sigma = 4°36'$ at the north gnomon of S. M. degli Angeli. Since h', the height of the north gnomon, is about 24.39 meters, the diurnal circuit of Polaris in Bianchini's time made an ellipse with major axis 4.4 meters and minor axis 3.0 meters. This is the outermost of the band of ellipses LM in figure 5.2. The diagram accurately shows the center of the external ellipse displaced slightly from the projection of the pole toward the north by $h'\sigma^2 \text{ctn } \phi \csc^2 \lambda$ $\approx$ a tenth of a meter.[74]

The inner ellipses correspond to a star closer to the pole than Polaris was in 1700. That star is—Polaris. The nest of ellipses indicates an evolution in time: the polestar has come closer to the pole since Bianchini lived. Their rapprochement derives from what the Greeks called the precession of the equinoxes. The equinoxes, we know, are imaginary points in the sky where the ecliptic cuts the celestial equator. It would be convenient if stars sat on these points. But it would not be convenient for long. The stars would slide off the equinoctial points, down along the ecliptic, from west to east, in the order of the zodiacal signs. Alternatively, one might suppose the stars fixed and the equinoxes in motion, in the direction of the daily rotation of the heavens. This is the solution that the Greeks preferred. Hence "precession" of the equinoxes, a very slight anticipation of the rotation of the fixed stars.

Figure 5.9 presents the relevant geometry. VE_1, AE_1, and P_1 signify the positions of the vernal and autumnal equinoxes and the north celestial pole around 1 A.D.; the same symbols with subscript 2, the same points when the equinoxes have precessed through a zodiacal sign. This represents the situation close to our time. It takes 26,000 years for the equinoxes to circle the ecliptic. In 2,000 years they cover 28°. Early in the next millennium they will have passed through Pisces to enter, and define, the Age of Aquarius.

Since the north celestial pole is on the axis of the equator, it must move among the stars as the equator shifts its orientation to the ecliptic. That is the geometrical reason that the pole has been approaching Polaris since Bianchini's time. For the physical reason the world is indebted to Newton. On Newtonian theory, the spinning earth sustains a gravitational force that tends to upset it; in consequence, like a top, the earth precesses around an axis perpendicular to the plane that "supports" it. The analogy to the vertical force of the earth's gravity, which causes the spinning top to precess, is the pull of the moon on the earth's

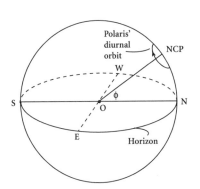

FIG. 5.8. Diurnal orbit of Polaris.

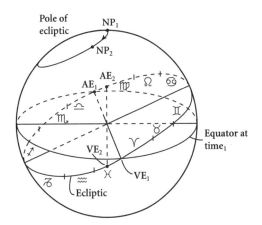

FIG. 5.9. The precession of the equinoxes.

equatorial bulge. The Copernicans of Bianchini's time therefore had to ascribe three motions to the earth: diurnal, annual, and precessional. The phenomenon of precession is included in the definition of the tropical or calendrical year, the interval between the sun's successive visits to the same equinox. The slightly shorter sidereal year is the interval between returns to the same place relative to the stars.

In twenty-five years the equinoxes slip forward some 20' of arc along the ecliptic. Likewise, the pole runs 20' along a circle parallel to the ecliptic (Figure 5.9). It will reach its closest approach to Polaris around 2100; the locus of the star's diurnal motion will then describe the innermost of Bianchini's ellipses. Since the closest approach will be around 26.5', the innermost ellipse has axes of 0.84 and 0.56 meters. After 2100, Polaris will decline from the pole and its diurnal locus will run through the sixteen ellipses one by one until, in 2500, it will regain the ellipse of 1700.[75]

With a little imagination, the construction can be construed liturgically. The ellipses, representing twenty-five-year intervals, make a grid of jubilee years. The *meridiana* itself indicates the current year and the place of Easter in it. Nature, God's creation, dictates the positions of the ellipses, the *meridiana,* and the equinoctial plaque once man exercises his freedom by opening a hole to admit the light from above. "And so [wrote Bianchini] in this single instrument not only astronomy but also sacred chronology and the Roman calendar may be seen and singled out and united by the rays of the celestial bodies." The theme was embellished to christianize the meridian in the former pagan baths by another protégé of Clement XI's, Giuseppe Piazza:

Now the astronomical terminology of planets, pole, meridian, horizontal, tangent, and parallel, of equinoxes, lunations, epochs, zodiac, ecliptic, vertical, horizontal, Arcturus, and Orion have found a wonderful and intimate connection with the imposing names and famous epithets of the Great Virgin Mary Mother of God: Aurora, sun, moon, star, noon, lamp, light, and luminary of the world . . . ; so that [at the Heliometer] we are stirred by exemplary instruments, and ingenious interventions, to praise God and his most holy Mother.[76]

· SOME RESULTS ·

In 1703 Clement XI wrote to the universities of Europe: "The grand installations that Gregory prepared for observing the sun by the mathematicians of his age, chiefly Father Egnatius Danti . . . are the greatest and most accurate of all instruments for fixing the equinoxes as the Gregorian system decreed." And, the Pope continued, the newest and best of these installations determined that Gregory's astronomers had arrived at a value for the year as close to the truth as needs be. They used $365^d5^h49^m12^s$; Bianchini got $365^d5^h49^m1.31^s$; or no difference at all to a gentleman unwilling to appear overexact. For the average lunation, Gregory had adopted $29^d12^h44^m3.11^s$, with which Bianchini concurred to within a hundredth of a second, "a most remarkable agreement."[77]

Cassini accepted the results, which agreed with his own observations. As we know, he had decided already that the Gregorian year was "as good as can be." The Pope's gnomon had not been superfluous, however. "A matter of this importance, in which masters of the art [he had in mind Bianchini and Maraldi] are engaged, requires a direct and careful examination of the heavens."[78] Leibniz concurred and, as he told Bianchini, hoped for more: "I read your book with satisfaction and approval . . . [I]n the Clementine gnomon you gave the church a perpetual index of time by which we can do without cycles. But I do not take ill the practice of using a cycle, nothing more apt yet being provided for the purpose. I hope that in more peaceful times I might bring Protestant mathematicians also to accept whatever in your exact investigation pleases the Pope."[79]

Bianchini remained resolutely on the side of cycles, on which he had become a master, and devised one of 1,184 years that would have returned the Paschal full moon to the same hour and minute until A.D. 4000, or later.[80] In this suggestion he agreed with Joannes Tidius, whose cycle of 592 years (one-half of Bianchini's) caused everything to repeat after 7,322 lunations. Leibniz advanced Tidius' cycle as the best going, even for those who rejected cycles, but admitted that its form and length gave it little chance of acceptance.[81] A fortiori Bianchini's, which practical computists rejected despite the celebrity of its author, "whose name will be immortalized by the observations made with the gnomon at the Baths of Diocletian," as unnecessarily fastidious for the uses of the church. Neither Tidius nor Bianchini, however, could compete with the magnificent Cassini, who derived a cycle of 11,600 years from a deep analysis of Indian astronomy.[82]

Like San Petronio, S. M. degli Angeli became a solar observatory in steady operation. Bianchini observed there for many years. The routine began in the summer of 1703, although the instrument was almost finished a year earlier; but Bianchini, being a court politician as well as an astronomer, had had to break off to accompany a papal legate sent to patch up relations between Rome and Neapolitan intellectuals.[83] Bianchini's voluminous measurements of solar images and stellar positions, most of them made at S. M. degli Angeli, and his meteorological records, extending over a quarter of a century, were later published by his admirer Eustachio Manfredi.[84] He was joined by other astronomers, Maraldi of course, Eustachio and Gabriele Manfredi, and Giovanni Bianchi, a busybody physician from Rimini who revolved in intellectual circles in Bologna and Rome.[85]

Many visitors came, especially around the vernal equinox and summer solstice, notably the pretender James III, with whom Bianchini was on close terms, and several lesser English aristocrats. The routine of observing the weather, the sunspots, and the sun was interrupted occasionally by an earthquake, like that of 2 February 1703, when the observers "fell on their knees and commended themselves to God." No harm came to them or the church. Bianchini loved the place. Even when fatally ill, he "would jump out of bed repeatedly to catch the sun as it crossed the *meridiana*."[86]

Among users of the instrument after Bianchini's death was Anders Celsius, who went to Rome in the spring of 1734 after six months of steady observation at San Petronio. "I doubt that I would have gone [to Rome]," he wrote in his travel diary, "were it not for the meridian line at the Certosa," that is, the Carthusian monastery of S. M. degli Angeli. But he also took to the study of the two most evident features of Rome, antiquities and clerics. The latter came in two kinds, he found, the cardinals, monsignors, and prelates, some of whom were cultivated, and their trains of ignorant "little priests, monks, and abbés, worthless men who walk and loaf in the streets, make love, and gossip in the coffeehouses."[87] When Manfredi was planning a visit to Rome in 1734, Bianchi warned that he would find very few people there apart from Antonio Leprotti, a physician in papal service who had observed at San Petronio, able to discuss mathematics and physics. "The men who live in Rome ordinarily are only waiting to litigate civil suits or to suck up ecclesiastical benefices or honors, for which our studies are no use, and therefore they neglect and despise our concerns, as we do theirs." It is a good if unfriendly reminder that few priests had the high literary culture of men like Bianchini, Noris, and Clement XI.[88]

At Bianchini's line Celsius made observations for comparison with ones Manfredi made simultaneously in Bologna. In the process, he detected that the Roman *meridiana* declined from true by two minutes of arc, an error of which he could not suppose Bianchini and Maraldi capable. He preferred to ascribe the deviation to a displacement of the earth's poles, which, if it occurred, would shift the *meridiana* in respect of the church and explain why ancient geographical coordinates disagreed so blatantly with modern ones. In fact the poles do not shift and the ancients were just wrong. Later observers, the Jesuits Boscovich and Maire, of whom more momentarily, found an even larger error at S. M. degli Angeli, almost 4.5 minutes, which meant a noon too late by 5 seconds at the summer solstice and by 17 seconds at the winter solstice.[89] But modern trials suggest that Bianchini built better than they measured and that the "universal applause of the learned," which greeted the completion of his work, was not misplaced. It appears that midsummer noon can be established at the Roman *meridiana* to within a second of time.[90]

Meridiane and Meridians

· FROM ONE TO THE OTHER ·

Bianchini returned from his trip to Northern Europe eager to carry out in Italy a trigonometric survey similar to the one the Cassinis had undertaken in France.[91] Their enterprise was to extend the survey of the short arc of longitude through the Paris Observatory measured by Picard from sea to sea, from the English Channel at Dunkirk to the Mediterranean near Perpignan. Their technique, simple in principle but troublesome in practice, was to set up a chain of virtual triangles from one end of the arc under investigation to the other; measure the angles of these triangles; lay out a level baseline, determine its length, and link it to one of the virtual triangles; calculate thereby the length a of the arc; observe the zenith distances z_1, z_2 of the same star at either end of the arc; and divide a by $(z_2 - z_1)$ to obtain the length of a degree of a meridian of longitude.

Figure 5.10 depicts the operations: A and K are the endpoints of the arc; B, C . . . J, intermediate stations, usually church towers or rocky prominences; PQ and RS, baselines. The process can be checked in two ways. One, measuring all the angles to be sure that those of every triangle sum to 180°, counseled a perfection seldom possible in the field, where a tall tree or other inaccessible height might have to serve as a vertex of a triangle. The other way, almost always applied, com-

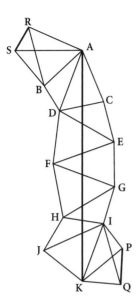

FIG. 5.10. A typical chain of geodetic triangles
linking the stations A and K.

pared the measured length of one of the base lines with its calculated value as determined from the chain of triangles and the other baseline. The desired arc AK
follows from the triangles by a calculation that is more tedious than informative.
The relation between the endpoints A and K and the zenith distances z_1 and z_2,
which caused Picard to remark that "we must seek in the heavens the measure of
the earth,"[92] will be understood from Figure 4.8: the thick lines point to a star that
crosses the meridian through A and B nearly overhead, z_1 and z_2 indicate its distances from the two zeniths. From Euclid I.32, $z_2 = z_1 + \angle AOB$. But $\angle AOB = \Delta\phi$,
the difference in latitude between A and B. Hence, having measured the arc AK (in
Figure 5.10) as some number a of royal toises, the Cassinis defined the length of a
degree of a meridian as $a/(z_2 - z_1)$ toise.

Bianchini undertook a similar determination, also from sea to shining sea, from
the Mediterranean near Rome to the Adriatic near Rimini. He had already drawn
a little piece of the arc, precisely, on the floor of Santa Maria degli Angeli; the
meridiana would anchor the meridian.[93] He thus realized literally Cassini's conceit
of interpreting his *meridiana* as a piece of the meridian running through it. It remained to specify the intermediate stations, determine the angles of the virtual
triangles, lay out baselines near Rome and Rimini, and find the difference in latitude between the endpoints of the arc. By 1724, with encouragement from
Clement XI, Bianchini had run a baseline down the Appian Way, made many measurements of latitude along his arc, and set out several virtual triangles. His preliminary results for the length of a degree agreed perfectly, he said, with the
number declared by Jacques Cassini in 1720.[94] The task was too great for a gimpy

old man to complete alone, however, and, for a few years yet, Italy had to "envy France what the Cassinis achieved there."[95]

It had not been easy for Jacques Cassini and Giacomo Maraldi to complete the extension of Picard's arc begun by Gian Domenico Cassini.[96] When they did finish, they announced their achievement with the family flair. Jacques described how he and his cousin had laid out a baseline along the beach near Perpignan with toise sticks aligned by telescope. Taking the baseline that Picard had measured near Amiens as their given, they calculated through the series of triangles what the base on the beach at Perpignan should be. The calculated and directly measured lengths of the distance between the tree trunks that they used as markers at the ends of the base—some 7,246 toises (ts)—agreed to within 3 toises before correction and precisely afterward. A similar result, to within a single toise, was achieved with a base laid out across the dunes near Dunkirk; a singular coincidence, since many of Jacques Cassini's angles differed by 10 minutes, and a few by whole degrees, from later measurements made from his stations. Having determined the length of the triangulated arc from Dunkirk to Perpignan to be 8°31'12", Cassini gave the weighted average value of a degree over the distance as 57,061 toises, astonishingly, suspiciously close to Picard's.[97]

As a sidelight, Cassini calculated that a degree of the arc north of Paris was slightly smaller than one taken to the south. The difference between consecutive degrees, amounting to some 31 toises, fell well within likely errors of measurement. Nevertheless, Jacques Cassini deemed it significant. Cassini *père* had reached a similar conclusion from measurements he had made toward extending Picard's arc. Cassini *fils* had no interest in overlooking what he thought were facts that confirmed his late father's opinions. If confirmed, the variation in the length of degrees along a meridian would indicate that the earth is not a perfect sphere. That agreed with deductions from Newton's principles and also with the principles of Descartes as developed by Huygens. But Newton and Huygens required the degrees to shorten toward the equator and Cassini had found them to grow.[98] The implied challenge stimulated much useful geodetic work for academicians who liked to travel.

What was involved appears from Figure 5.11, which displays a section of an ellipsoid of revolution that, on Newton's theory, represents the shape of the earth. Here C is the earth's center, O an observer at about 40° latitude, and NTP the north terrestrial pole. (To preserve ϕ as the height of the pole, latitude is not measured from the center of an ellipsoidal earth, a concept that gave many early analysts trouble.) The curvature at the equator is less than that of a circle of radius a (the semi-major axis); calculation shows that the radius of curvature in the first case is a^2/b, in the second b^2/a (Figure 5.12). Since the distance traveled a short way along an arc is proportional to the radius of curvature at the point of departure, one

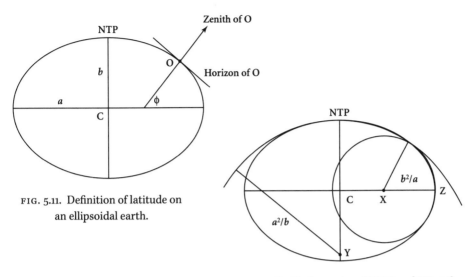

FIG. 5.11. Definition of latitude on an ellipsoidal earth.

FIG. 5.12. Radii of curvature NTP·Y and XZ at the terrestrial poles and equator, respectively.

must walk further along a meridian to raise a degree of latitude at high than at low latitudes. A degree along the meridian at Dunkirk exceeds one at Perpignan by about 100 toises, far too little to have been discerned by the methods of Jacques Cassini.[99]

The greater the distance in latitude between the endpoints, the better the chances of finding evidence of a deviation of the earth's figure from perfect sphericity. If the upper station were at the Arctic Circle and the lower at the equator, the difference in degrees north and south would amount to 650 toises. That was detectable. To detect it, the Paris Academy of Sciences dispatched two expeditions during the 1730s, one to Peru and the other to Lapland. These expeditions are among the great adventures of civilized men. The northern surveyors, who included Celsius, braved freezing cold, glaciers, and mosquitoes; the southern, raging heat, withering cold, huge mountains, and mosquitoes; and all to learn whether a degree of longitude contains a few more feet at the Arctic Circle than at the equator.

The northerners had the better of it. They had a frozen river on which to lay out a baseline, six stout soldiers and fifteen boats to carry their instruments, and only one winter to survive. They returned to Paris in 1737, after an absence of sixteen months, with a result that, as Voltaire put it, "simultaneously flattened the poles and the Cassinis." In fact, their best value was out by 200 toises, which, had it fallen in the other direction, would have confirmed a spherical or other non-Newtonian earth. The southerners got a better number after much suffering from prolonged exposure to the Andean climate and to one another. They arrived in the New

World in 1735 and began straggling back in 1744. Among the many causes of delay were the weather, fatigue, disease, earthquakes, adverse terrain, and the hostility of the Indians, who thought that grown men who spent years lining up sticks end to end on the ground must be sorcerers or lunatics. By combining the good Peruvian degree with the poor Arctic one the French academicians managed to obtain a value for the ellipticity of the earth (call it the Perlap value) not far from the one Newton had calculated without leaving Cambridge.

Too much information is a dangerous thing. A recalculation of Picard's arc around Paris gave a new datum that, when combined with either the equatorial or the Arctic degree, yielded an ellipticity different from Newton's and from Perlap's. To bring everything into harmony, the French academicians remeasured Picard's arc. Their result could not be reconciled with the others. The Academy proposed to resolve the matter by judicial inquiry and named commissioners for the purpose; but they could not agree about anything except that the numbers allowed no firm decision about the degree of the earth's ellipticity. At this point the distressing possibility that the earth might not be an ellipsoid of revolution gained currency. This possibility was considered a certainty by the busy Jesuit mathematician Roger Boscovich, then a professor at the University of Rome, the Sapientia.

Boscovich grasped the opportunity to confirm his conviction during a conversation with Cardinal Silvio de' Marchesi Valenti, who had risen through the Vatican hierarchy by learning and diplomacy to become Secretary of State under Pope Benedict XIV. Boscovich described the problem; Valenti asked if anything could be done; Boscovich replied that completion of Bianchini's project of running a meridianal arc from Rome to Rimini would probably settle the matter.[100] He had in mind that the Rome–Rimini arc had the same average latitude as the Cassinis' line from Dunkirk to Perpignan; consequently, its measurement would allow a test whether at the same latitudes the length of a degree along all meridians is the same. Boscovich expected the test to fail, which it did; "prejudice for regularity and simplicity," he wrote, "is a source of error that only too often has infected philosophy."[101]

Valenti consulted Benedict, a strong supporter of experimental natural science, which he patronized both at the Sapientia and at the Academy of Sciences in Bologna. (Benedict bought up most of the Campani lenses available in Italy for his favorite philosophers.)[102] He commissioned Boscovich to complete Bianchini's work and, into the bargain, to map the Papal States; but, so Boscovich insisted, the primary purpose of the business, "of much greater interest to the Republic of Letters," was the meridian, not the map.[103] Even Boscovich could not perform the exacting task of running a meridian alone. He enlisted the support of a fellow Jesuit, Christopher Maire, a good astronomer with a taste for geography, "and, moreover, healthy enough to withstand the fatigue of travel."[104]

They began by checking Bianchini's determination of the latitude of S. M. degli Angeli, which, perhaps, they thought to take as the anchor of their net of triangles. They found errors of a few seconds and, what was worse, a slight deviation of the line from the meridian; and, what was still worse, a serious error in division, where Bianchini, apparently losing count, had chopped a part of a scale into 900 instead of 1,000 parts. So Boscovich and Maire decided to anchor their net a little to the west, in the dome of Saint Peter's, from which they could obtain wide views to other stations along the meridian. They also followed Bianchini's lead in choosing a northern station near Rimini. They found it in the villa of Francesco Garampi, a gentleman whose grasp of astronomy had earned the praise of his teacher Manfredi and whose home contained much useful astronomical apparatus, including a meridian line.[105] It was time to get on the road.

Boscovich and Maire started their work in rains the like of which had not been seen since the time of Noah. Twice marooned in villages, they improved their time by preaching, "thinking we could fill the void in our geometrical occupations in no better way than by works of zeal." They were attacked by dogs and, worse, by peasants, who knew that no one could be idiot enough to do what Boscovich and Maire claimed to be doing; the Jesuits must be looking for buried treasure or otherwise irritating the hydraulic spirits. Reasoning thus and needing nails, the peasants tore down the signal towers that Boscovich and Maire had built on the summits of the Apennines. The French Jesuit translator of the report of the expedition, which Boscovich and Maire drew up in Latin, was astonished. "Who could have believed that there would be so strong a resemblance between the peasants of the Apennines and the Indians of the mountains of Quito?"[106]

Using old ship's masts as rulers, Boscovich and Maire laid out baselines of about 12 kilometers along the Appian Way and on the beach near Rimini; they set the masts with small gaps in between, which they measured with dividers, so as to avoid inadvertent displacements; two measurements of the base differed from one another by 1/36 of a toise in over 6,037 toises, or, in accordance with a nineteenth-century remeasurement, by 28 cetimeters in over 3000 meters. They anchored their angles more accurately than the French had done in Lapland; they determined the number of toises in their masts from a toise standard sent from Paris; and they ended with a value for a degree along the meridian at a latitude of 42° that did not agree well with the latest French degree at an average latitude of 43.5°. Boscovich was pleased. "Look where you will, you will see nothing regular, nothing fixed or constant."[107]

To push this anarchistic finding farther, Boscovich pulled upon the strings available to a well-placed Jesuit savant. He observed that the surveyors in Peru had found that the gravitational attraction of the Andes had drawn their plumb bobs aside by a few seconds of arc, falsifying the vertical and the orientation of instruments in its neighborhood. Similarly, the Pyrenees might have influenced observations taken at the southern end of the Dunkirk–Perpignan line and the Apennines the labors of Boscovich himself in the Papal States. To obtain measures without mountains, Boscovich proposed to the Royal Society of London and the Austro-Hungarian Empress that they commission determinations of a degree across the plains in America and in Eastern Europe; and to the King of the Two Sicilies that, for comparison, he do the same, in the foothills of the Alps around Turin.[108]

The Royal Society commissioned Charles Mason and Jeremiah Dixon, already surveying in the New World, to measure an arc in Maryland; the Holy Roman Empress sent off a Jesuit astronomer, Joseph Xaver Liesganig, to the prairies of Hungary, where he measured out baselines using rafters from the Jesuit college in Vienna; and the King of the Two Sicilies asked the professor of physics in the University of Turin, Giambattista Beccaria, to try his hand at a degree in Piedmont. The results only increased confusion.

Mason and Dixon did not triangulate, but measured their degree directly on the ground; they had the advantage of a level terrain and the best English instruments; they corrected for temperature and compared their measuring rods carefully with the "toise de Pérou"; "yet [it is the criticism of the late eighteenth century] the result was no better than the rest." The critic, a prolix promoter of astronomy in central Europe, Franz Xaver von Zach, could not grant even that much to Beccaria or Liesganig. Beccaria had bobbled his arc by 900 toises ("an altogether intolerable error") and missed an angle by 13 minutes, according to measurements Zach took on the spot; whereas Liesganig, according to his manuscripts, which Zach acquired, had falsified his angles, taken one church tower for another, misplaced one endpoint by 4,500 toises, and misidentified the star by which he fixed his latitudes.[109]

The confusion and contradiction created by extending *meridiane* from the church into the field grew worse as data accumulated. Boscovich turned out to be right. The shape of the earth is too irregular to be mimicked or modeled accurately by an ellipsoid of revolution. All meridians do not have equal length. The shape of the earth is just that—the geoid, to use a term introduced in the nineteenth century.[110] The grand conception of Cassini and Bianchini, that the *meridiane* of San Petronio and Santa Maria degli Angeli occupied well-defined, though perhaps not perfectly known, fractions of an unproblematic, useful, universal quantity, the length of a meridian of longitude, was thus exploded. Progress has its price.

As we know, Gian Domenico Cassini fought with the architect of the Paris Observatory almost from the moment of his arrival in France. His most pressing concern was to create a space large enough for a *meridiana* to rival San Petronio's. This much he achieved, that a great hall was set aside, which, with its adjoining tower to the north, allowed the implanting of a full *meridiana* extending 96'10" Paris measure (31.45 meters).[111] To give this dimension a higher authority than the recalcitrance of the architect, Cassini worked out that it stretched one second plus one third (that is, $1/60^2 + 1/60^3$) of the circumference C of the earth. This *jeu d'esprit* came very close, for, with Picard's value of 57,060 toises/degree, it amounts to 96'8" or 31.38 meters. Had Cassini lowered the hole a little further (he had put it too high at first to fit the winter solstice into the tower and had to remeasure with Picard's help), he could have made his conceit a reality. Putting off construction to allow the building to settle and taking up new projects, like the Dunkirk – Perpignan line, he deferred too long and died before he could install his trademark in the great hall of the Observatory.[112] No doubt its diminutive size in comparison to the heliometer in San Petronio ($h = 9.94$ m against 27.1 m) condemned it to perpetual second priority.

It rose to the top of the Cassini agenda when Jacques Cassini finished the survey of the Dunkirk – Perpignan line. "Having extended the meridian line that passes through the Paris Observatory in both directions, north and south, to the confines of the kingdom, it appeared necessary to the full perfection of the work to trace within the Observatory itself a meridian line that would be part of the one crossing the kingdom and, at the same time, would serve for the astronomical observations pursued vigorously there ever since its foundation." Thus Jacques Cassini justified the expenditure of 6,307 livres from the treasury of the Paris Academy, a little over half of its annual budget for ordinary expenses, for a fit monument for his father.[113]

The location of the *meridiana* on the second floor of the Observatory will be clear from Figure 3.1, which shows the south façade of the building and the entrance, through which the virtual meridian of France ran. (The Observatory's *meridiana* goes at right angles to a line joining the centers of the towers.) Jacques Cassini modestly improved the laying of meridian lines. Number 1 of Figure 5.13 shows the copper plate (2' × 18") holding the hole (with diameter equal, as usual, to $h/1000$); the parallel wooden sticks hanging from it measured the height. Number 2 shows the plumb line for determining the vertex of the meridian; the disk at the top fits the hole in Number 1 precisely, so that the suspended weight shown at F dips into the water-filled box beneath it to damp its oscillations.[114]

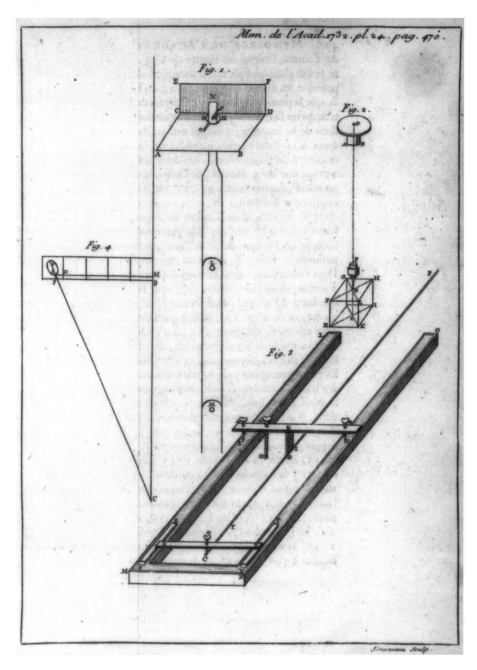

FIG. 5.13. Leveler for the *meridiana* in the Paris Observatory. From Cassini, *MAS*, 1732, 470.

Number 3 depicts the leveler, Jacques Cassini's most original contribution to the meridianal art. He divided the brass rod constituting the line into thirty strips, each exactly 3'0"8''' long, and cemented each piece precisely between, and on a level with, two correctly dressed marble slabs each exactly 3'0"8''' long (Jacques' favorite word was "précisément"). Having dug a trench along the north–south line through the vertex to the depth of the marbles, he began leveling the first segment by placing its southern end "exactly" at the *meridiana's* vertex C and lowering the tip of a screw on the leveler so as just to touch the segment beginning at C. The screw was carried by a metal link between wooden floats in the water-filled frame LMNO, 36' long, which straddled the direction of the meridian (LMNO is the shaded U-shaped receptacle). Once he had settled the first segment at C, he moved the floats to the segment's north end, which he raised or lowered so as to bring it into contact with the point of the leveling screw. Proceeding thus, he laid all thirty of his brass-and-marble sandwiches, "precisely."[115]

In much of the work, and particularly in the determination of the direction of the meridian through C, Cassini had the help of Maraldi. They found the direction in 1729, by observations of the sun at equal distances on either side of the solstices. They then approached the moment of truth. "All the pieces of the *meridiana* being thus disposed, we checked to see if it was precisely in the direction of the meridian line running to the furthest limits of the kingdom." They hung a plumb line over the north end of the *meridiana,* centered the line in the eyepiece of a telescope set up over the south end, and looked beyond it, through the tower window. They were no doubt relieved to see, smack in the center of the eyepiece, the post that Picard had planted on Montmartre as a marker on the meridian he had traced to Amiens.[116]

It remained to add the usual decoration, marbles with the zodiacal signs, and to make use of the instrument. Comparing its orientation with that of two large mural quadrants elsewhere in the building, Cassini confirmed what had long been suspected, that they deviated from their meridians. And, comparing the value of the obliquity of the ecliptic that he deduced from his observations at his new *meridiana* with his father's made sixty years earlier at San Petronio, he found a discrepancy of 27". Did that mean that the Cassinis had at last answered the question of the decline of the obliquity and fixed it at 45" per century? That would have been a dramatic and fitting climax to their construction of *meridiane.* Alas! Jacques was not his father. He declined to pronounce, continued his observations, and lost his opportunity.[117]

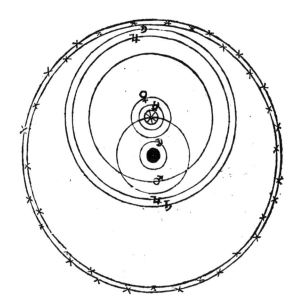

6: The Accommodation of Copernicus

Neither Cassini nor his Jesuit collaborators had in mind subverting the Church's condemnation of the concept of a moving earth and stationary sun when they built and worked their meridian lines in Bologna, even though Cassini expected to confirm that notorious consequence of the elliptico-heliocentric system, the bi-section of the eccentricity of the sun's orbit. No more no doubt did the *fabbricieri* of San Petronio expect that their cathedral would provide information about the heavens opposed to the teachings of their church. Nonetheless, there was good reason to expect the worst of observations made at meridian lines.

Heliometers and Heliocentrism

· GALILEAN CONNECTIONS ·

Galileo met Cesare Marsili, a prominent Senator in Bologna and one-time super-intendent of its waters, in Rome in 1624 when both were there to congratulate the

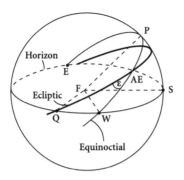

FIG. 6.1. The ecliptic with the summer solstice setting to illustrate Galileo's thought experiment for detecting a change in the obliquity.

new Pope, Urban VIII, on his election. A few years later, as Galileo was finishing his ill-fated *Dialogo,* Marsili sent some interesting news, which Galileo was pleased to insert in the book, "as an indication to the world of my respect for your powers and your esteem for my researches." This news had to do with Danti's *meridiana* in San Petronio, which, after thorough examination, Marsili decided was further off true than it had been in Danti's time.[1] Marsili asked Galileo to check whether Danti's instruments on the façade of S. M. Novella in Florence showed the same decline.[2] Galileo was interested: if the meridian had shifted in relation to the church, the axis of the earth must have moved relative to the axis of the world. An earth-centered person would think that the earth's axis had held steady and that the direction of rotation of the entire heavens had changed; whereas the sensible and modest Copernican would attribute whatever motion occurred to the mobile earth.

Galileo had had an idea for checking for a change in the obliquity of the ecliptic that could be adapted to detecting a shift in the meridian. With a very good telescope he would observe the direction of sunset on the day of the summer solstice over a sharp ridge some sixty miles distant. "If God grants me life for four or six years and clear skies at the solstices, I do not doubt that in that time, although short, I will see some sensible change."[3] Galileo's plan is indicated in Figure 6.1, which shows the setting midsummer sun at Q as seen from Florence (F). Since the arc Q·AE is 90°, $\angle Q \cdot AE \cdot W = \varepsilon$, and $\angle PFS = 90° - \phi$, the solution of the spherical triangle Q·AE·W gives ε in terms of the measurable angle QFW and the latitude ϕ.[4]

Galileo expected to see a change in $\angle QFW$ owing to a diminution in the obliquity ε. In response to Marsili's inquiry, he recognized that a change in $\angle QFW$ might also arise from motion of W (the west point of the horizon) consequent on a shift of the meridian. Observation of midwinter sunsets should distinguish between the two possibilities: a decrease in ε would move the midwinter sunset, like the midsummer, toward the west; a fixed ε and moving W would bring the ridge that marked the west in the first year of observation toward the direction of midsummer, and away from that of midwinter, sunset. Nothing useful could be learned, however, from S. M. Novella. "I judge it to be a difficult and uncertain matter because [Danti's] instruments are small and because the pavement, being very uneven, is not suitable for a new *meridiana*." Although Galileo had no faith in the accuracy of Danti's instruments, he thought that it might be worthwhile to compare whatever meridian they had indicated with a *meridiana* that could be drawn in a friend's house on the same square as the church.[5]

Marsili tried to measure the height of the hole Danti had put in San Petronio and to find the direction of the meridian by the usual trick of bisecting the angle between equivalent solar images before and after noon, but the inequality of the pavement, the general "malevolence large material objects have for exact observations," and the "trembling of the rays and the blurring of the shadow" conspired against him. Nonetheless he had found clear indications of a decline in ε and reinforcement for his belief that big instruments are better than little ones. He would have the pavement of San Petronio redone where necessary and, if all went well, provide strong evidence for the motion of the earth.[6]

Galileo judged the chances for success good enough that he honored Marsili by including his demonstration of the motion of the meridian among the five best observational evidences for the motion of the earth with which he ended his great polemic on the world systems. According to Salviati, Galileo's spokesman in the *Dialogo*, Marsili had observed a very slow but continuous change in the meridian and had written out the results in a learned paper, "which I have seen recently with amazement and which I hope he will have copied for all students of the wonders of nature."[7] This advertisement piqued the curiosity of many students of wonders. One of them, writing to Galileo to ask when Marsili's paper would appear, made clear what a definitive detection might mean. "If it is found to agree with Copernicus, *a Dio Thomaici*, good-bye to the Thomists!"[8] Bolognese astronomers seem to have believed in the reality of the effect down to the time of Manfredi.[9] In the end nothing definitive did emerge, except the recognition that meridian lines could be used to support dangerous doctrines. Marsili's failed effort at San Petronio makes Cassini's work there the more admirable in its execution and the more problematic in its purpose.

Marsili's argument in aid of Galileo naturally interested the Jesuits. One of their most able and eclectic natural philosophers, Juan Caramuel, investigated, and found no evidence for a shift of the meridian.[10] Riccioli treats it fairly, acknowledging its relevance in principle but denying it observational authority owing to the substantial errors in Danti's *meridiana*.[11] He took greater interest in another argument, or, rather, thought experiment, in favor of the earth's motion, this one concocted by Galileo himself, which was to frame the technical debate about Copernicanism in Northern Italy during the 1660s. Galileo's argument, which he

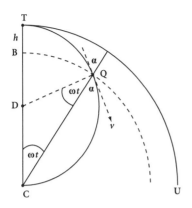

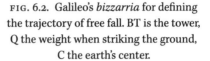

FIG. 6.2. Galileo's *bizzarria* for defining
the trajectory of free fall. BT is the tower,
Q the weight when striking the ground,
C the earth's center.

characterized as more whimsical *(bizzarria)* than demonstrative, appears in the *Dialogo*.[12]

Suppose the earth divided into two parts by its equatorial plane so that a weight dropped from a tower at the equator could fall to the earth's center. What path would the weight follow? Galileo suggested that it might follow a semicircle centered at D with diameter equal to the distance from the center C to the top of the earth's tower T (Figure 6.2). The reason he gave was characteristic of his thought. Since $\triangle QDC$ is isosceles, $\angle TDQ = 2\omega t$, where ω is the angular velocity of the earth's suppositious rotation. The velocity of the falling weight along the tangent to its path around D is therefore $2\omega(r + h)/2$, where r is the earth's radius and h the tower's height. But before its release the weight moved along the arc TU with a velocity $\omega(r + h)$. Hence dropping an object would not change the magnitude, but only the direction, of its velocity. To Galileo this consequence had the advantage of giving a plausible trajectory without an acceleration and, hence, freeing him from the obligation of offering an explanation.

This same kinematics, or description of motion without invoking causes, underlay Galileo's fundamental contribution to mechanics, his rule of free fall. His final formulation of this rule was that, when dropped without constraints of any kind, a heavy object will increase its velocity in proportion to the time elapsed since its release; and that, consequently, the distances it covers will be as the squares of the elapsed times. Now this rule, which Galileo put forward as a generalization of experience, cannot be independent of the weight's semicircular orbit to the center if we take his *bizzarria* seriously. When the weight strikes the ground at time t, its velocity along the tower (and hence perpendicular to the horizon) is about equal to $\omega^2(r + h)t$.[13] The result agrees with Galileo's rule in making the velocity of free fall proportional to the elapsed time; but it is much stronger than the rule in giving a value for the constant of proportionality. Since the earth turns once in 24 hours, $\omega = 2\pi/(24 \cdot 3600)$ rad/sec; $r + h \approx (4 \cdot 10^7/2\pi)$ m; hence $\omega^2 r = 3.4$

cm/sec^2. That came nowhere near the mark. The constant of free fall is 980 cm/sec^2.

· THE DEMANDS OF DUTY ·

Riccioli made the first good measurements of the constant of free fall. He did so expecting to defeat Galileo's rule. He began his experiments in 1640, when teaching philosophy at the Jesuit college in Bologna. Building on Galileo's observation of the regularity of pendulum beats, Riccioli used a chain and weight as a clock. But how to find, precisely, the number of seconds in each beat of the pendulum? Riccioli's answer, which raised his determination of the constant of acceleration to the "first sustained attempt of an experimental measurement," was to choose his pendulum of such a length that its bob took exactly one second to make one swing.[14] He proposed to find this convenient length by experiment.

Riccioli and Grimaldi chose a pendulum 3'4" long Roman measure, set it going, pushed it when it grew languid, and counted, for six hours by astronomical measure, as it swung, back and forth, 21,706 times. That came close to the number desired: 24·60·60/4 = 21,600. But it did not satisfy Riccioli. He tried again, this time for an entire 24 hours, enlisting nine of his brethren including Grimaldi; the result, 87,998 swings against the desired 86,400. Riccioli lengthened the pendulum to 3'4.2" and repeated the count, with the same team: this time they got 86,999. That was close enough for them, but not for him. Going in the wrong direction, he shortened to 3'2.67" and, with only Grimaldi and one other staunch counter to keep the vigil with him, obtained, on three different nights, 3,212 swings for the time between the meridianal crossings of the stars Spica and Arcturus. He should have found 3,192. He estimated that the length he required was 3'3.27", which — such is the confidence of faith — he accepted without trying.[15] It was a good choice, only a little further out than his initial one, as it implies a value of 955 cm/sec^2 for the constant of gravity.

Armed with this information, a smaller, faster pendulum calibrated by it, and balls of wood and lead, and accompanied by a chorus of musical brethren to complete their clock, Riccioli and Grimaldi repaired to the Torre degli Asinelli. (The musical brethren chanted "do," "re," etc., as the pendulum beat so that Riccioli needed only to keep track of units of eight, rather than of individual, swings.) As everyone had expected, Galileo was disproved. The lead ball always hit the ground before the wooden one when they fell from the same height. The discrepancy between the experiment and Galileo's claim that they reached the bottom simultaneously was so great that Grimaldi supposed that Galileo must have known about it, but suppressed his knowledge in order to secure a proposition dearer to him

PLATE 1. Toscanelli's *meridiana,* S. M. del Fiore, Florence, showing the solstitial mark of 1510.

PLATE 2. San Petronio, Bologna, interior. The *meridiana* touches the piers on the left. From Bellosi et al., *Basilica* (1983).

PLATE 3. Danti's *meridiana* in the Torre dei Venti, the Vatican. The beam has been enhanced.

PLATE 4. The sun on the *meridiana* of San Petronio.
From Heilbron, in Shea, *Scienze, 2* (1992), 349.

PLATE 5. Santa Maria degli Angeli, Rome. From Alberti-Poja, *Meridiana* (1949), after p. 36.

(A)

(B)

PLATE 6. The midsummer sun at Santa Maria del Fiore, Florence, at the brief reopening of the *meridiana* in 1997: (A) the light falling on Toscanelli's gnomon; (B) the sun's image a little before noon a few days from the summer solstice.

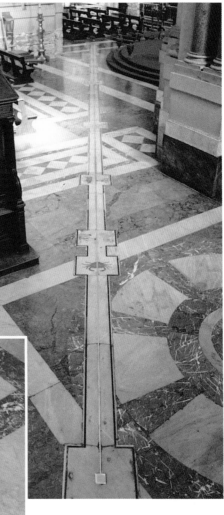

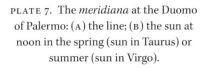

PLATE 7. The *meridiana* at the Duomo of Palermo: (A) the line; (B) the sun at noon in the spring (sun in Taurus) or summer (sun in Virgo).

(A)

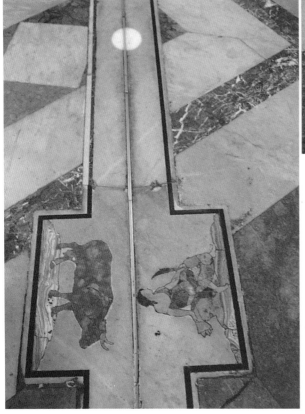

(B)

PLATE 8. The Shrine of Remembrance, Melbourne.

than truth. According to this Jesuitical interpretation, the proposition that had to be saved was the *bizzarria,* the semicircular descent of freely falling bodies to the center of the earth. Riccioli explained Grimaldi's reasoning: according to the *bizzarria,* bodies freely falling from the same height should move along their semicircles with one and the same velocity, that of the place from which they fell; whereas, in fact, experiments at the Torre degli Asinelli by a chorus of priests demonstrated exactly and irrefragably that different bodies descend at different speeds.[16]

Riccioli was more indulgent. Although Galileo's rule did not hold universally with the same constant of proportionality, it did hold — as Riccioli had found by releasing weights from different heights along the tower — for the lead ball and the wooden ball taken separately. But that concession provided still another argument against the semicircular trajectory. From his own measurements he calculated that, if Galileo's rule of fall held all the way to the earth's center, it would take a lead ball about twenty minutes to get there. Proceeding along their semicircles, however, all bodies, independent of the height from which they fell, would need six hours to reach the center since the distance to be covered was a quarter of the circumference of the circle described by the top of the tower in twenty-four hours. That appeared to make no sense. Worse still, the consequence that so pleased Galileo, that the falling body had the same velocity when free as when sitting in the tower, implied that, no matter what the height or place of the tower, the weight would always strike the ground at the same speed and hence with the same impulse. But that violated the most evident experience.[17]

Riccioli gave great weight to this argument because he thought it applied not only to Galileo's special semicircular path but to any curve derived from the diurnal motion of the falling body. He reasoned that, on the Copernican theory, the motion of the body along the tower was merely apparent and hence (the "hence" does not follow) could not create the observed impacts. According to him, the real motion, the motion with the force of motion, of a free body on a spinning earth would be a curve in space; relative motions constructed from this curve and the circumstances of the observer would have little or no physical effect. In this conception he was wrong, as Cassini and Montanari told him; but his eagerness to subvert heliocentrism, his faithfulness to scholastic physics, and, perhaps, an inability to grasp the properties of relative motion held him obdurate to the end of his days.[18]

Riccioli first published this, his best physical argument against heliocentrism, in his *Almagestum* of 1651. It produced no public polemic. When he repeated it fourteen years later in his *Astronomia reformata,* however, he untied the tongues and pens of several Italian mathematicians. Cardinal Leopold de' Medici assigned the best geometer of his Accademia del Cimento, Giovanni Alfonso Borelli, professor of mathematics of the University of Pisa, to answer Riccioli; Borelli pointed out that the old man had misunderstood relative motion. Then the Jesuat (not Jesuit) priest Stefano degli Angeli, professor of mathematics at the University of Padua, who esteemed Galileo's works as "so many miracles," delivered the same message in several prolix pamphlets. Both Borelli and degli Angeli made mistakes of their own, however, which gave rise to further tracts by them, by Riccioli, and by their students.[19]

This long dispute was the last in which mathematicians tried to refute heliocentrism on physical grounds. These refutations derived what force they had from common sense ("where is the wind blowing from the East?") or Aristotelian physics ("bodies cannot undergo two distinct motions at the same time"). Therefore they told, if at all, only against a literal interpretation of Copernicus' theory. That did not confound mathematicians who read Copernicus in the instrumentalist manner recommended by Osiander and abominated by Galileo. These free spirits could hold the traditional theory or none at all and compute heliocentrically whenever they deemed it advantageous. It was particularly advantageous for the calculation of planetary distances. Indeed, there was no other reasonable choice since Ptolemy's theory cannot specify the order, let alone the distances, of the planets without special assumptions. Traditionalist astronomers had only to know how to suspend disbelief to be able to adapt information deducible only in the heliocentric system for application in a geocentric cosmology.[20]

To find the heliocentric distance r_i of an inferior planet, pick a time when it appears at maximum elongation at P_{max} (Figure 6.3), when $\angle SP_{max}E$ is 90°; then, by measuring this elongation β (the angle between the sun and the planet as seen from the earth E), you have r_i in terms of the solar distance s and β.[21] Finding the heliocentric distance r_s of a superior planet is only a little more difficult. Begin at opposition (Figure 6.4), when the sun and planet appear diametrically opposite as seen from the earth (P_1E_1S). Wait a few months and measure the angle $\gamma = SE_2P_2$ between the sun and the new direction of the planet E_2P_2. All the angles in ΔSE_2P_2 are then known, since $\angle E_2SP_2$ is the difference between θ_E, the angular motion of the earth around the sun, and θ_P, the angular motion of the planet, during the time elapsed between opposition and the measurement.[22]

Minds subtle enough to mix Copernican distances and traditional cosmology spied other loopholes in the general proscription to which Galileo's violation of his personal injunction gave rise. A leading example is Gassendi, who, in his *Institutio astronomica* of 1647, justified including the Copernican as well as the Ptolemaic and Tychonic "hypotheses" on the quibble that the Holy Office had silenced only Galileo. No doubt, Gassendi added, should the Church properly proscribe heliocentrism in general, Copernicans would be happy to recognize and renounce their error. Gassendi's often reprinted *Institutio* was the most important early source of

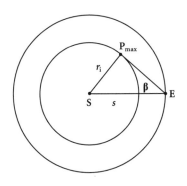

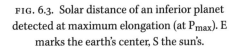

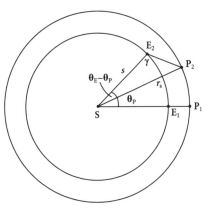

FIG. 6.3. Solar distance of an inferior planet detected at maximum elongation (at P_{max}). E marks the earth's center, S the sun's.

FIG. 6.4. Solar distance of a superior planet from two observed positions (P_1, P_2).

Copernicus' ideas in France.[23] Again in France, the Minim priest Marin Mersenne, who for many years acted as Descartes' mouthpiece to the republic of letters, publicized Galileo's results, blamed his trouble on his disobedience, favored heliocentrism, recommended the Tychonic system, and offered planetary distances according to the Copernican hypothesis.[24] Even in Italy during the 1630s and 1640s some bold souls taught heliocentrism hypothetically, protected by the arguments that the prohibition applied only to Galileo or, if more general, then only to physics, not astronomy.[25]

The latter was Riccioli's preferred position. He admired Copernicus enormously and also his theory, taken hypothetically. "The loftiness of Copernicus' mind, the depth of his understanding, and the keenness of his intellect have not been, nor ever will be, sufficiently admired; for he, by the triple motion of one little globule (for such is the earth compared with the heavens) demonstrated what most astronomers before him could not even represent without the crazy machinery of the celestial spheres." At one stroke, Copernicus had eliminated half-a-dozen unnecessary circles, explained the stations and retrogradations of the planets, done away with the ponderus process of the precession, and much more. "The deeper one digs into the Copernican hypothesis, the more ingenuity and precious subtlety one unearths." Unfortunately, before entering into the mathematical astronomy of *De revolutionibus*, and occasionally even within it, Copernicus wrote as if he believed in the truth of the great structure he was building. He slipped from the straight but narrow path of the hypothetical. Riccioli again: "What before so many Atlases could not support, this Hercules has dared to carry. Would that he had kept himself within the limits of his hypothesis!"[26] The same went for Galileo, "a mathematician of immense power wonderfully skilled in astronomy; he would have been greater still if he had put forward the opinion of Copernicus as a mere hypothesis."[27]

Under the shield of hypothesis, Riccioli discussed sympathetically all the chief astronomies — not just the Ptolemaic and Copernican, to which Galileo limited his world systems, but also Kepler's and Tycho's. As typical among the Jesuits, he favored the Tychonic system or, rather, his modification of it, which left Jupiter and Saturn circling the earth. Had it not been for Galileo's intervention, Riccioli might have been able to spare himself the labor of assembling the 126 arguments philosophical, mathematical, and theological, 49 *pro* and 77 *contra* heliocentrism, with which he filled the second book of his *Almagestum.* [28]

He regarded this exercise not only as a duty to his superiors but also as a favor to himself. He had received the grace to be a helpmate to God and the Sacred College of Cardinals. He wrote, doubtless sincerely, that efforts to read the Bible as Galileo had tried to do, as neutral toward or even in favor of proscribed cosmologies, irritated him. "It is almost impossible to believe how much they provoke me to contradict and to answer with all my strength, especially since in this Copernican struggle the literal sense of Scripture is at stake." And yet — knocking at the door that Bellarmine had opened a crack — literalness must not be driven to absurdity. In the unlikely event that the Copernican system, now apparently ruled out by the plain meaning of the holy writings, acquired arguments so strong that clinging to the obvious reading would be absurd, the church would have to reinterpret its gospel.[29]

Galileo's disciples did not deride Riccioli's conclusion. They respected the man and understood his circumstances. Rather, they criticized his insertion of Galileo's condemnation and retraction in his big book and wondered that he had bothered with all his arguments. As Vincenzo Viviani, Galileo's last direct disciple, wrote to Montanari, there was no reason for all that artillery. Riccioli had only to say that the Copernican theory "had been prohibited by something that cannot err in these matters."[30] That, in fact, is exactly where Riccioli came out. The corrosive fictionalism represented by Osiander dissolved all his mathematical and astronomical arguments: a purely hypothetical statement can no more be proved false than true. In the end, there was only the decree of the Holy Office:

> That the sun is revolved by diurnal and annual motion, and that the earth is at rest I firmly hold, infallibly believe, and openly confess, not because of mathematical reasons, but solely at the command of the faith, by the authority of Scripture, and by the intimation [*nutu*] of the Roman See, whose rules, laid down at the dictation of the spirit of truth, may I, as everyone should, uphold as law.[31]

No utterance of the oracle of San Petronio could shake this faith. God had not intended Riccioli or anyone else to unravel the labyrinthine motions of the heavenly bodies far enough to succeed, by the light of nature, in fixing any point as definitively at rest.[32] Unafraid, therefore, to find further evidence in favor of heliocentrism and willing to use the best data and geometry available to save the phenomena, Riccioli could encourage Cassini's installation of the *meridiana*, accept the results obtained with its aid, adopt an ellipse for the sun's orbit, and attack the heliocentrism that underlay elliptical astronomy without confusing his judgment or risking disobedience.

· THE BEST OF BOTH WORLDS ·

Cassini had no obligation to publish whether or not he accepted the Copernican point of view. At the outset of his career, however, when under Riccioli's constant influence, he composed several tracts explicitly based on a geocentric world view. Thus he wrote out a system of his own, in which the planets describe spiral paths around a stationary earth, for presentation to Pope Alexander VII, for whom he subsequently constructed a planisphere that showed the revolutions of the superior planets "circa terram." Later on, when he had no reason to curry favor with Jesuits or popes, he made another earth-centered planisphere; and he would always refer to the astronomy of Copernicans not as "his," but as "their" system.[33] "They" gathered that he "allows not the Motion of ye earth at least the annual."[34] Of course, Cassini understood that for many purposes heliocentrism had important advantages over the Ptolemaic picture.

He gained access to these advantages when necessary by adopting the Tychonic system, which, he thought, was "precisely equivalent" to the Copernican. Once, distinguishing between satellites and planets (the former orbiting moving, the latter stationary, centers), he observed that the phases of Venus and Mercury, discovered through the telescope, proved them to be satellites; from which it appears that he favored Tycho, for only on that system would they orbit a moving center (the sun). Huygens, who worked with Cassini for many years, supposed him to be Tychonic.[35] Later, however, having reasserted the equivalence of the systems, he waffled: the Copernican system "represents the universe as it is in fact" and the Tychonic as we see it. But seeing, he said, was better than imagining, and he left, as his last gift to astronomers, a set of beautifully engraved plates showing the spiral orbits of the planets as viewed from earth. "More concrete models are always more persuasive than less concrete ones." Cassini remained Tychonic, agnostic, or indifferent toward astronomical systems; in any case, he never declared for Copernicus.[36]

Nonetheless, Cassini made several discoveries supportive of heliocentrism. We already know his demonstration of the bisection of the eccentricity and his discovery, in 1666, of the rotation of Jupiter. This second detection seemed almost demonstrative to the editors of the *Journal des sçavans*. "It is one of the most beautiful discoveries ever made in the heavens, and those who believe in the motion of the earth will find a perfect analogy in it." The argument turned on a triple analogy. Before finding the spot on Jupiter that appeared fixed in its body, Cassini had identified other, fleeting spots as shadows of its moons. He thus answered a Jesuitical objection against assimilating the Medici stars to moons, that is, that they might go back-and-forth behind, rather than circle about, the body of Jupiter. "The satellites revolve around Jupiter as the moon does around the earth." And more: the relationship between the speeds of the satellites and their distances from Jupiter obeyed the law Kepler had found to hold for the sun and the planets. Who could miss the parallel between the earth and Jupiter (both spin and have moons) and between the Jovian and the solar systems (both primaries spin and have a swarm of satellites that move in similar ways).[37]

The discovery of Jupiter's rotation and its inferences were widely discussed. Adrien Auzout, "the most famous and accurate of [French] astronomers," thought they would make all astronomers Copernican. "Since it is the truth, the prohibitions inspired by the scandal of novelty will be lifted, as [certain Jesuitical writings] give us reason to hope."[38] We shall review these encouraging writings soon. The first few issues of the *Philosophical transactions* gave full play to the discovery of Jupiter's spin and its consequences. The very first article in the *Transactions*, a review of Campani's account of Jovian discoveries made with his lenses, referred to the support they gave to "the opinion of the Copernicans." Later it quoted Auzout and the *Journal des sçavans*.[39] Willy-nilly, Cassini had become a major contributor to the Copernican cause.

He made another significant contribution in 1686 by showing that the moons of Saturn also obeyed Kepler's law. In reporting this finding, Cassini did not mention its obvious bearing on the geometry of the solar system. Flamsteed thought he knew why: "To be thought a good Catholick he says nothing of it but conceales it."[40] On the contrary. To be thought a good Catholic he interpreted his demonstration as one more proof of the harmony of part and whole throughout the fabric of nature. "Thus the satellites of one order agree with those of another according to their rank in a perpetual concert, in praise of the author of this admirable harmony of the universe, and mankind's progress in the study of these marvels justifies more and more every day the truth of the divine words, *Dies diei eructat verbum, & nox nocti indicat scientiam*."[41] The showing that the satellites of Saturn and Jupiter obey the same numerical rule in their velocities and distances from their primaries as do the planets with respect to the sun was a powerful talisman. "Certainly," according to the uncertain Copernican J. B. Duhamel, writing as secretary to the Paris Academy, "it demonstrates the wonderful consonance of the systems of Jupiter and Saturn with the great system of the universe."[42]

Cassini's conservatism or vacillation in the matter of world systems does not appear to have hampered his capacity for discoveries or kept him from the vanguard of astronomers or forced him to self-censorship. He sometimes wrote in Copernican terminology in his contributions to the Academy's publications to ease composition or to avoid unflattering comparisons, not to express a preference.[43] He criticized Riccioli's argument against Galileo and proposed an experiment that, he thought, would settle a point of relative motion at issue in favor of Borelli, that is, against Riccioli. He did not thereby fall into disfavor. Riccioli mentioned in his *Apologia* that some good friends of his, including "the most excellent professors in the University of Bologna, G. Domenico Cassini and Geminiano Montanari, who are widely known both in Italy and abroad by their published works and to me also well known and highly esteemed through intimate conversations," had tried to dissuade him.[44] We are led to the unappealing, unromantic, counterintuitive conclusion that Cassini led the astronomers of his time without commitment to any cosmology or world system and that he lived amicably with the persistent traditionalists of Bologna and the convinced Copernicans of Paris.

Protective Measures

The silence with which Riccioli's argument against Galileo's *bizzarria* met when first published in 1651 has been ascribed to fear of persecution. The noise that greeted its republication in 1665 would therefore indicate an abatement of fear, or an increase in courage, or a reassessment of the Church's interest in the matter. Still, Riccioli's opponents hesitated. Both Borelli and degli Angeli began by declaring the falsity of the doctrine of a mobile earth. Borelli sent a draft of his reply to degli Angeli for comment by Cassini, Mengoli, and Riccioli before publishing it and asked for help in revising it so as to avoid offending the censors. Montanari wished to enter the polemic, but decided that it would not be prudent: "I regret [he wrote Cardinal Leopold] not being in a country where I too could have my say; for here [in Bologna] one must keep one's mouth shut even about things that admit no doubt." And Borelli declined to answer a second booklet by degli Angeli, who lived safely in Padua, from concern that "this mixing too much in Copernican matters, even if hypothetically," would displease the authorities.[45]

People became bolder as the Jesuits moved in to occupy and tidy up the expandable space between defending an opinion and asserting it hypothetically. For once textbook writers were in the van. André Tacquet, who spent his career between the Jesuit colleges of Angers and Louvain, a region no less conservative than Rome, formulated the standard Catholic teaching about Copernicus while following an instruction issued directly to him by the Jesuit General. It happened this way. Tacquet acquired a reputation as a mathematician by a tract on cycloids and other difficult curves published in 1651. News of his success reached the General in Rome, who supposed that a mathematician could write something more useful than mathematics. He sent congratulations and an order: "I would like to see you complete the project that others have suggested to you of writing a course of mathematics for classroom use. That would be a great contribution to learning. Let God give you enough strength."[46]

God so disposed. In 1654 Tacquet published an *Elementa geometriae* that went through many editions and, with a later addition of trigonometry, marked a permanent change from the slavish following of Euclid in Continental geometry texts. From geometry Tacquet proceeded through the various branches of mathematics, including astronomy (in which he declared his admiration of Galileo), before his strength gave out, permanently, in 1660. His excellent pedagogy is preserved, however, in his *Opera mathematica*, published posthumously in 1669, reviewed favorably even in England as one of the best books available on mathematics, and reissued as late as 1727.[47] Most of the portion on astronomy (220 pages of 350) concerns the sun, moon, and stars, and so has no need of the heliocentric hypothesis. "Both in the common, and true, opinion and in the Copernican . . . the earth is apparently [*quoad sensum*] at rest in the center of the firmament." One need assume only that in the false opinion the diameter of the earth's orbit is insensible in comparison to the distance to the stars.[48]

When the planets came into the picture, however, Tacquet insisted on bringing in the heliocentric hypothesis since it allowed useful simplifications and calculations, "and is clear and easily comprehended." Many great astronomers had favored it; "indeed, it is remarkable how great a hankering [*ingeniorum pruritus*] there was everywhere for this old novelty. One can take it in a poetical sense, as in Virgil's verse, *Provehimur portu, terraque urbesque recedunt,* 'We set out from port, and the land and cities recede.'" Can it be taken as true? "For my part, since I see no arguments adduced on either side that are anything more than probable, I am not going to worry about them." However, where logic, mathematics and astronomy could not decide, Scripture spoke plainly. Tacquet quoted the usual verses:

"The sun rises and the sun goes down"; "the earth is established immovably"; "you fixed the earth on its foundation so that it will never be moved"; "stand still, you sun, at Gideon."[49]

Tacquet allowed that the words of the Bible had to be taken literally unless they conflicted with a clearer part of Scripture, or were reinterpreted by the Church, or were decisively confronted "by the very light of nature." The response of the Copernicans, that Scripture expressed popular notions in the manner of Virgil's verse, could not alter its plain meaning. Hence their remedy had not been applied. On the contrary, in the matter of Galileo the Sacred Congregation of Cardinals has found the quiescence of the sun and the annual motion of the earth expressly contrary to faith, and the diurnal motion of the earth erroneous in faith at the very least.[50] That settled the matter for Tacquet. No doubt he held his position sincerely. He was an open and engaging man, and a strong believer in the truth of the Catholic Church.[51]

Claude-François-Milliet Deschales, S.J., was a more worldly person than Tacquet. At the outset of his career, at his request, he was sent to convert the Turks; but he only converted himself, from a missionary into a mathematician, as he developed a greater interest in how his ship navigated the Mediterranean than in his occupation at its destination. Returned to France, he became a royal professor of hydrography at Marseilles, one of seven such positions in the kingdom, all entrusted to Jesuits. Thence he was ordered to the great Jesuit college in Lyons, where he spent most of the rest of his life teaching mathematics. His pedagogical writings followed the style of Tacquet's. He began in 1660 with an edition of Euclid, which had a long life, and then worked his way through the syllabus, which he pursued down even unto mathematical astrology, "although it contains not a single demonstration, but continuous nonsense, and nothing else." The full course, first published in 1674 in three unwieldy volumes, reappeared in 1690, edited by a disciple, who increased the compass to four volumes by adding a history of mathematics and a refutation of Descartes.[52]

Deschales took his reader from the first elements of mathematical sciences through their most abstruse and recondite parts, never (said he) losing track either of persons of average attainments or of born mathematicians, "[who], once they have fully understood what is taught here, can go through all the enigmas and asperities of other authors with no guide but themselves." Whoever made it to the third volume would learn that "Copernicus explains [the retrogradations of the planets] more simply than all the other systems so that, if his hypothesis were not contrary to Scripture, it could be called *divina prorsus,* utterly divine."[53]

Deschales discussed the *systema terrae motae,* the theory of a moving earth, in a separate essay, as did Tacquet, and he emphasized that its physical truth, whether it could be proved or not, did not interest astronomy. That was a job for philosophy. "It is shameful to mathematicians to play the philosopher and mix assertions that do not go beyond the probable with sure and certain demonstrations."[54] Deschales' take-home message: the Copernican is the best hypothesis, but cannot be held as true; which should not bother the astronomer, who blushes to hear speculations about the truth of things. With this understanding, his students could indeed read other astronomers and the writings of those who would censor astronomical hypotheses with no guide but themselves.

A further step into the space between hypothesis and assertion was taken by Deschales' predecessor at the Jesuit college at Lyons, Honoré Fabri, who flirted so openly with modern physical ideas, including the more frightful ones of Descartes, that his superiors transferred him to Rome, to a bureaucratic post in which he could not contaminate young minds. Undeterred, he published a *Physica* in 1670 that praised Galileo and took a soft line toward Copernicus. At the same time he published a heterodox theological treatise. He promptly found himself in the jail of the Inquisition. It was not his bold physics, however, but his defense of the overly flexible Jesuit doctrine of probabilism, that got him into trouble. Indeed, his physics extracted him from the hole into which his theology had plunged him; Cardinal Leopold, who had been impressed by Fabri's contributions to the Accademia del Cimento despite his polemic against Huygens in the affair of Saturn's ring, procured his release from jail and his restoration to his office and reputation. "As for his books on philosophy [physics and cosmology], don't worry about them," wrote the Jesuit General to the Order's Provincial in Lyons. "One can indulge a man of such parts."[55]

Fabri's contribution to the evolving Jesuit position on Copernicus was to prepare the way to explain away the scriptural passages that, taken literally, opposed Copernican theory. Fabri interpreted the condemnations of 1633 as merely provisional, enacted to meet the surprise and challenge set by Galileo in a way that would not upset and confuse the faithful. It was expected, he said, that after careful and prudent review and the evaluation of subsequent progress in astronomy, the authorities would relax or remove their strictures. Fabri first published this view in the second pamphlet he wrote under Divini's name in the Saturnian wars. Huygens had asserted that the true reason for their attack on him was his invocation of the ring hypothesis in favor of the Copernican system; which, he claimed, was held by many Catholics, and, in France, even by ecclesiastics, openly, not as hypothesis but as truth. Fabri replied through Divini that the decree of the Pope and cardinals against holding or teaching Copernicus bound all Catholics.[56]

But that was not to say everything. Following up on Riccioli, Fabri insisted that Scripture must be interpreted literally only as long as no absurdity resulted. Should it become absurd, the church, which understood better than Huygens the true character of science, would know what to do:

You [Huygens] and your colleagues do not ask whether you have a demonstration for establishing the earth's motion; nor do you dare assert that you do; nothing therefore stands in the way of the Church's interpreting those famous passages of Scripture literally, and in declaring that they are to be so interpreted as long as no demonstration proves the contrary; but if you should find a definitive proof, which I scarcely believe you will, then the church will not hesitate to declare that those passages are to be understood in a figurative and nonliteral sense, like the poet's *Terraque urbesque recedunt.*[57]

Tacquet had taught that no physical evidence opposed the hypothesis of heliocentrism; Deschales, that no other hypothesis nearly so good existed; Fabri, that the disabilities against it raised from Scripture could be set aside. Their opinions held sway among the mathematicians of the Order, who, after 1670, seldom bothered to try to refute Copernicanism, taught mathematics and mechanics in the style of the Galilean school, and made useful contributions to geometry, positional and physical astronomy, and applied mathematics.[58] This generalization applies also to Jesuits in Rome. We know the views of Fabri. At the Roman College, Athanasius Kircher, acting as an advisor to censors, recommended books that used the Copernican system hypothetically, as he did himself in computing planetary distances, and his colleague at the Roman College, Gaspar Schott, treated the solar system Copernically and hypothetically in his *Cursus mathematicus* of 1661. Francesco Eschinardi, the recording secretary of Ciampini's Accademia Fisicomatematica but not an advanced thinker (he would defend Jesuit-Aristotelian natural philosophy at the Academy), put the standard fictionalism at the head of his elementary textbook of 1668: "However things really are, mathematicians usually suppose various and multiple circles in the heavens . . . ; nor does this fiction hurt, for, as Aristotle says, truth is not required in examples used only for explanation." In a more serious *Cursus physico-mathematicus* of 1689, Eschinardi laughed with and at mathematicians for playing with the earth as Baliani had in proposing a moon-centered system. As for Copernicus, Eschinardi praised him for eliminating superfluous circles and for being philosopher enough to know that astronomy has nothing to do with truth. The troublemakers were Galileo and his followers, who, "led astray by contrived fancies, go beyond an hypothesis usefully framed by mathematicians to [a claim of] physical demonstration [*conclusionem*]."[59]

Fabri returned to this theme in 1665, in a book modeled on Galileo's *Dialogo.* The characters in his *Dialogi physici* are Antimus, who represents Fabri; Augustinus, a Copernican; and Chrysomachus, a thick-headed scholastic like Galileo's Simplicio. Augustinus opens by criticizing Antimus for being hard on Copernicus;

Antimus answers by requesting correction, "for I am first of all a lover of truth." He brings forth against the earth's motion none of the old objections, like the missing wind from the East, "made by people less experienced in mathematics and physics," but a central attack on Galileo's ideas about free fall and the tides. Augustinus answers as best he can. Antimus turns to Huygens, accepting Saturn's ring but interpreting it against Copernicus. Chrysomachus enters. "To all this you could add those perspicuous passages in Scripture, which confirm that the earth stands and the sun moves according to the literal sense, and which must be fully retained if they can be so without absurdity." With that he reached the bottom of the barrel: "Finally, since the divine power in the sacred text wanted to be praised and magnified by the very fast motions of the stars, who rightly can deprive Him of that praise? No one, certainly, who accepts the holy scripture."[60]

Outside of Italy Jesuits could go further. In 1685 Adam Adamandus Kochansky, S. J., then the librarian and mathematician of the King of Poland, proposed to all of learned Europe through the *Acta eruditorum* that they seek phenomena explicable only on the assumption of a moving earth. He began about where Fabri had left off. "Since through too much overheated study many minds of our time had embraced the hypotheses of Philolaus as improved by our fellow countryman Copernicus, the Congregation of the Holy Inquisition prudently ordered (on the occasion given by the Florentine Galileo) that this opinion not be set up as a dogmatic thesis and bandied about as a pure truth." The injunction caused the followers of Galileo to search for a proof that would prompt the reinterpretation of Scripture. They have found nothing but mere probabilities. They were right in this, however, Kochansky allowed, that "a firm physico-mathematical demonstration showing the revolution of the earth would not only permit, but make necessary, a reinterpretation."[61]

Kochansky offered two possible demonstrations. For one, if, as his investigations made likely, the ground always begins to vibrate from east to west during an earthquake, it would show that the earth spins from west to east. (The argument, which shows rather the deep difficulty that even the best mathematicians experienced in grasping the principles of relative motion, does not deserve a rehearsal.)[62] The second possible demonstration was inspired by reading in the proscribed works of Descartes that a cannon ball fired vertically upward hard enough does not return to earth. Being a philosopher, Descartes explained the effect without, perhaps, believing in it. Kochansky did not believe in it. In fact, he said, the shot would fall to the west of the cannon: entering into the upper atmosphere where the air is too tenuous to drag it to the east, it would return to the earth at a distance behind its point of departure. The experiment would be a costly one. Kochansky advised lining up a Maecenas to pay for the gun and its transport to the top of a high mountain at the eastern edge of a wide desert; and an army, to station its soldiers in a long line to the west of the mountain, to await the descent of the shot. If the shot always fell in the desert, "it will be a proof that the earth has a diurnal motion."[63]

The approach to Copernicus and heliocentrism worked out by the Jesuits during the 1660s and 1670s was not peculiar to them. One preeminent example, *Philosophia vetus et nova ad usum scholae acommodata* (1678) by the Oratorian Jean-Baptiste Duhamel, can stand for many. Duhamel had a particular authority because he served as the secretary of the Paris Academy of Science when he adapted his views about astronomy to the level of the students at the Oratorian Collège de Bourgogne. Since Duhamel had made a niche for himself in the republic of letters by balancing, harmonizing, compromising, and trimming ancient and modern opinions on controversial subjects, he was a perfect expositor of a liberal-conservative, geo-heliocentric, Ricciolian-Copernican astronomy. His *Astronomia physica* (1660) opens with an exhortation to modesty: "A philosopher is not too enthusiastic an admirer of antiquity nor does he condemn completely the discoveries of the moderns."[64] Duhamel put his substantive message in the mouths of three philosophers who, in their enthusiasm, sometimes displayed greater attachment to one or another system than good taste and manners allowed: the traditionalist Theophilus, who thought that the only difficulty astronomy faced was the unnecessary complication cooked up by Copernicus; Menander, the Copernican; and Simplicio, a conservative calculator.

Menander delivers himself of a long account of heliocentrism and its excellencies; Theophilus agrees that it works, especially for planetary motions, but objects that it violates physics and Scripture. Menander says that Scripture can be reinterpreted. This distresses Simplicio. "I beg you, my dear Menander, put the sun and earth back in their places." Otherwise all the theologians and philosophers will be after you. And why bother? Fabri's Tychonic system, says Simplicio, gives you all the advantages of Copernicus', without the negatives. More wrangling ensues. This time Menander calls a halt and requests a placid discussion of celestial hypotheses, none of which can be said to be more than probable. He then describes the Cartesian system. There the dialogue about world hypotheses ends, having settled nothing, giving the moderns the last, but neither side the decisive, word.[65]

In his school text, *Philosophia vetus et nova*, Duhamel delivered astronomy as a part of physics. First comes the Ptolemaic system, which would have offered great simplicity and facility, "if it agreed with the phenomena." Then there is Copernicus' hypothesis, "which almost all contemporary astronomers follow." These propositions might appear to settle the matter. That was not, however, the method of Duhamel. He reviewed the various arguments, arrived at the usual impasse, and jumped to the usual safety net: "No argument demonstrates the stability or the motion of the earth, still it is safer to defend its quiescence than its mobility."[66] But then, it is no less safe, and more in keeping with the nature of the subject, to defend nothing at all: "Whether the Copernican system is true or false is not for an astronomer to say; for he draws conclusions hypothetically, not absolutely. It is enough for him if he can explain the celestial motions in any way at all; the physicist investigates their causes."[67] Duhamel's even-handed formulations had a strong appeal around 1680 and enjoyed several editions. The Jesuits used it in their missions in Asia and translated it into the Tartar language for the benefit of the Emperor of China, that he might thereby learn the diverse opinions of the mandarins of Europe.[68]

· PHILOSOPHICAL FIG LEAVES ·

Leibniz, who noticed everything, kept track of the development of the astronomical apologetics of the Jesuits. "They are beginning to come around," he wrote early in 1681 to the Landgraf Ernst von Hessen-Rheinfels, a converted Catholic with influential connections in Rome, who had sent him a speech by the Jesuit General Oliva. Leibniz respected Oliva's Order for its many excellent savants. "[By] making its members work together, it could establish propositions [in science] as certain as those of Euclid's Elements."[69] A few years later, referring to the writings of Fabri and Deschales, he asked Landgraf Ernst to sound out a few cardinals about the prospect of lifting the censorship against Copernican ideas. Very likely Leibniz had in mind not only enabling Italian Catholics to play a larger role in the search for true propositions, but also removing a barrier to the realization of a modest project that he had discussed with the Count. This was nothing less than the union of the Catholic and Protestant churches.[70]

In keeping with his high purpose, Leibniz supplied Landgraf Ernst with a beguiling message for the cardinals: there was no reason to go as far as Galileo, and lately Fabri, had proposed and reinterpret Scripture in favor of heliocentrism. Not at all. Had Joshua been a Copernican, his address to the sun would have been the same. "Otherwise, he would have shocked the people as well as common sense. All Copernicans, in their ordinary speech and even among themselves, when the issue is not scientific, will always say that the sun has risen or set, and will never say the same of the earth." So it would be easy to patch up the dispute. The Roman church had only to allow people to hold the hypothesis as truth and the Copernicans had only to acknowledge that Scripture could not have spoken more appropriately.[71] Each party would retain authority in its domain.

In 1689 Leibniz took the matter directly in hand. He went to Italy for a year to begin the rapprochement between Catholics and Protestants by arranging a marriage between the daughter of his employer, the Elector of Hannover, and the Duke of Modena.[72] He managed the marriage and also—almost, or so he thought at the time—brought the Church to accept his hermeneutics. He arrived in Rome on 14 April, discussed his ideas with influential Jesuits, "who did not completely reject them," and learned that the Pope, Alexander VIII (Bianchini's patron Pietro Ottoboni), was intelligent and open-minded. At the very heart of Catholic assertiveness, in the Congregatio de Propaganda Fidei, Leibniz found "the light of enlightenment." There he met regularly with learned men for "discussions about all problems of scientific work."[73] He had the further satisfaction of discussions at Ciampini's Accademia Fisicomatematica, where he met Bianchini, through whom he hoped to gain access to the Pope. Also at Ciampini's he cultivated a person of much consequence for his quest, Antonio Baldigiani, S.J., professor of mathematics at the Roman College and an advisor to the Congregation of the Index.[74] "If there are more like [him] in ability and authority, I would hope that the old liberty can be regained, whose loss much restricts the lively minds of the Italians."[75]

With this encouragement, Leibniz composed an essay on the relativity of motion, which, as he later wrote Huygens, he had deemed "capable of persuading those at Rome to allow the opinion of Copernicus." Here is his argument. No one can know which, if any, of several bodies in relative motion truly rests. The analyst chooses one or another as stationary as best suits his purpose, which is to make the motions intelligible: "The truth of an hypothesis is nothing but its intelligibility." The apparent motion of the sun and stars is more intelligible on the geocentric than on the heliocentric hypothesis; Copernicus does better than Ptolemy with planetary motions. Hence Joshua spoke correctly; but had he been talking about Mars instead of the sun, he would have spoken badly.[76]

Deschales and responsible clerics who agreed with him would not have to resort to circumlocutions and qualifications for fear of offending the censors if the truth of a hypothesis were regarded as its intelligibility. "[Then] there would be no more distinction between those who prefer the Copernican system as the hypothesis more in agreement with the intellect and those who defend it as the truth. For the nature of the matter is that the two claims are identical. . . . And since it is permissible to present the Copernican system as the simpler hypothesis, it would also be possible to teach it as the truth in this particular sense. This would preserve the authority of the censors, so that a retraction would never be needed in the future . . . , while, at the same time, there would be no violence done to the distinguished discoveries of our age through the outward appearance of official condemnation."[77]

Having proposed this neat paradox, Leibniz began his return journey, stopping off in Florence, where he had long talks with Viviani and met Enrico Noris, who would head Clement XI's committee on calendar reform; and in Bologna, where he met Guglielmini, with the enduring benefit to Italian mathematics already noticed.[78] Perhaps in consequence of his talks with Viviani, Leibniz wrote out another memorandum for the conversion of the popes addressed to "R. P. B." The mysterious reverend father B has been identified as Baldigiani, whom Viviani rated as the best intermediary available for advancing the Copernican cause.[79]

Another candidate, also praised by Viviani, is Cardinal Gregorio Barbarigo, a man of great piety and learning, who organized the meetings Leibniz attended at the Congregatio de Propaganda Fidei. Barbarigo had studied mathematics at the University of Padua before becoming a priest and entering the service of Pope Alexander VII. He had time for self-culture as he climbed the ladder of preferment. When he left Rome to take up his duties as Bishop of Bergamo in 1657, he owned copies of Kepler's *Mysterium cosmographicum* and Ismael Boulliau's *Astronomia philolaica*, both Copernican works. As bishop he reserved an hour after dinner for mathematics and other diversions and engaged Cosimo Galilei, a nephew of the sinner, as his secretary. Cosimo helped him with Apollonius and gave him a copy of the proscribed *Dialogo* annotated by Galileo himself. Later, as Cardinal Archbishop of Padua, Barbarigo reformed the local seminary into a center of learning, gave it a printing press (which promptly reprinted Nicholas Mercator's Copernican *Institutiones astronomicae*), a professor of mathematics, an observatory enriched with a meridian line laid out by Montanari, and generally performed miracles. He came within an ace of being elected pope and ended up a saint. Baldigiani was better placed to further Leibniz's plans, Barbarigo freer to do so.[80] But neither father B undertook to upset the condemnation of heliocentrism with the lever supplied by Leibniz: not even an angel can determine which bodies rest and which move; a historian would make himself ridiculous by writing that the earth rises and sets; both the Ptolemic and the Copernican doctrines are true in their proper domains.[81]

These missiles missed their mark owing in part to a change in the variable Roman intellectual climate during the reign of Innocent XII, who became pope in 1691. Innocent was devout, charitable, economical, and literal. He decreed that popes should never grant estates, offices, or income to relatives, and generally made war on nepotism. He believed in discipline. There were always zealous and jealous people around the Curia eager to charge savants with going beyond the bounds of licensed expression. Often the charges were ignored. In 1693, however, they resonated. That January Baldigiani sent Viviani the breathless news that "all of Rome is in arms against the mathematicians and physico-mathematicians, extraordinary meetings are held by the cardinals of the Holy Office, and with the Pope, and there is talk of a general prohibition against all writers on modern physics, they are making very long lists of them, and at the top Galileo, Gassendi, and Descartes as most pernicious to the republic of letters and the candor of religion."[82] The interests that might have been in play here will engage us momentarily.

When Innocent's successor, Clement XI, showed himself supportive of astronomy, Leibniz tried to revive his project through his and the Pope's friend Bianchini. Soon after the successful completion of the *meridiana* at Santa Maria degli Angeli had strengthened Bianchini's position in the Vatican, Leibniz wrote to express his amazement that people so enlightened thought that they could know for certain whether a body rests or moves. He had no objection, he said, to restraining expression, or prohibiting books opposed to received religious opinion; but pretending to know indubitable truths about motion and condemning heliocentrism in consequence showed ignorance of both philosophy and nature. The Pope should come to the aid of truth. At the same time, he would free the great minds of Italy from the "chains by which, in science, especially astronomy, they are tied to the ground."[83] As we know, Leibniz had advertised the missing premise in this harangue on many other occasions: the simplest hypotheses that save the phenomena are, ipso facto, the true ones.[84]

Bianchini had no wish to interrupt his quiet life to promote a doctrine, or an outcome, that meant nothing, practically speaking, to him. As an observational astronomer, he recorded his data in terms of times and angles measured on the earth. During his most extensive and original astronomical work, the mapping of the features of Venus, he pinpointed the orientation of the planet's axis of rotation and the magnitude of its period and drew it circling the sun. He devised illustrative armillaries on the Ptolemaic and Tychonic, but not on the Copernican, scheme. Fontenelle took him gently to task for "his great care always to indicate how everything can be made to accord with Tycho." But Bianchini dwelt peacefully with Copernicans, commissioned a clockwork heliocentric armillary for his great patron, King John V of Portugal, praised Galileo extravagantly for throwing open the heavens to scrutiny by the telescope, and pursued his own telescopic investigations successfully without feeling obliged to declare his allegiance or opposition to one or another system of the world.[85] Had he been asked, he doubtless would have replied in the sense of his teacher Montanari: "I hold many opinions for probable, many for improbable, and none for absolutely true."[86]

Book Banning

Italian savants who wanted to speak and write openly about the *systema terrae motae* faced the uncertain and inconsistent operation of the complicated Roman book-banning apparatus. The inconsistencies could be exploited to advantage. The situation of Montanari, Bianchini, Manfredi, and other leading Italian astronomers during the century after Riccioli's death may be inferred from a review of the Roman apparatus and its application to the "system of the moved earth." Two lists of prohibited books bracketed the period. The earlier, published in 1664, was commissioned by Alexander VII, the Pope who patronized Cassini; the later, published in 1758, was the work of Benedict XIV, a generous and enlightened supporter of the arts and sciences.

· THE ROMAN CENSORSHIP ·

All books printed in Italy had to be approved by officials authorized by the Church. This decentralized prospective censorship did not provide permanent protection. To take a notorious case, Galileo's *Dialogo* came out with the high imprimatur of the Master of the Sacred Palace; nonetheless it was censured, along with the official who had licensed it. Complaints against it had arisen from outside the review system. That also was part of the system. Retrospective complaints about licensed books and denunciations of works published in Protestant countries triggered the central bureaucracy presided over by congregations, or standing committees, of cardinals. Complaints could be submitted to several of these committees, which had the power to propose actions to the Pope, who in routine cases accepted their recommendations. The congregations most involved in the censorship were those of the Inquisition (the Holy Office), the Index, and, after its establishment in 1621, Propaganda Fidei. In addition, the Master of the Sacred Palace, acting usually on orders from the Pope, and the Pope himself, acting by brief or bull, could prohibit books.[87]

The procedures of the Holy Office as reorganized by Sixtus V in 1578 endured with little change. Its business was prepared by an administrative staff headed by a prefect, always a Dominican, and his deputy. Every Monday morning they would meet with their permanent consultors, perhaps ten in all, most of them bishops, and, ex officio, the General of the Dominicans, a Franciscan, and the Master of the Sacred Palace, to decide the order in which complaints should be investigated and to draw up recommendations for submission to the cardinals. Expert advisors or qualificators prepared the reports substantiating the recommendations. The staffs met with the Congregation on Wednesday morning for routine business and Thursday morning (when the Pope might preside at the Inquisition) for difficult cases; a Thursday condemnation by the Holy Office indicated a very serious offence.[88]

The Congregation of the Index, organized independently in 1571 as a spin-off from the Inquisition, to which it returned 346 years later, may at first have operated similarly. But by the eighteenth century it had come to rely more even than the Holy Office on staff reports. The Index's secretary decided whether to proceed against a book after reading it and investigating the motives of its denouncers. If he decided to go ahead, he assigned it to two consultors, whose report, if unanimous, went to a larger committee of consultors that met about once a month. The committee submitted its recommendations to the cardinals, who might call for other opinions, including a defense by the author of the offending work. The Master of the Sacred Palace also belonged to the Congregation of the Index ex officio.[89]

It appears that the cardinals did not attend to their business as faithfully as their staff work would have permitted. Some were ignorant, others indifferent, and all overwhelmed by long reports on the large tomes that entombed the learning of the day. The author of a big folio might have to wait until long after his death to see his work allowed or condemned.[90] Or left somewhere between. The outcomes of the deliberations of the congregations included lesser penalties than burning the offending book in every accessible copy. It might be removed from circulation but permitted to individuals who could justify their need to read it; generally it was not difficult to obtain this indulgence for books on cosmology and natural philosophy. As an extreme example, the famous blue stocking of Bologna, Laura Bassi, received a license to read Descartes' and others' bad books sometime before 1740 although she suffered under the double disadvantage of being female and underage.[91] Finally, as we know, a book containing a few incorrect passages but otherwise meritorious could be indexed *donec corrigatur,* until corrected, either by a new edition or, in rare cases like Copernicus' *De revolutionibus,* for which the church itself stipulated the necessary changes, by marking one's copy.

As the traditional practice was codified and used under Benedict XIV, the congregations were to give books the benefit of the doubt if written by Catholics of good moral or scholarly reputation. Consultors should interpret favorably expressions capable of diverse interpretations; should judge professionally, as unprejudiced experts; and (it had to be added) should read the entire book before rendering an opinion about it.[92] So, in theory, did the retrospective censorship work—fairly, impartially, prudently, deliberately. In practice the fate of a book depended upon the tenacity of its denouncers, the connections of its author, battles among the orders represented in the congregations, and, above all, the relationship between the challenged doctrines and larger matters then claiming attention in Rome. An instructive example is the persecution of Enrico Noris, the very learned Augustinian who served as secretary of Clement XI's committee on the calendar, the exemplar, according to Fontenelle, of a cardinal appointed for his talent and erudition.[93] The process, which continued long after Noris' death in 1704, spanned almost the entire period from Alexander VII to Benedict XIV.

The trouble centered on Noris' account of Augustine's battles over the efficacy of grace and the freedom of the will. That embroiled him in the affair left unfinished by the Council of Trent, postponed indefinitely by Paul V, and reopened by Jansen's *Augustinus* in 1640. The Jesuits opposed licensing Noris' book, *Historia palagiana* (1673), which is now considered a milestone in the history of dogmatics. Noris got around them (everything depended on whom you knew) by arranging to have his manuscript reviewed by the Holy Office, where he had supporters, rather than by the Congregation of the Index.[94]

His enemies could not allow the matter to rest there. Urban VIII's successor one removed, Alexander VII, had forbidden five theses supposedly contained in Jansen's book. Alexander's bulls met with resistance in France, where Jansenism had some strength among the parliamentarians and the pious. Their defense included an effective attack on the Jesuits' moral code, from which Alexander took some 50 of the 110 laxist propositions that he anathematized in 1665 and 1666. A fitful peace was established in 1669 by Alexander's successor, who took the appropriate name of Clement IX. He allowed the Jansenists to comply with Alexander's bulls with the reservation that the five theses, though condemnable in themselves, could not be found in Jansen's tremendous book.[95] It was an administrative solution reminiscent of the agreement between the censorship and the astronomers. In the one case, a Jansenist could accept the condemnation of the core of his master's doctrine while following him faithfully, merely by saying that the prohibited theses could not be found in the *Augustinus*. In the other, astronomers could accept the decrees against Galileo while adhering to the theory for which he was condemned merely by calling the theory a hypothesis.

The Jesuits first denounced Noris' book during the pontificate of Clement X, who understood that they were energizing the censorship not to protect the faithful but to attack their enemies. The Pope knew the system well. He had been a consultor to the Holy Office under Alexander VII and had made Noris, whose career he favored, a qualificator to the Congregation of the Index. Papal understanding and support did not clear Noris of suspicion, however. Three times the Jesuits brought charges from which three separate committees of theologians, stacked by the popes, exonerated him. Each victory brought preferment: from qualificator to consultor in the Congregation of the Index, then overseer of the Vatican Library, and, in 1695, cardinal. Noris remained in the business of banning books. He intervened decisively in favor of the Bollandists when the Spanish Inquisition indexed fourteen volumes of their famous series of *Acta sanctorum*.[96] That was generous of him, since the Bollandists (after Jean Bolland, one of their early leaders), were Jesuits. "You would laugh [he wrote an old friend] if you heard how I vote on torture, imprisonment, and galley sentences and saw how I have been transformed from a chronicler into a criminologist."[97]

While Noris climbed the ecclesiastical ladder, Jansenism gained strength in

France as a political rather than a theological movement. Louis XIV regarded it as a threat. He pressured Clement XI for a new condemnation of Jansenist opinions. Clement reluctantly complied in 1713 with the famous bull *Unigenitus* that anathematized 101 propositions drawn from a book by the leader of French Jansenism. Many of these propositions read like statements from Augustine and other revered fathers. Louis' effort to force the bull through the clergy, the parliaments, and the Sorbonne gave rise to a protracted paper war. The Jansenists gained ground and the Jesuits, alarmed at the progress of their adversaries, countered at random.[98] They revived the old charges against Noris and managed in 1748 to persuade the Grand Inquisitor of Spain to ban his books.[99]

The General of the Augustinian order appealed to the Pope, Benedict XIV, who was sympathetic to Augustine's teachings, cool toward the Jesuits, and annoyed by the capriciousness of the censors.[100] The result was a decisive rebuke to the Spanish inquisitor and a flash of light on the workings of the censorship. Benedict observed that Noris' books had been cleared on several occasions, but that, even if there had been no previous examination and they did contain condemnable propositions, they would not have to be prohibited on that ground alone. That is because, according to the Pope, the Church should not ban books when the prohibition would do more harm than good. Furthermore, he admonished, the censorship process must not be invoked to resolve conflicts in the schools. Thomists, Augustinians, and Jesuits can all say what they please about predestination, grace, and free will since none of their opinions has been condemned by the popes. "In a word, bishops and inquisitors should not notice the arguments that the doctors throw at one another in their mortal combat."[101]

The Pope said much more to the inquisitor: "It will not have escaped your erudition that church history offers examples of prudent economy, in which, to curb scandal and avoid present dangers, our elders thought to draw back from the rigor of the law." For example, Clement XI silenced attacks on the Bollandists, and Clement XII denunciations of the polymath Ludovico Antonio Muratori, although their writings contained important errors. "How often in [Muratori's] books are things deserving of censure found! How much of this material have I not encountered myself in reading them! How many have been pointed out to me by rivals and denouncers!" Some of these offenses were serious. Muratori sailed close to Jansenism; taught a political philosophy opposed to the interests of the Holy See; and praised Galileo, Gassendi, and Descartes, all of whom had been indexed, as models for "extracting truth from the deep mine of mind and things." But Benedict was not going to do anything about it. "Instructed by the examples of my predecessors, who for the love of peace and concord ceased from proscribing what merited proscription when, manifestly, they thought more evil than good would come from it."[102]

It may be gathered that the charges against a book might not be the reasons for attacking it; that even popes felt obliged to let the censorship machinery run once it had been triggered and to seek consensus before issuing or revising serious condemnations; that this process could take many years, even decades; that a book's fate could be fixed by the chance presence of a champion, enemy, or idiot on the staffs or among the cardinals of the Holy Office or the Index; and that, as in everything else in Italy, the author's status and connections could determine the difference between condemnation and approval.

No doubt the established censorship could be burdensome, oppressive, threatening, capricious, and ridiculous. Fellow scholars and zealous informants outside the regular system could be dangerous. Noris ascribed the Holy Office's censure of one of his books to the influence of the Jesuits on a censor jealous of his success — and to the inconvenient deaths of two of his well-placed supporters.[103] Only in unusual circumstances did this irksome and intricate process bother with books on mathematics or natural knowledge. Most of the 1,500 items banned in the seventeenth century concerned central theological questions, like Jansenism, and most of the 1,200 banned in the eighteenth century, the opinions of Jansenists and free thinkers.[104] The works of Copernicus and Galileo occupied an exceptional place among indexed books.

· SYSTEMA TERRAE MOTAE ·

Heliocentrism

The charge against Galileo rested primarily on his disobedience, which made him, according to the Inquisition, "vehemently suspected of heresy." The fatal word recurs, as we know, in the finding of the Holy Office, included in the proscription disseminated by the legates and nuncios, that the assertion of a fixed sun is formally heretical because contrary to Scripture. Because the Pope approved the charge and the finding, it might appear that the Church condemned Galileo as a heretic and Copernicanism as a heresy. That was not the case.

Galileo's heresy, according to the standard distinction used by the Holy Office, was "inquisitorial" rather than "theological." This distinction allowed it to proceed against people for disobeying orders or creating scandals, although neither offense violated an article of faith defined and promulgated by a pope and a general council. Galileo's teaching of propositions opposed to the literal reading of Scripture fell into this category of lesser offenses. "Formally heretical," as applied to heliocentrism, meant opposed to the obvious meaning of the Divine Word. Since, however, the church had never declared that the biblical passages implying a moving sun had to be interpreted in favor of a Ptolemaic universe as an article of faith, optimistic commentators, like Gassendi and Fabri, could understand "formally heretical" to mean "provisionally not accepted."[105]

Informed contemporaries appreciated that the reference to heresy in connection with Galileo or Copernicus had no general or theological significance. Gassendi, in 1642, observed that the decision of the cardinals, though important for the faithful, did not amount to an article of faith; Riccioli, in 1651, that heliocentrism was not a heresy; Mengoli, in 1675, that interpretations of Scripture can only bind Catholics if agreed to at a general council; and Baldigiani, in 1678, that everyone knew all that. "Galileo is no longer condemned for his teachings, nor even is it said that they are a heresy against Scripture, [or] of doubtful faith; people only dispute over the way in which he wrote, which is a very different matter."[106]

Despite these subtleties, Urban VIII made a bad mistake in associating himself with Galileo's condemnation and with the characterization of heliocentrism as heretical in any sense. That made it difficult for his successors to reject or revise his prohibitions when they came to understand that the choice of celestial geometry had little to do with faith and morals. The best they could do was to avoid doing more. In return, mathematicians practiced the prophylaxis of hypothesis and wrote what they pleased about astronomical systems, suffering, perhaps, some vexation or delay in accordance with their and the system's circumstances. Sometimes a mere change of venue made the difference between ease and hassle, as when Borelli, encountering trouble in licensing a book in Florence, sent the same manuscript to Bologna, where it passed without incident.[107]

Manfredi's experiences indicate the situation during the first few decades of the eighteenth century. We can guess at the pressures he felt at the outset of his career from a draft statute of 1702 for the prospective Accademia delle Scienze of Bologna that was to incorporate and succeed his Inquieti. The statute required the academicians to swear "never to fight with Copernicans, but to convince them with physical reasons, and in general to promise as much for astronomy as for any physical principle in experimental philosophy, viz., to bring everything into conformity with the Holy Roman Church."[108] Although this excessive self-censorship did not survive into the definitive statute, it lived on in Manfredi, as appears from his response to the trouble he had in obtaining a license for his book on stellar parallax of 1729 and the whimsical depiction of Galileo's experiments that he included, for no apparent reason, in his later book on the *meridiana* of San Petronio (Figure 6.5).

The book on parallax sets out the theory of the annual motion of the fixed stars and evaluates the observational evidence for it. Manfredi did not hide that should the evidence prove conclusive it would be an unanswerable demonstration of Copernican truth. "So I propose in this book to show what sorts of observations of the fixed stars we would see, supposing the earth to revolve around the sun, if

FIG. 6.5 An angelic research assistant confirms Galileo's law of falling bodies.
From Manfredi, *Gnomone* (1736), 95.

only [their motions] are large enough for us to perceive." Although Manfredi
cagily and correctly found that existing observations, by Picard and by Jacques
Cassini, were not decisive, he more than hinted that he expected that conclusive
positive evidence would be found using his methods. After a short delay the censor, who worried that Manfredi devoted so much attention to the earth's motion,
approved the manuscript. The book appeared from the Inquisition's printer with
a dedication to Giovantonio Davia, the cardinal-president of the Congregation of
the Index.[109]

Davia was a native of Bologna, a former student of Montanari's and friend of
Bianchini's, who began his career, in 1681, as an official in the city's water service.
Then he fought the Turks as a military engineer, collected coins and medals in
Rome, and, at the request of Innocent XI, to whom he had applied for a military appointment, entered the diplomatic service of the Curia. He settled a Jansenist problem to general applause and was made a bishop; he bobbled Clement XI's policy of
neutrality during the War of the Spanish Succession and was retired to his bishopric in Rimini. He thereupon renewed his ties with Bologna, drew close to Man-

fredi and Bianchini, patronized the Bologna Academy of Sciences, and worked his way back into favor with Clement, who made him a cardinal. That was in 1712. Returned to Rome, he served the Congregations of the Inquisition, Index, and Propaganda, and twice came within a few votes of winning the papal tiara. He died in 1740, the year his great friend Prospero Lambertini became Pope Benedict XIV.[110]

When Manfredi discussed the dedication of his book on parallax with Davia, he proposed to put in the usual ritual censure of heliocentrism, which the censor had recommended. Davia did not want it. His reason: the ritual might "make an article of faith of what certainly is not one." He materialized this viewpoint in a Copernican armillary, which he gave to the Bologna Academy; whose members, as timid as Manfredi, commissioned two others, a Ptolemaic and a Tychonic, to flank Davia's unwelcome gift and to demonstrate that they did not know the true system of the world. This caution was not misplaced.[111]

While procuring a license for his piece on parallax, Manfredi was following up James Bradley's discovery of the aberration of starlight, of which more later. Manfredi confirmed the aberration and explained it on Copernican principles. His lengthy report on his work, in the form of a letter to the Pope's physician, Antonio Leprotti, was intended for the first volume of the Academy's *Commentarii*. In keeping with the custom set by Fontenelle in Paris, the secretary of the Bologna Academy, Francesco Maria Zanotti, planned a historical preface to the *Commentarii* to review the articles that followed. He would have to describe Manfredi's compromising demonstration of aberration and say a good word about Copernicus. He feared that the censor would make trouble. He wrote Leprotti for advice. "I am very much afraid that the censors will require that . . . where I say what Copernicus thought, I must immediately add that I detest his system as a heresy." And what to say about Davia's gift? "I would not want them to oblige me to say that both His Eminence and I regard the machine as a *jeu d'esprit,* knowing as we do that it is contrary to the faith. You see very well how these protestations would be received by Catholics, especially learned ones, in the first book published by this academy of sciences, which supposes that it contains the flower of the literature of Bologna."[112]

Eventually the censor approved the volume, including Manfredi's dissertation, which characterized the assumption of a moving earth as a "principle" in Bradley's unreflective usage and a "hypothesis" ("nothing prohibits us from making them") in the correct interpretation made by the Church.[113] This stale subterfuge, or solid epistemology, may have been necessary. Leprotti wrote Manfredi in November 1730, before the license came: "It will not be possible to publish the dissertation except in the way you thought; and the showing that [certain earlier observations] do not demonstrate the annual parallax will go far to allay suspicions." The thick first volume of the Bologna *Commentarii* appeared in 1731, less than two years after Zanotti had a final manuscript. Although the delay caused by the censorship could not have been very long, it was exasperating. As Zanotti pointed out, it could cause Italians to lose priority in discovery and to publish material already superseded or disproved by better-favored academies elsewhere.[114]

Compliance with the continuing obligation to insert the standard disclaimer had become routine and perfunctory. Here is an example from 1712: "Dear reader, I want you to know that I hold the Copernican system to be completely false and, with due veneration, accept the decree by which, most fairly, the system was condemned." The author then goes about his business, discussing gravity and a spinning earth, without further qualification. Here is another, from the important edition of Newton's *Principia* published by two Minim monks around 1740. "In this third book, Newton assumes the hypothesis of the moving earth. The author's propositions can be explained in no other way.... Hence we are forced to play another person's part. Otherwise, we openly declare that we comply with the Popes' decrees against the earth's motion."[115]

Few, perhaps, credited such disclaimers. Words that seemed sincere when written by a Baliani, Riccioli, or Tacquet rang hollow half a century later.[116] But freed from the constraint that may now seem its rationale, the hedge about the truth was not unsound. In one guise or another, the view that mathematical theories have only an instrumental value has recurred in Western thought without the guidance of the Catholic Church. It developed in antiquity with particular reference to Ptolemy's devices and dominated the epistemology of physics at the end of the eighteenth and the nineteenth centuries.[117]

It is likely that Manfredi, like Montanari, Bianchini, and, perhaps, Cassini, shared this instrumentalist epistemology. His posthumously published lectures on astronomy set its goal as the discovery of precise laws confirmed by observation; hypotheses about the structure and nature of the heavens are at best instruments for seeking regularities, and, if taken too literally, as by the Cartesians, stymie the hunt. Where did Newton's theory belong? It certainly agreed best with celestial phenomena. Still, Manfredi could not raise it above hypothesis, not even its underlying principle of universal gravity, which Newton had offered as a fact derived from experience. Astronomers did not have to believe in a world system to write about astronomy. The division of Manfredi's text on astronomy conveys this wisdom. It begins with a Ptolemaic-elliptic description of the sun, moon, and stars in 225 pages; goes on to a Copernican-elliptic account of the motions of the planets in 160 pages; and ends with 15 pages of directions for translating Copernicus into the language of Ptolemy and Tycho.[118]

A last example will suggest how matters stood before the censorship of learned books eased under Benedict XIV. One of Manfredi's former students, who had not adopted his precautions, wrote an account of the excellences of English philosophy. Manfredi warned that such carelessness could bring down upon an errant author and his friends "the sort of hatefulness that cannot be regarded lightly by people who live in Italy, rather, it is the greatest and most dreaded there is." The book in question was Francesco Algarotti's famous *Neutonianismo per le dame.*[119] Manfredi and Davia intervened during prepublication review but the Congregation of the Index, alarmed by Algarotti's association of Newtonian physics with Lockean philosophy, which had been forbidden three years earlier, could not let the ladies have their Newton so contaminated. Even Algarotti's supporters had trouble tolerating his enthusiasm for foreign thought. "That devil of an Algarotti appears like an aurora borealis that lasts but briefly and smells a lot of the North; he is so full of France and England that he seems a barbarian." After much negotiation the business was composed. The prohibition, which did not prevent publication, softened to *donec corrigatur.* To "correct" for the Italian market, Algarotti dropped his ardent praise of the transalpine and his irrelevant psychologizing, changed the title of his book, and saw the new edition, of 1746, approved with the inclusion of a "Notice from the Printer" rehearsing the usual disclaimer. A newer edition, of 1750, came through without the epistemology of the printer although, of course, it was thoroughly heliocentric. The original edition remained upon the Index.[120]

Galilaeus Sanctificatus

Meanwhile the juridical situation was changing. In 1710 Galileo's *Dialogo,* while remaining firmly indexed, was republished in Naples, with a false Florentine imprint and no named publisher. It included Galileo's hermeneutics (in the form of his letter to the Grand Duchess Christina) and, as an offset and warning, his condemnation and abjuration. The Inquisition knew perfectly well that the editor and printer were Lorenzo Ciccarelli, a Neopolitan lawyer who liked to publish controversial books.[121] In 1712 the Holy Office decided, without telling anyone, that it would not make objections to the teaching of heliocentrism as a hypothesis.[122] In 1741 it sanctioned the first printing of Galileo's works to include the *Dialogo.* It came out in 1744, printed from the copy annotated by Galileo once owned by Saint Gregorio Barbarigo and preserved where he had left it, in the Seminary of Padua. The printing was done at the Seminary's press, established by Barbarigo. "It was not a commercial venture, but a scientific and philosophical battle fought and won by the Seminario di Pavia." The new *Dialogo* appeared intact apart from 53 postils (comments printed in the margins), 13 of which were dropped and 40 reworked to insert "supposed" before "motion of the earth."[123]

Two additions enriched the text. One was a dissertation by a theologian, Agostino Calmet, on the bearing of Scripture on astronomy. Calmet reached the balanced judgment that no world system could be known to be true. "It seems that God, being jealous, so to speak, of the beauty and magnificence of his work, has reserved to Himself the perfect understanding of its structure, and the secrets of its motions." The second addition, a preface by the editor, Giuseppe Toaldo, later professor of astronomy at the University of Padua, who represented the "fullest fusion of the ecclesiastical, scientific, and literary ideas . . . of Gregorio Barbariga," laid down the established line. "As far as the principal question of the motion of the earth is concerned, we too conform to the retraction and declaration of the author, stating in the most solemn manner that it cannot and must not be admitted except as a pure mathematical hypothesis for the easier explanation of certain phenomena. For this reason, we have removed or reduced to a hypothetical form the marginal postils that were not, or did not appear, completely indeterminate."[124] Among those who helped to obtain the assent of the Holy Office to this violation of its own injunction was the Papal Legate to Bologna, the future Benedict XIV.[125] Only in Rome, according to Roger Boscovich, Jesuit and Copernican, could Copernicus not be taught; that, anyway, is what he published openly in Rome in 1747, forgetting, perhaps, his hypothetical treatment of the earth's motion in a book on comets published the previous year.[126]

In 1758 Benedict issued his Index. It omitted the general proscription against Copernican works made by the Inquisition in 1620 but left Galileo in place. The decision of the Holy Office, taken in 1757, ran, "After discussion with the Holy Father, the decree is omitted prohibiting all books teaching the immobility of the sun and the motion of the earth." Why was Galileo not also reprieved? According to the French astronomer Jérôme Lalande, who talked with the cardinal-president of the Inquisition in 1765, the problem was the old judgment of the Holy Office, which had to be set aside before the prohibition could be removed; Benedict, who died in 1758, had no opportunity to see the negotiations through, nor did his successor, Clement XIII, although, according to Lalande's informant, he had intended to do so.[127]

The opening made by Benedict stimulated Italian mathematicians to seek the clinching argument. Giambattista Guglielmini of Bologna responded. After finishing his *laurea* he went to Rome to teach mathematics to Cardinal Ignazio Boncompagni, the Papal Secretary of State, and his nephews. While holding this sinecure he developed the old idea of proving the diurnal motion by measuring the distance east of the vertical at which a weight dropped from a high tower strikes the ground. His reasoning is easy to follow. Owing to the rotation, the weight on top of the tower has an easterly velocity $(r + h)\omega$, where r is the earth's radius, h the tower's height, and ω the speed of the diurnal rotation. The point V vertically underneath the weight has a corresponding velocity $r\omega$. Hence the displacement from V should be $h\omega t$, t being the time of fall. Since, according to Galileo's rule, $t^2 = 2h/g$, g being the known constant of gravity, everything was in place for a dramatic test. All Guglielmini needed was a great height. He went to the top. He asked

Boncompagni for permission to drop balls from the dome of Saint Peter's into the Saint's sepulchre at the base of the Bernini altar. That would be a perfect place to exonerate Galileo. The cardinal agreed.[128]

While Guglielmini wondered why no one, not even, apparently, the all-knowing Newton, had tried this simple experiment, his cardinal lost his place. Guglielmini took his experiment home, to the Torre degli Asinelli. There he found out why no one had succeeded earlier. The calculations had to be developed to account for air friction and the experiment had to be managed without giving the balls any lateral displacement or spin on their release. Competing mathematicians pointed out that Guglielmini's elementary theory did not take into account that the direction of the vertical changed as the rock fell, that the deflection had a component to the south, that the Alps and the tower exerted gravitational forces, and so on. Guglielmini solved his main problem, releasing the balls, by burning rather than cutting the strings that suspended them, and a secondary problem, finding the point V, by employing the technique described in Manfredi's book on the *meridiana* of San Petronio. The result, published in 1792, could be interpreted as agreeing with the theory, more or less.

Thereafter, for a hundred years and more, mathematicians and physicists, including Laplace and Gauss, worked to improve the experiment and its theory. Around 1900 physicists dropped, or proposed to drop, weights from the lantern of the Florentine Duomo, the cupola of the Pantheon, and the top of the Washington Monument. The first objection-free demonstration of the eastern deflection was made by an American professor, Edwin Herbert Hall, famous for the discovery of an obscure electrical effect named after him. Having been refused the Washington Monument, he made do with lesser heights at Harvard. In 1912 the Jesuits at the Roman College confirmed and extended his work, thus concluding after some interruption the Society's investigation of free fall begun by Riccioli and Grimaldi.[129]

Galileo did not have to await these results to lose his place on the Index. That occurred in 1822, in consequence of a protest from Giuseppe Settele, a professor of mathematics at the Roman College, to the Pope. Settele had been refused permission to publish his *Elementi di ottica e di astronomia* by the Master of the Sacred Palace because he had not followed the formula. The Pope referred the protest to the Congregation of the Index, which gave permission. The Master again objected, on the ground that the decree of 1633 had made the quiescence of the earth a matter of faith, thus showing his own ignorance and the aptness of the concern Davia had expressed to Manfredi. The matter next went to the Holy Office, which reviewed everything and decided that belief in what stands where was never a matter of faith; by "heretical" the old inquisitors had meant only "against the traditional reading of Scripture." In its meeting of 16 August 1820, the cardinals of the Congregation of the Inquisition resolved as follows: "Nothing is opposed to defending Copernicus' opinion about the motion of the earth in the manner in which it customarily is now held by Catholic writers." The Master still objected. He held up matters, to his own increasing peril, for another two years.[130] Had the business dragged out a little longer, the Church would have removed the disabilities from the Copernican system just as definitive evidence of the existence of stellar parallax was found.

The removal of Galileo's *Dialogo* from the Index canceled a black mark against the book but not against its author. The official rehabilitation of Galileo took another century and a half. It began around 1940 in connection with the three hundredth anniversary of Galileo's death. That was not a good time for a party. Nevertheless Pope Pius XII approved a campaign to demonstrate the Church's openness to science. As an indication of this openness and on the recommendation of his Pontifical Academy of Sciences, he commissioned an unrestricted biography of Galileo. The assignment went to Monsignore Pio Paschini, rector of the Pontifical Lateran University, a historian known for his balanced account of the Church during the Reformation. It was a bold choice, since Paschini tended to be liberal and judicious. These virtues worried some of the Pope's senior advisors, who had the satisfaction of being proved right. Paschini took Galileo's part, admitted that the condemnation had been an error, and lost no opportunity for criticizing the Jesuits, on whom he blamed the entire affair.[131]

The Jesuits objected. Paschini's two-volume work disappeared in the review mechanism, much as academic articles submitted to scholarly journals do today. The anniversary date, 1942, was long past when the Vatican journal, *Civiltà cattolica,* got around to the subject. It admitted that the Church, or, rather, intemperate and ill-informed churchmen, had erred in condemning Galileo; and it recommended that the best use that could be made of this fact was to forget it. Paschini understood and fell silent. As a reward he was made a bishop two months before he died and an honorary member of the Pontifical Academy of Sciences, which had opposed the publication of his work.[132] With this unimaginative solution, the Church shut up a work that it had commissioned to demonstrate its openness. No good administrator could have wished to leave the matter there. What to do? Wait. The celebration of the tercentenary of Galileo's death had not worked out; the four hundredth anniversary of his birth might do as well, or, as it turned out, better. It fell during the Vatican Council, ten years after Paschini's death had removed the party most interested in seeing his work appear as he wrote it.

Paschini's biography of Galileo appeared in 1968, heralded as an indication of the Pope's program for the peaceful coexistence of religion and science. The general scholarly press reviewed the book favorably. Lay and clerical critics commended its balance. Very few knew that, to use the old expression of the censorship, the book had in effect been indexed *donec corrigatur,* and then corrected, before publication, by the keeper of the Jesuits' archives.[133] Since the publication of this collaborative work, efforts at rapprochement between science and religion have intensified, particularly under the auspices of John Paul II. In 1979, on the occasion of the centenary of Einstein's death, the Pope told his Pontifical Academy of Sciences that he wanted theologians, scientists, and historians to work together on a reassessment of the Galileo affair. He endorsed Galileo's principles of biblical exegesis. He gave as an earnest of the project the Church's sponsorship of Paschini's biography, without knowing (let us hope) that it had been censored more crudely than the old books on heliocentric astronomy.[134]

Thirteen years later, in 1992, having received the reports of the study committees he had appointed, John Paul announced that the theologians who had condemned Galileo had erred. By not recognizing the proper distinction between the Bible and its interpretation, they had "transpose[d] into the realm of the doctrine of faith, a question which in fact pertained to scientific investigation." The Pope then exonerated Galileo and confirmed the historical framework in which he had placed the problem of reconciliation. This framework identified the theologians of 1633 with the operations as well as with the edicts of the Inquisition, and the scientists of our time with a correct and timeless epistemology. The administrators, including some of John Paul's predecessors, who had found ways to render the edicts of 1633 dead letters long before Galileo was removed from the Index dropped from sight. That is the fate of good bureaucrats.[135]

As for the ultimate fate of Galileo, it is too early to tell. The business has been going on for less than four hundred years. But there are signs for those who can read them. Some forty years ago a theologian trying to extract something positive from the contretemps of 1633 worked out that its providential purpose was to teach the lesson that divine assistance to the Church does not include reliable advice about science. Galileo had to suffer to bring light to the Church. The cardinal appointed by John Paul to head the study teams repeated and endorsed this view. He may not have known that Galileo occasionally referred to himself as a saint in his self-appointed mission to enlighten the Church.[136] No doubt shortly—within a hundred years or so—Galileo will be canonized.

More Dangerous Doctrines

The principal and significant exception to the generalization that, after Galileo, the Church did not condemn books on world systems was the work of Descartes, indexed by the Index *donec corrigatur* in 1663 primarily for his radical materialism. The Church considered atomism a threat to dogma regarding the human soul; and, since the materialism to which the censors objected lay at the core of Cartesian philosophy, the Congregation of the Index could never deliver the corrections that would have made it readable.[137] Consequently, during most of the rest of the seventeenth century a good tactic in attacking an enemy's book was to allege a connection between it and the philosophy of Descartes.

The tactic may have been used against Galileo even before the publication of Descartes' damnable books. In this rendering, the atomism of Galileo's lampoon of Jesuit natural philosophy, the *Saggiatore* or *Assayer* (1624), and not the lampoon itself or his advocacy of Copernicus, aroused the vigilance of the Jesuits. The ringleader, Orazio Grassi, one of those weighed and found wanting in Galileo's balance, was an old and unforgiving enemy. The two had fought hotly over comets, producing much smoke, error, and ill-will. Grassi was not an unworthy opponent: he taught mathematics at the Roman College, acted as architect on many important Jesuit building projects, and ended, after a period of ostracism following Galileo's trial, as an administrator in the Jesuits' school system. Grassi answered the *Assayer* by pointing out, anonymously, its advocacy of a materialism incompatible with the standard account of the mystery of the Eucharist. Even the Pope was unable to protect his old friend Galileo from prosecution on this serious charge. The Holy Office, with uncustomary philanthropy, engineered a plea bargain: Galileo would acknowledge guilt on the lesser charge of disobedience in a matter that, in contrast to materialism and its threat to eucharistic mechanics, did not menace faith and morals.[138]

It is not necessary to accept this extravagant reading to arrive at a connection between corpuscularism and Copernicus. Descartes defended heliocentrism. He did so with his usual cleverness. On his system, a vortex centered on the sun sweeps around all the planets, including the earth, and so, he said, escapes the general proscription since the apparently moving bodies are in fact at rest in the swirl. Few were gulled by this attempt to make a quiescent earth the consequence of heliocentrism. The Copernican cause suffered a further burden of guilt by association. As Leibniz put it, "Descartes' astronomy is at root nothing but that of Copernicus and Kepler."[139]

The Jesuits anticipated the Index in attacking Descartes. During the 1630s and, especially, the 1640s, as its more conservative elements gained the upper hand, the Society issued one directive after another against teaching anything outside the consensus of the schools. Furthermore, it ordered its professors to stick to their lasts, theologians to theology, mathematicians to mathematics. That was too much for Riccioli. "It is preposterous and iniquitous." An instruction of 1651 forbade teaching new ideas. That was almost too much for Grassi. "In the last general councils [the ninth and tenth, 1650 and 1652], the teaching of many opinions, some of which form the basis of my work [on optics], have been prohibited, not because they are judged evil or false, but because they are new and out of the ordinary; so that I shall have to sacrifice my work to holy obedience, by which, no doubt, I will gain more than by publishing it."[140]

Many of the condemnations that deprived the world of Grassi's optics concerned Descartes. The ninth general congregation proscribed fifteen Cartesian propositions. Subsequent congregations repeated the prohibition against these and all other novelties; the fifteenth congregation, held in 1706, pointed out thirty obnoxious statements in Cartesian physics alone, which the next congregation, in 1732, reduced to an economical ten.[141] But Descartes crept into the Society's teachings nonetheless. The edict against his writings of 1663 was drawn up with the help of Fabri, himself suspected, rightly, of harboring Cartesian ideas.[142] In 1672, an ornament of the Order, Gaston Pardies, in a *Lettre d'un philosophe à un Cartésien*, declared a curious compromise: "Just as formerly God allowed the Hebrews to marry their captives after many purifications, so after having washed and purified the philosophy of Descartes, I could very well embrace his opinions." And did. The liberty he demanded, and perhaps more, was conceded by the general congregation of 1706; the last of its "prohibitions" allowed "the defense of the Cartesian system as a hypothesis."[143] And the Cartesian system was a carrier of the Copernican.[144]

The Jesuits' internal censorship could weigh heavily on its authors with a modernist bent. But not necessarily. The cases of Tommaso Ceva and Yves-Marie André will illustrate the leeway available to independent thinkers in good and bad odor, respectively, in the Society during the first decade or so of the eighteenth century.

Ceva was a prolific Latin poet and a mediocre mathematician at the Jesuit College in Milan. His long poems gained him a wide reputation and admission to the Arcadia, the leading Italian literary academy. His device for dividing angles in any odd number of parts, described in the *Acta eruditorum* in 1695, won him the attention of Leibniz and an intense correspondence with Guido Grandi, a professor at the University of Pisa, a Galilean in mechanics, a Cartesian in physics, and a Leibnizian in mathematics.[145] Ceva readily acknowledged Grandi's superiority in mathematics and good-naturedly advised him not to commit himself immoderately to it: "You run the risk of ruining your complexion, as happened to my brother and to [one of] my closest friends, who thanks to geometry came close to drawing his last line."[146] The brother had suffered from a second intellectual affliction from which, happily, Ceva's poetry rescued him: he had been a convinced Cartesian.[147]

Ceva's anti-Cartesian verses sang the proposition that gravity is innate to bodies. It would be tedious to report how he employed it to prove the existence of God, and vice versa; but it is easy to see that, once secured, it destroyed the physical systems of Gassendi and Descartes. Ceva's demonstration, in a tract *De natura gravium*, captivated Grandi. "It is a grand system, most ingenious, and full of the most beautiful thoughts." Grandi rhapsodized not over the argument, which he thought easily refutable by any good Cartesian, but over Ceva's way of dealing with his opponents. "The praise you give these philosophers is handsome and considerable.... Oh! I scarcely expected this from a Jesuit!"[148] Grandi stayed with Descartes, but not slavishly. It is best to be eclectic, he said, to take what is best from Descartes, Gassendi, Aristotle, but not innate gravity, not from anyone. "Padre, no, since I cannot conceive of it and have the same repugnance for it that the philosophers of the last century must have felt on hearing for the first time from Copernicus that the earth goes around the sun. Perhaps some time I too will give up this prejudice just as by now the most sensible philosophers, I think, hold the system of the moving earth, which, although false, is not impossible, nor opposed to sense, as it appeared to be at the beginning."[149]

Ceva responded in a *Philosophia nova-antiqua* (1704), written in verse in emulation of Lucretius and dedicated to the Pope's nephew to ensure it an easy passage through the censorship. It opens with the proof of the existence of God from the principle of gravity; extols Descartes as a mathematician and destroys his system of vortices; accepts Galileo's treatment of motion and attacks his cosmology; insists on the necessity of Aristotelian philosophy, since matter in motion cannot explain everything, but rejects much of it; and explains the adherence of northerners to heliocentrism as an expression of Protestant opposition to the Pope.[150] *Philosophia nova-antiqua* was a great success among eclectics who took more from the old than from the new. "The poem is written with that sweetness and delicacy of style, and with that subtlety and strength of reasoning, of which Father Ceva has given samples in so many of his writings."[151] It became an important text in Jesuit schools. Five editions and one translation were called for between 1704 and 1732, although, to be sure, the last ones gave comfort to rival teaching orders eager to demonstrate the backwardness of Jesuit physics.[152]

Ceva hit the conservative eclectic center of the Italian Jesuits of his time and prospered. André missed the consensus of French Jesuits and suffered. Around 1705, while still a student of theology in Paris, André adopted the views of Nicolas Malebranche, a Cartesian who had got himself indexed for a tract about Jansenism. On discovering this doubly compromising connection, André's superiors sent him to the provinces to complete his studies; there he converted a fellow Jesuit, who had thought Descartes the "mortal enemy of the faith," so completely that the convert left the Order.[153] André was a dangerous man. His superiors sent him to teach in Amiens. There he proposed the thesis, "We defend the Copernican system as an ingenious, if not a true [*si non vera*] hypothesis"; his censor required, "*etsi non vera* [although untrue]"; André insisted and was removed to Rouen.[154]

He persevered in his obnoxious ways, teaching and, as his manuscripts show, also holding to Descartes ("his physics could be made perfect with only a very few changes") and Copernicus; and, worse, mixing theology into his philosophy. "I speak too much of God and his gospel in my philosophical writings," he told Malebranche, "not to be suspected of novelty, fanaticism, and heresy." The authorities required a retraction. André refused to sign anything declaring the opinion of Descartes or Malebranche contrary to faith; what right have we Jesuits, he asked, to throw suspicions of heresy on good Catholics or use our schools to root out opinions we do not like? He was shipped off to Alençon. That was in 1713, the year of Clement's *Unigenitus*. André showed himself too little zealous against Jansenism. He was removed from teaching and sent to Arras. There he wrote manifestos on the need to revamp the Jesuit curriculum and completed a biography of Malebranche. By then his superiors had run out of patience and unpleasant little colleges. In 1721 they accused him of Jansenism, confiscated his biography of his master, in which they found much praise of Descartes and much criticism of themselves, and arranged a stay for him in the Bastille.[155]

André soon saw the unimportance of knowing whether the earth moved or not. He apologized and returned to his teaching, at the respectable college of Caen. There he played by the rules. He taught astronomy hypothetically, invited his students to choose among the systems, and offered the hint that the experts had preferred the Copernican for over a century. One student who had the opportunity to follow this advice became the greatest mathematical astronomer the world had yet produced, Pierre Simon de Laplace. For his excellent teaching and good judgment, as tempered in the Bastille, André was made rector of his college. He died in 1768, a few months before his Order was expelled from France.[156]

André's story is not the quashing of free thought by unreasoning oppression, but the mutual adjustment of novelty and conservatism in an autocratic organization required to teach and defend old values. Both parties, having overstepped the territory they could control, drew back. As a Jesuit, André owed obedience to his superiors; he tested them again and again, pushing out the boundary of allowed expression until he insisted too much on the excellences of Cartesian philosophy and the shortcomings of his confrères. For its part, the Society, which had the right to control its own schools, had gone too far extramurally to curb ideas it did not like; its victories over Galileo and Descartes brought it into territory it could not hold. There were too many ways for good Catholics to learn obnoxious ideas.

In Rome itself, during the second half of the seventeenth century, Cartesian physics was frequently discussed in Ciampini's academy, in the pages of the *Giornale de' letterati d'Italia,* and in books and soirées patronized by Queen Christina. Her very presence in Italy recommended the philosophy of Descartes, for it had opened the road to her conversion to Catholicism. Ideas that worked such wonders were not to be despised.[157] In Bologna, the intellectual center of the Papal States, the exemplary Catholic layman Montanari interpreted his experiments in the manner of Descartes, Galileo, and Gassendi, though without metaphysical commitment. He disseminated these "innocent speculations," as he called them, without molestation in published books and unpublished lectures, some of which have survived in notes by Bianchini.[158] Even the timid Manfredi wrote favorably of Cartesianism, in his poetry if not in his astronomy.[159]

Print was only one form of publication. Some manuscripts on condemned physics circulated so widely as to have amounted to public documents. A good example is the Italian translation of Lucretius' poem on atomism completed in 1670 by Alessandro Marchetti, Borelli's successor as professor of mathematics at the University of Pisa. Cardinal Leopold, concerned not to escalate tension at the university, where some professors were denouncing others for introducing unprotected youth to atomism and impiety, ordered Marchetti not to publish his work. When the tension eased, Leopold approved sending the manuscript to the censors. The Holy Office would not approve it. So it went about as it was, two copies to Florence, two to Rome, two to Brussels (to Davia), several to Naples, and so on. Marchetti became a famous author without publishing a word. The Roman censorship vainly indexed his Lucretius when it finally came from the press in London in 1717.[160]

The growing absurdity of the censorship system could not have been expressed more clearly. What was the use of condemning a translation, which had circulated freely for over forty years, of an exemplar of Latin poetry that had been available in print, and used in schools, for more than two centuries? No doubt many high officials of the Church recognized the ridiculousness of a censorship they could neither change nor enforce. Their solution was to ignore offending physics books unless some zealot, literalist, or rival triggered the mechanism that made them take notice.

· MODERATION AND BALANCE ·

When the Papist King of England, James II, was visiting Paris in 1690, he asked to visit the Academy of Sciences. Cassini arranged most of the entertainment. The King heard a lecture on reckoning the longitude from eclipses of Jupiter's satellites, saw the positions of places thus located on the world map inscribed on a floor in the Observatory, and, the high point of the session, examined a silver planisphere, which Cassini had had made as a gift for Louis XIV. This machine displayed the world systems of Ptolemy, Copernicus, and Tycho in exquisite and exact detail. James, who had impressed his hosts with his knowledge of geography and astronomy, took an informed pleasure in the silver simulacrum of the heavens. "His Majesty saw how it worked and observed with delight the precision of the relationship among the three systems whose hypotheses seemed to be so different."[161] His Majesty did not need, and may not have cared, to know which if any of the three systems God had chosen. James had no interest in trying to change the opinions of English astronomers, all of whom were elliptical Copernicans. Heavenly geometry would only become an affair of state if he tried to impose the Catholic view.[162] Wise rulers were indifferent about world systems.

James could have read as much in a little book on mathematics designed for emperors, kings, and princes published in 1693. Its author was Samuel Reyher, the professor in Kiel whose calendrical computations Leibniz later recommended to Noris. Rehyer gave examples of the use of mathematics in its twenty-seven applied branches by rulers from Nimrod down. They had not bothered much about geometrical astronomy; according to Reyher, all kings needed to know about planetary orbits was how to hire competent astronomers to calculate them. "As for cosmography, or the science [*doctrina*] of the various systems of the universe, it certainly does not seem to be very necessary; however, since it is very easy, and also very pleasant, it will not be amiss [to study it] — provided the subtler controversies are omitted."[163]

A similar message came from picture books on geography and astronomy, which capitalized on the superfluity of world systems to multiply illustrations, and from works on cosmology and universal history, which exploited the squabbles among mathematicians to underscore the relative civility of other sorts of savants. All leave the choice of world system, of which they customarily presented three or four, to their readers, and the final judgment to posterity.[164]

Among the worldly wise happy to leave the decision to posterity were senior administrators of the Church. Take that hot spot Naples in the early 1690s, where the doctors and lawyers were full of the books of Descartes and Gassendi then under attack in Rome. The agent of the Holy Office dispatched to compel readers of modern philosophy to give up their evil ways was a former inquisitor, Jacopo Cantelmo, who employed his weekends in self-flagellation. He jailed a few prominent citizens. The Viceroy, acting in agreement with the Pope, threw him out of town and restored the doctors and lawyers to their proscribed books and printing presses.[165]

Almost simultaneously, in the very Catholic University of Louvain, the papal nuncio stifled a fight between the faculty and a professor of mathematics, Martin van Velden, who wanted to dispute the thesis "the Copernican system of the motion of the planets is indubitable: and with good reason, the earth is considered a planet." The faculty had expelled van Velden; the nuncio arranged his reinstatement; he resumed where he had left off, with Copernicus and Descartes.[166] Throughout the nuncio acted reasonably and pragmatically in inhibiting the faculty from using the machinery of the censorship to silence a dissident colleague. All of which will be unpleasantly familiar to observers of the operation of political correctness in contemporary universities.

The Viceroy of Naples acted to protect the citizens on whom the prosperity of his realm depended; the nuncio of Louvain, to contain an academic squabble. A similar pragmatic motive lay behind the Jesuits' teaching of Copernicus and eventually of Descartes: to retain their position as schoolmasters to the governing and professional classes in Catholic countries. These classes also made up the audience for the coffee-table books on world systems, the manuals for princes, and the Jesuits' general cultural magazine, the *Journal de Trévoux,* begun in 1701, which explicitly excluded works of piety and devotion, reviewed books in all other respectable subjects, and managed to maintain a large readership against many competing periodicals.[167] For this audience, the burning question of their grandfathers' day, whether the sun goes around the earth or vice versa, was scarcely lukewarm.

The instrumentalist epistemology that allowed administrators weary of turning the ineffectual machinery of the censorship to ignore books in mathematics and physics does not appear to have diminished the production of mathematical works in Italy during the seventeenth century.[168] In this matter, however, quality may not be proportional to quantity. It is not the army of sophisticates like Bianchini, Manfredi, and, in his way, Ceva, or wafflers like Cassini, but true believers like Galileo, Kepler, and Descartes, who change the ideas of people other than themselves. In any case, by 1700 savants inside and outside the Church had accepted the superiority of the Copernican scheme in Kepler's form for describing the motions of the planets. The Church's objections to it came to look far-fetched and irrelevant. Competent censors, like Davia, found their task increasingly disagreeable and uncomfortable. The more honest and informed they were, the quicker they were disarmed. "For how," as Fontenelle wrote in 1723, anent Jacques Cassini's paper on determining "the apogee and perigree, or the aphelion and perihelion of the planets," "how defend oneself from the system of Copernicus?"[169]

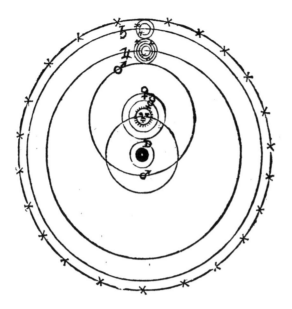

7: The Last Cathedral Observatories

The Things Themselves

During 1734–35, on his return journey to Uppsala from Italy, Celsius spent some time in Paris fraternizing with the academicians and observing with the astronomers. With Jacques Cassini he studied the sun's image at the then newly completed *meridiana* at the Observatory. With P. C. Lemonnier he did the same at the equally recent line in the vast church of Saint Sulpice, the only installation north of the Alps worthy of comparison with the heliometers of Bologna and Rome.[1] It was the penultimate such *meridiana*. One more was built twenty years after Celsius observed at Saint Sulpice, in Italy, of course, at the Florentine cathedral, which thus became the first and the last Catholic church to serve as an instrument of science. These late *meridiane* faced stiff competition from improving telescopes, to which they soon succumbed. Like today's instruments of big science, cathedral observatories died suddenly; unlike modern instruments, they are buried in hallowed ground.

About the time that the *fabbricieri* of San Petronio went into debt to enlarge their cathedral, the custodians of the church of Saint Sulpice developed even more grandiose plans for the accommodation of the faithful. Their parish, which included wealthy enclaves around the Palais du Luxembourg and stretched into the rapidly growing suburbs to the west, had become too big for its church. It was decided to raze and rebuild on a scale to rival Notre Dame, the vast medieval cathedral of Paris. Operations began in 1646 and went forward apace until 1678, when, with the transept only just begun, a crippling fault was found: a huge undisclosed debt. Everything stopped for forty years as the parishioners paid for the excess of zeal of their contractors. Then in 1719, Jean-Baptiste-Joseph Languet de Gergy, who had been named curé of Saint Sulpice five years earlier, decided to push his church to completion. He was a gifted raiser of gifts. An indefatigable philanthropist himself, who spent most of the income from his benefices on good works, he shamed his parishioners, who included some of the richest people in Paris, into giving; and his vigilant opposition to Jansenism recommended his projects to a regime happy to help so sound a shepherd house his large and influential flock. It is said of him that he never returned from dinner at an aristocratic house without his table setting and that the result of these pious pilferings was a six-foot solid silver statue of the Virgin Mary.[2]

In the mid-1720s, before the completion of the nave, Languet had the idea of running a meridian line through the transept, which lies almost north–south and has an internal length of about 180 royal feet (57 m). He may have had this purpose in view in arranging the underpinnings of the pavement of the nave, one of whose pillars, free of the others and anchored in rock, stands directly under the noon image of the midsummer sun; a plaque placed there, supported independently of the walls, would not settle with them and so would serve as a fiducial mark for years of solstitial observations.[3] Languet's stated purpose in installing a *meridiana* as part of the furnishing of his new church was entirely liturgical: the by then unnecessary redetermination of better dates for movable feasts. In 1727 he engaged an English clockmaker resident in Paris, Henri Sully, to make the *meridiana*.[4]

Sully had an agenda of his own. It annoyed him that the clocks of Paris did not ring the hour in unison. The near completion of Saint Sulpice suggested to him how to preserve Parisians from the inconvenience of multiple noons; he needed only to lay down a *meridiana* in the church and attach its attendant to a noisemaker to alert Languet's parishoners that midday had arrived. The application of technical innovations for the public good was the objective of a small group to which Sully belonged, which called itself the Société des Arts. Its financial backer, the Comte de Clermont, was one of the richest men in Languet's domain. Clermont might have been an intermediary, even a sponsor, in the construction of the *meridiana* if, as Sully wrote, he had approached Languet, not Languet him, with the project.[5]

Sully proposed an elaborate construction beginning with a hole in a plaque sealed into the wall of the main window of the south transept of the church, 75 feet (24.34 m) above the ground. The line would run 175 feet across the transept, which deviates 11° from north–south. There were to be four scales on either side of the *meridiana*, giving, to the east, the sun's declination, its place in the ecliptic (ascending signs), the time of sunset, and—a new one this—the time taken by the sun's image to cross the line; and, to the west, the time of sunrise, the sun's place in the ecliptic (descending signs), the tangent of the sun's zenith distance, and the distance from the vertex in seconds and thirds of the earth's circumference. In fact, the full *meridiana* would have occupied too much of the earth's circumference (some 240 feet) to fit within the 170-foot transept. Sully planned to bend it at right angles to itself at the north wall of the transept and—another innovation in major *meridiane*—run it up an obelisk far enough to catch the midwinter sun.[6]

Sully's time came before he could complete his elaborate project. No one knows how far he carried it; only a few traces of his work survive, 45 centimeters to the west of the definitive line implanted in 1742 at Languet's request by Pierre-Charles Lemonnier. The new man had an agenda that did not coincide with the curé's or the clockmaker's. He built not for religious observance or for public welfare but for science. Although then not yet thirty, Lemonnier was already a seasoned member of the Paris Academy. He had entered there as a junior mathematician *(adjoint géomètre)* in 1736, at the prodigiously young age of twenty, no doubt for his accomplishments as an observational astronomer and also because his vigor fitted him perfectly for the then imminent expedition to measure the meridian in Lapland. On his return to Paris, Lemonnier started a detailed study of the elements of the solar orbit.[7] He therefore welcomed the task of making Saint Sulpice into a heliometer to help determine the exact equinox, the perigee of the sun's orbit, the effect of winter weather on refraction at low altitudes (that of the midwinter sun in Paris is 18°), and how and whether the obliquity changes.[8] By the 1740s this project stood, in up-to-dateness, between Languet's old-fashioned chronology and Sully's new-fangled chronometry.[9]

Lemonnier did not do the exacting engineering work himself. He directed the work of Claude Langlois, regarded as the best instrument maker in France, who had supplied the sectors, quadrants, and standard toise sticks for the expeditions to Lapland and Peru. The result of their collaboration, which brought several new elements to the design of *meridiane*, earned high marks for both execution and application from the author of the article on meridian lines in Diderot and d'Alem-

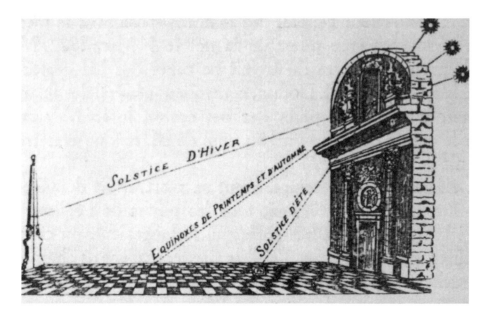

FIG. 7.1. The *meridiana* in Saint Sulpice, Paris. From Lemesle, *Saint-Sulpice* (1931), 32.

bert's avant-garde *Encyclopédie:* "[Their] precautions, combined with so many new means of accuracy, made the méridienne of Saint Sulpice a unique instrument, and one of the most useful ever furnished to astronomy."[10] The chief precaution the encyclopedist had in mind was the independent suspension of the pavement. The new sources of accuracy came to two: the obelisk for viewing the winter sun and a lens for focusing the rays of the midsummer sun (Figure 7.1).

The form of the vertical portion of the *meridiana* suggested ancient Rome, where obelisks had had a connection with time telling since Augustus erected one he had stolen from Egypt to serve as the style of a huge sundial expressive of his power and glory. Nothing visible remains of Augustus' complex dial today except the obelisk, now displaced from its original position, and the *ara pacis,* the altar of peace, through which the obelisk's shadow ran on the afternoon of the equinox (Figure 7.2).[11] To continue the parallel, Lemonnier's *meridiana* no longer works. No discernible sun disk crosses the floor because the window, which Lemonnier had covered over apart from the hole in order to create an image, has long since been returned to its original transparency.

Like Augustus' obelisk, Lemonnier's carried an extensive text. On the left in Figure 7.3 we read of Languet's pious pre-academic project for "An astronomical gnomon for the certain study of the Paschal equinox," for the fulfillment of the program begun by the Council of Nicaea and carried forward by popes Gregory XIII and Clement "with unbelievable zeal." On the right, between the lamb of God

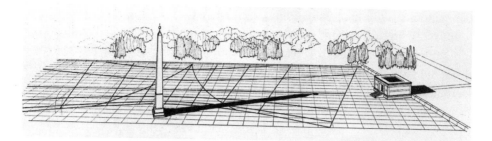

FIG. 7.2. Artist's reconstruction of the obelisk-sun dial of Augustus.
From Buchner, *Sonnenuhr* (1982), 43.

FIG. 7.3. Inscription on the obelisk in Saint Sulpice; the vertical line is the section of the
meridiana for receiving the sun's image during the winter. Photograph by author.

and the credit to the Academy, comes a blank that adds much to the inscription.
It once memorialized two of Louis XV's senior administrators responsible for public buildings, the Comte de Maurepas and Philibert Orry.[12] The removal of their
names, accomplished under the French Revolution in its purgings of vestiges of
the bad old régime, anticipated the wholesome modern practice of ignoring the
contributions of administrators to the advancement of science.

The effacement of the names of Maurepas and Orry was the slightest of the depredations suffered by Saint Sulpice at the hands of revolutionaries who made it into a "Temple of Reason," stole its treasures, burnt its statues, and removed the monument to its builder Languet. In November 1799, three days before the coup d'état of 18 Brumaire, when a great banquet in honor of Napoleon took place in the church, a statue of victory stood where the high altar had been. The pavement and balustrade of the choir containing the altar survived, however, because the sanctuary overlaps the course of the *meridiana*. Lemonnier and Langlois had had to put a piece of the line in the choir, some 36 centimeters above the floor of the church, just behind the balustrade; as luck had it, this raised piece included the equinox, so the brass plate that Langlois cut to receive its image in 1742 rested beyond the destructive feet of visitors. This circumstance in turn preserved the entire choir. Two pharmacists of the parish convinced the revolutionary Committee on Public Safety that the destruction of the choir would interrupt the *meridiana* that ran through it to the great injury of French science.[13] And so the *meridiana* of Saint Sulpice, conceived for religious purposes but used for secular ones, helped to preserve the most sacred precincts of the church that housed it from the vandalism of the state.

The vandals did not object to the pagan obelisk. The contrivance had much to recommend it besides merely capturing the winter sun. It allowed Lemonnier to put the gnomon as high as he could without enlargement of the image and it made the midwinter image almost circular. These geometrical facts may be deduced from Figure 7.4, where O represents the hole, QPR the meridianal diameter of the noon image of the midwinter sun as it would appear outside the church if it had no north wall, and KLM the same diameter intercepted by the obelisk XY. From the similar triangles QKY and QOV, $KY = h[1 - w/(x + \Delta x)]$, where $x = PV$, $\Delta x = QP = PR$, and $w = VY$, the distance from the vertex to the base of the obelisk. From the triangles RMY and ROV, $MY = h[1 - w/(x - \Delta x)]$. We have for the vertical diameter KM of the solar image, $KM = KY - MY \approx 2wh\Delta x/x^2$. Since x and Δx are proportional to h, the height cancels out of the expression for the size of the image.[14] Hence by going higher, Lemonnier could expand the image on the pavement, where it was useful to do so, without making the winter images any larger. Recent measurements make w 57.64 meters, whence KM = 55 cm at midwinter.[15] The diameter IJ perpendicular to KM is almost as large. The geometry gives $IJ = 2PT(OL/OP) = 2OP(PT/OP)(OL/OP)$, which, when $\alpha = 17°40''$, is 52.3 centimeters. The image of the noon sun at winter solstice formed almost a perfect circle on Lemonnier's obelisk.[16]

The winter image moved about a sixth of an inch a second as it came to its rendezvous with the obelisk. That made its dichotomy easy to spot; Lemonnier estimated that, by taking the mean time between the passages of the leading and trailing limb across the vertical *meridiana*, he could specify exact local noon to

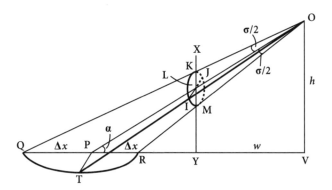

FIG. 7.4. The vertical diameter KM of the image on the obelisk.

within a half, and even a quarter, second. That was the sort of feat possible "with a gnomon of so prodigious a size."[17] The midsummer sun did not behave so well. Its image, a bit more elliptical than that of the midwinter sun, sauntered along at less than a twelfth of an inch a second; or would have, had it not been blocked altogether by a cornice. To cope, Lemonnier put in a second hole, just where Sully had placed his, at 75 feet, and—following a suggestion made earlier by other astronomers—inserted a lens in it to throw the image onto the *meridiana* between the vertex and the position for the first point of Cancer. (The lower hole was opened, and its lens fitted, only on midsummer day.) The device enabled Lemonnier to read the progress of the midsummer image as if it moved as quickly as the midwinter one; Lalande judged that so equipped Saint Sulpice performed better at the summer solstice than San Petronio.[18]

The lens had a focal length of 80 feet. So what? Well, as Lemonnier admitted and an easy calculation shows, the correct length would have been between 82 and 83 feet. Why should the Paris Academy of Science and the King's administrators settle for a lens that made an image two feet above the desired focus? The reason was the difficulty, even impossibility, of making long-focus objectives to specifications.[19] As will appear, the elimination of the need for such long lenses in standard telescopic astronomy marked the end of cathedral observatories.

Lemonnier observed the solstices regularly at Saint Sulpice for almost half a century, from 1743 until his death in 1791. The most significant of these observations, because they played a part in the study of the obliquity, took place during the first twenty years in collaboration with other academicians.[20] The instrument died with its maker. A recent thorough examination could not locate the hole at 80 feet or the lower one supposed to throw the image at the summer solstice; a hole does exist at 75 feet, but it works, badly, only between the equinoxes and the winter solstice; and a second hole, at 21.25 meters (around 65 feet) does not seem to have any purpose at all. An external cornice blocks rays from the sun through the upper hole for eighteen days on both sides of the summer solstice. Something is missing. Unfortunately, something unintended is present: many perforations in the glass window that throw superfluous images on the floor and the obelisk, washing out the weak signal through the one purposeful hole. Friends of French sundials have submitted a proposal to the current curé to repair the window (but not, alas, to darken it) and to place a hole so as to admit rays from the midsummer sun into the church as in times gone by.[21] May the spirit of Languet descend upon his successor.

· SANTA MARIA DEL FIORE, FLORENCE ·

"[It is] the most beautiful monument to astronomy in the world." Thus lavishly did Jérôme Lalande, who, as a former student of Lemonnier's, knew his *meridiane,* compliment Leonardo Ximenes, S.J., the restorer of Toscanelli's old line, just after completion of the work in 1755. "[It is] the greatest instrument of astronomy." Thus Algarotti extolled the new line to the reverent caretaker of the superseded heliometer of San Petronio.[22] Ximenes matched his meridian. According to his countryman Giuseppi Piazzi, a good astronomer and meridian maker in his own right, Ximenes was the only Sicilian genius in astronomy and, what was more, "one of the most meritorious and courageous astronomers of our age."[23]

Ximenes' career differed only in distinction from those of many mathematical Jesuits of eighteenth-century Italy. Born to a noble family in Trapani in 1716, he was educated by the Jesuits, to whom, at the age of fifteen, he presented himself as a novice. Having shown his abilities, he was sent to finish his studies at the Roman College. That done, his General assigned him to teach mathematics to the sons of an important Florentine nobleman, the marchese Vincenzo Riccardi. The assignment left Ximenes ample time to study mathematics in all its branches in the libraries in the Jesuit college and the marchese's palace.[24]

As the life of Bianchini suggests, acquaintances formed in the libraries of the lay and the clerical nobility could be platforms for launching distinguished careers. Anatole France hit the mark in forming his fictional eighteenth-century abbé, Jérome Coignard, the learned and dissolute trencherman of the *Rôtisserie de la reine pédauque,* as a bishop's librarian. Riccardi's librarian was a writing machine named Giovanni Lami. As editor of a news magazine, *Novelle letterarie,* which covered all the arts and sciences, Lami was able to promote Ximenes' career by publishing indications of the young tutor's competence in mathematics.[25]

It was thus, launched by a librarian, that Ximenes came to the attention of Count Emanuele di Richecourt, who ran the Grand Duchy of Tuscany for the Holy Roman Emperor. (The duchy had been under new management since 1737, when the death of the last male Medici turned it over to the Duke of Lorraine, who rose to be Emperor.) The new régime naturally desired an inventory. Several different plans for mapping and surveying Tuscany had been proposed but not accepted when, in 1750, Richecourt asked Ximenes for his opinion. Ximenes recommended beginning with a serious geodetic inquiry of the kind that Bianchini had started, and Boscovich and Maire would soon complete, in the Papal States. The inquiry would be centered in Florence on a point whose geographical position had to be determined with exemplary exactness. Ximenes suggested as the origin of Tuscan coordinates the *meridiana* of Toscanelli, entirely refurbished and modernized. In December 1750, as part of his reply to Richecourt's commission, he set out the requirements for converting Santa Maria del Fiore into a heliometer.[26]

In the end, neither the hopeful "contractor general of astronomy in Tuscany" nor anyone else made a geodetic survey of the Grand Duchy during Ximenes' lifetime, probably because the imperial treasury refused the funds for the cadastre the survey was to orient. For a moment, in 1777, it appeared that the survey would be entrusted to the last sprig of the Cassini tree, named Jean Dominique after his great grandfather. That annoyed Ximenes. "Your subjects are not Americans, Lapps, or Africans," he protested to the Emperor, "to need foreign astronomers to draw up a map and measure a degree of a meridian." By then, having acquired extensive experience as a surveyor, Ximenes was ready and willing to direct the work himself, including the cadastre. But again Vienna declined to pay.[27]

After drawing up his fruitless plans for triangulating Tuscany, Ximenes returned to astronomical observations and mathematical exercises. In 1752 he issued an abbreviated almanac, containing in addition to the usual information about the sun, moon, and planets, the longitudes and latitudes of important cities, the effects of atmospheric refraction according to the latest Cassini tables, standards of length from around Europe, his own recent observations, an explanation of his calculations, and a censure of philosophical systems. He allowed that much progress had been made in astronomy compared with the achievement of the ancients, but very little compared with what modern instruments made possible to minds unprejudiced by cosmological systems. Even the Newtonian had become a hindrance to advance for those who believed that "universal attraction [was] the great secret unveiled at last from nature and confided through authentic writings to the happy philosophers of England."[28]

Despite this unfriendly remark, Ximenes relied on Newton's mechanics and the heliocentric world picture in a thesis on the ocean tides. Ximenes' manuscript went for review to Maire in Rome, who counseled caution; Ximenes complied in the familiar way, by calling his assertions hypotheses. In his big book on the *meridiana* of the Duomo, he expressed himself more subtly and firmly. He took universal gravity ("whatever its cause may be") as established fact, affecting the earth and everything on it just as it regulated the motions of all the (here we are to understand "other") planets. The circulation of the planets around the sun was to him "un fatto innegabile," "an undeniable fact."[29]

In 1755 Ximenes decided that he would like to be a professor. Nothing easier. He requested appointment to a position in applied mathematics at the University of Florence. Through Richecourt he received not only the professorship but also the resounding title of Imperial Geographer and a useful sum for the purchase of books and instruments. He had little in salary, however, and complained about it for years: the Emperor was not nearly so generous with his mathematicians, whom he treated as civil servants, as the popes were with Bianchini, whom they treated as an ornament.[30] Ximenes added to his income through the usual business of an applied mathematician in Northern Italy: water. From 1756 on he canalized rivers, drained swamps, opened ditches, and argued with his opposite numbers, including Boscovich, Zanotti, and Gabriele Manfredi, employed by neighboring territories. His philosophy for managing wetlands, which limited his success in his time, would win him high marks in ours. He insisted that they not be reclaimed entirely for agriculture, but preserved in part for wildlife and waterborne transportation.[31]

Despite his frequent fruitless requests for substantial increases in salary, Ximenes died wealthy enough to endow two professorships with funds for salary and instruments without touching the capital he had inherited in Sicily. One chair holder had responsibility for hydraulics, including fieldwork; his legacy included Ximenes' level, compass, and relevant books. The other chair holder received all of Ximenes' astronomical books and instruments. He was to teach astronomy, keep up the observatory Ximenes had set up in the Jesuit college, and make observations, "particularly of the summer solstice, every year, diligently, at the testator's *meridiana* in the Cathedral, according to the method he has published."[32] The publication, which had appeared in 1757, was Ximenes' masterwork, the best and most comprehensive meridian book, *Del vecchio e nuovo gnomone fiorentino*, that is, *On the Florentine gnomon, new and old*.

Ximenes dug out the history of the old gnomon at the urging and with the support of Richecourt, and, very likely, with the help of the learned Lami, who excerpted it liberally in his *Novelle*.[33] Ximenes' clever guess that the history of the old gnomon began in 1468 fell short by seven years. As we know, Toscanelli pierced the lantern in 1475 to place a small segment of a meridian line in the north arm of the transept (Figure 2.30). Ximenes' impressive research turned up a reference to the use of the old gnomon in 1502. That year an astrologer answered a technical point in Pico della Mirandola's famous attack on astrology by claiming that observations at S. M. del Fiore showed a change in the obliquity. In 1510 the *fabbricieri* ordered a bronze plate placed where the sun's image fell at noon on the summer solstice, a precious benchmark for settling the moving obliquity (Plate 1).[34] The following year they had the hole reset. That did not inspire many recorded solar observations, however, and nothing of interest before Ximenes started his operations in 1754. He then found that Toscanelli's line ran about one degree off true, which would not impugn the placement of the bronze plate of 1510.[35]

When Ximenes reached this conclusion, Lalande was in town. They agreed that the apparent reliability of the old bronze made the next several summer solstices "extremely important for astronomy." For, as Ximenes continued in his proposal to the *fabbricieri* of the Duomo, redoing the old gnomon would give Florence the chance, "unique in the entire world," for deciding the vexed and vital question of the inconstancy of the obliquity. To make the measurements adequately, "in a way that will satisfy the Royal Academy of France," very considerable work would be required. Ximenes specified raising scaffolding to determine the height of the hole, leveling the marble floor below, bringing in implements foreign to churches, and erecting walls to keep people from interfering with the observations. "I or some other dependable person must keep the key to this enclosure, so that no one can damage the marks that have to be made on the marble on the days near the solstice." The *fabbricieri* approved this honest proposal, which came without the traditional open-sesame of calendar reform, on 9 May 1755. According to Lalande, Richecourt got the Emperor to pay for it.[36]

During the installation, the cathedral functioned as a physics laboratory. Like his predecessors, Ximenes leveled his line against a water basin using a device more convenient and exact, he claimed, than those employed at Bologna or Rome. Its business end, the point R (just above K in Figure 7.5), terminates a micrometer screw one turn of which advanced or withdrew R by about one Parisian ligne. A tongue EH holds the screw, the stand DFG holds the tongue, and the rest of the apparatus maintains the stand horizontally with the help of plumb bobs a and b (below C). Ximenes obtained the exact level with respect to the water of the mark C by lowering R until a droplet of water sprang to it. When illuminated the droplet shone like a little halo, "a delightful, easy, and trouble-free sight." He estimated that by watching these haloes he could determine the level to within 0.2 ligne or half a millimeter.[37]

The leveling was anything but trouble-free. The two points to be referred to the same water surface, the vertex and the solstitial mark, fell in the choir and the north transept, respectively. Between them stood (and stands) a balustrade five feet high. Ximenes had to connect the two points by canals. His ingenious contrivance (Figure 7.6): a series of tin troughs connected by siphons running from the vertex β (near the center of the altar precinct) out the entrance to the choir at pavement level (thus avoiding the balustrade) and back to join a series of wooden troughs beginning near the solstitial mark ρ (just to the north of the altar precinct). The large water surface suffered appreciable evaporation. After finishing his leveling,

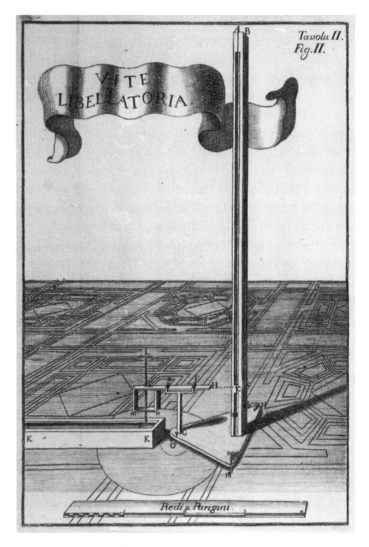

FIG. 7.5. Ximenes' instrument for leveling the line in S. M. del Fiore,
Florence. The circular plate in the floor records the observation of 1510.
From Ximenes, *Gnomone* (1755), plate ii, fig. ii.

Ximenes used his apparatus for a "precise experiment in physics," that is, the determination of the rate of evaporation of water as a function of temperature.[38]

The leveling also provided an opportunity to measure the morals of measurement. When the leveling had been completed with exquisite accuracy, a workman
slipped and lowered a spot. Ximenes called attention to the minute and insignifi-
cant depression lest posterity discover it and impugn his honor. "It is true that the
depression is very small, and insensible for this great gnomon, but nonetheless it

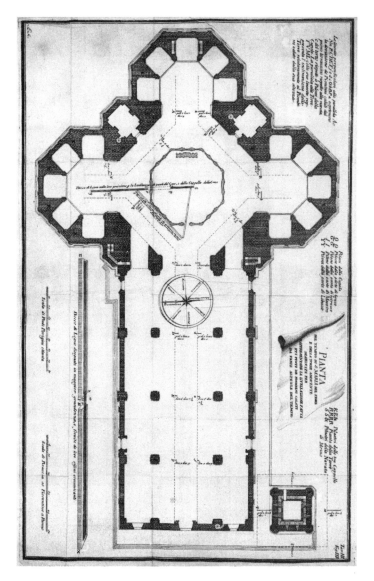

FIG. 7.6. Canals for leveling the line in S. M. del Fiore and avoiding the octagonal balustrade. From Ximenes, *Gnomone* (1755), plate iii, fig. iii.

exists, and could have been, and was to be, passed by entirely unnoticed; there will always be seen there an irregularity, which, though small, is visible to the eye, and which would be blamed on me if the true cause were not known."[39]

Another necessary preparation for the installation of the *meridiana* was an exact measurement of the height of the hole above the choir. The obvious and usual method, using a chain suspended from a fitting in the gnomon, had the difficulty that the chain stretched under its own weight. Earlier designers of cathedral ob-

FIG. 7.7. Apparatus for measuring the height of the gnomon at S. M. del Fiore, Florence. AB is the standard toise. From Ximenes, *Gnomone* (1755), plate v, fig. vi.

servatories had ignored or slighted the difficulty; Ximenes could not be equally cavalier because of the great length (and thus weight) of the chain needed to span the space. But how to measure the stretch? "When it actually comes to measuring great and inaccessible heights, especially if an exact and precise measure is wanted, such great obstacles are encountered that it seems almost impossible to overcome them." Ximenes measured the chain under strain by pulling it up through the hole in twelve-link increments via a winch with a hexagonal drum abcd (Figure 7.7).

Each surface of the drum had a width, and each link of the chain a length, equal to half a Parisian foot; hence two turns of the winch brought up about six feet of chain. Ximenes stopped the turning via the stays l, m; tied a silk ribbon to the chain opposite the lower end B of the fixed toise standard AB; released the stays and turned the winch twice; tightened the stays again; and then found the difference in length between the stretched piece of chain just brought up and six Parisian feet by lowering the drum with the screws L and M until the ribbon came opposite A. (Since the chain shrinks during the winding as the weight on it is removed, the ribbon must go beyond A to provide six unstretched feet.) He thus obtained not only the exact measurement wanted but also data about the lengthening of metal chains as a function of weight and temperature. To this gift to experimental physics Ximenes added a precise correlation of the fall of air pressure with height by carrying a barometer as well as his chain when he climbed around the cathedral. He thus made the installation of his *meridiana* an occasion to improve the hypsometric formula that had enchanted mathematicians for a century.[40]

The problem of finding the vertex of the instrument likewise extended an opportunity for advanced physics. The chain, fitted with a weight, gave only an indication of the place of the vertex, around which it oscillated like a pendulum. Ximenes tried bobs of various shapes and sizes to damp the oscillations and, adopting Cassini's technique, immersed them in water. But always they trembled. Ximenes subjected the roving to careful study, recognized that it had a definite cycle and a diurnal period, and conjectured that it might have owed something to gravity. Eventually he blamed the irreducible oscillations on air currents. Very probably his long chain acted as a Foucault pendulum. If it did, he had in his hands one of the best demonstrations available of the earth's spin.[41]

Ximenes investigated whether his *meridiana,* though short, was sufficiently precise to require correction for the curvature of the earth (no) or for diffraction of light at the hole (yes). He tried to apply Lemonnier's idea of focusing by lining up a telescope with the hole, but the allowable margin of error was so little owing to the great distance between them that even Ximenes could not do it reliably.[42] He expressed all his linear measurements in terms of a standard brought from Paris, equal to precisely one-half of the toise de Peru; and, for good measure, he provided a graphical means for the inter-conversion of the main Italian units of length. The bottom line: the height of the hole was 277'4"9.18''' of the toise de Peru, about half an inch higher than the perforation for the old gnomon; or, if you please, precisely equal to the sum of the heights of Gassendi's *meridiana* at Marseilles (52'), plus San Petronio's (83'), S. M. degli Angeli's (62'), and Saint Sulpice's (80'). The segment of *meridiana* on the floor sufficed only to receive the sun's image for about thirty-five days on either side of the summer tropic.[43]

A *meridiana* is not complete without an inscription. Ximenes put one up on the great pier of the tribune of the cross in the south transept of the cathedral. It gives the vital statistics, the height (the Parisian equivalent of 91.05 m) and the length from vertex to solstice in 1755 (33.69 m); and the reference data, the same length in 1510 (33.65 m) and the deviation of Toscanelli's line from true (56'41"). The small difference in solstitial distances, 4 centimeters, was the grand prize, the great fruit of the Florentine gnomon, "the greatest in all the world, which the Grand Duke of Tuscany, most solicitous of sacred matters and wholesome arts, gave to his people in the year of our Lord 1756 for finding the slightest variation in the ecliptic."[44]

The *fabbricieri* covered the *meridiana* in 1894. Only a determined inquirer can discover it now, under the chairs and pews that hide and protect it. Plate 2 suggests what Ximenes saw as he waited for the image of the solsticial sun, 90 centimeters in diameter, to reveal, as it crossed the beautiful old marble floor, how far the summer sun's rendezvous with the *meridiana* in the Duomo had slipped in 250 years.[45] Another 250 years have elapsed since Ximenes' vigils. Their passing almost coincided with the sixth centennial of the foundation of the Florentine cathedral. As part of the celebration, the *meridiana* and its gnomon were uncovered and, for a few weeks around midsummer day, 1997, the sun's golden image again flitted across the floor of Santa Maria del Fiore. Plate 6 shows the show.

Their Results

The Jesuit college in Florence, San Giovanni Evangelista, which stood across the street from the Duomo, had a small *meridiana,* about 20 feet high in Paris measure. There, even before completion of the line segment in S. M. del Fiore, Ximenes devoted many hours to determining the latitude of Florence. Likewise, Lemonnier tried over and over again to obtain a precise value for the latitude of his little observatory in his rooms before he began at Saint Sulpice.[46] Their compulsion derived not only from the standard conviction of observational astronomers that they could measure more accurately than their predecessors, but also from the certain knowledge that all stellar positions obtained before 1727 had need of correction. The herald of this affliction was the Savilian professor of astronomy at Oxford, James Bradley. In 1742, in recognition of his standing as the greatest astronomer of his time — indeed, according to Delambre, as one of the three greatest astronomers of all time — Bradley succeeded Edmund Halley, who had succeeded John Flamsteed, as Astronomer Royal of England.[47]

Bradley had faced up to the great challenge that had defeated generations of his predecessors. Armed with a better instrument, he had hoped to detect the parallax of a star close to the zenith that would afford a direct proof of the earth's annual motion. He succeeded only too well: his star moved about in what he eventually saw as a circle almost forty seconds of arc in diameter. He declined to interpret the little circle as the long-sought desideratum of Copernicans, however. Not only did the diameter seem improbably large, but the circumference was described in a way that ruled it out as evidence of parallax.[48] Figure 7.8, in which A and Π represent the aphelion and perihelion of the earth's orbit, X and Y midpoints between them, and Z Bradley's star, indicates what was at stake. For ease of representation, the plane ZAΠ is supposed to be perpendicular to the ecliptic; if we neglect the eccentricity of the orbit, the representation is generally valid.

Let the earth proceed from A to Π via X; if its parallax were discernible, Z would appear to move in the little ellipse A'X'Π'Y' of angular diameter proportional to a/d, where a is the radius of the earth's orbit and d the distance SZ from the sun to the star (see Appendix I). A capital point is that the displacement of the star owing to parallax occurs in the plane defined by the sun, the star, and the instantaneous *position* of the earth. In the motion discovered by Bradley, however, the displacement occurs in the plane defined by the star and the instantaneous *velocity* of the earth relative to the star.

Figure 7.9 represents Bradley's observations. When the earth is at aphelion (or perihelion) its velocity is perpendicular to the line of apsides AΠ and the star Z appears displaced to A' (or Π') in the plane through ZS perpendicular to the plane ASZ. When the earth is at X or Y, that is, out of AZS, Z appears displaced in AZS. The explanation of this bizarre behavior lies in the finite speed of light. Starlight entering a telescope at P (Figure 7.10a) takes a little (a very little) time to reach its

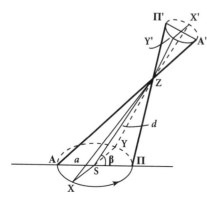

FIG. 7.8. The parallactic circle A'X'Π'Y' of a star Z occasioned by the earth's annual motion.

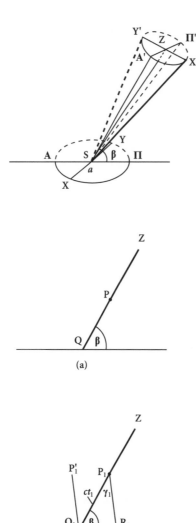

FIG. 7.9. The circle of aberration A'X'Π'Y' deduced by Bradley.

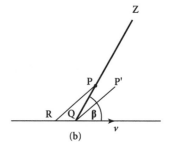

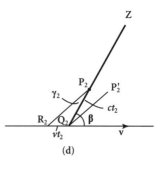

FIG. 7.10. Bradley's explanation of the circle of aberration. (a) direction to a northern star Z from a resting earth at Q; (b) and (d), the same, from an earth moving south to north; (c) the same, from an earth moving north to south; (e) the same, from an earth moving east to west. The P's indicate the objective, and the R's the eyepiece, of the telescope.

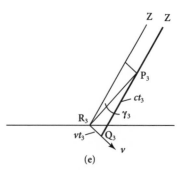

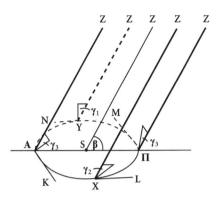

FIG. 7.11. The circumstances in Figure 7.10 combined; for simplicity, Z is shown in the plane perpendicular to the earth's orbit through the line of apsides.

focus at Q. If the telescope remains at rest relative to the star, it must be pointed along the ray ZPQ; but if it moves (Figure 7.10b), it must be tilted by an amount that brings the center of the eyepiece to Q to receive the light that entered at P. Figure 7.10c shows the situation (case 1) when the telescope moves away from the light source; Figure 7.10d, that when it moves toward the light (case 2). The angles of tilt, γ_1 and γ_2, are very slightly different in the two cases (Figures 7.10c, d); to a good approximation, however, they are equal and proportional to v/c, to the ratio of the velocities of the earth and of light. If the telescope moves at right angles to the light ray (case 3), the tilt equals v/c (Figure 7.10e). (See Appendix I.)

Bradley's explanation of the observations reproduced over-tidily in Figure 7.9 is that at A (or Π), the velocity being along AK (or ΠM) perpendicular to AΠ (Figure 7.11), the telescope must tilt by an angle (v/c) in the plane AKZ (or ΠMZ); at X (or Y), the velocity being along XL (or YN) parallel to AΠ, it must tilt by an angle $(v/c)\sin \beta$ in the plane XLZ (or YNZ). And, since on this analysis parallax is undetectable, plane AKZ coincides with ΠMZ, and plane XLZ with YNZ. Every star appears to describe a little ellipse, identical for all stars at the same elevation β; the figure increases in ellipticity from a circle of radius v/c near the pole of the ecliptic to a degenerate line segment at the ecliptic. Since $v/c \approx 20$ seconds of arc, the seasonal variations in the apparent places of stars within 30° of the pole of the ecliptic, which includes the polestar, came well within the precision of measurement claimed by astronomers from the time of Cassini. Indeed, earlier observers, Picard for example, had noticed seasonal fluctuations in the places of a few stars but had not tracked down their cause or regularity.[49]

Bradley had another big surprise for astronomers. His systematic investigations showed that stars moved about in yet another manner. This motion, which has a longer period than the aberration (19 years versus 1 year) and only half the amplitude, affects the zenith distances of all stars equally. It may be represented therefore by a periodic rise and fall of the earth's axis. Newton had foreseen such a nutation, to use the name Bradley gave it, as a consequence of gravitational forces acting on the earth's midriff bulge.[50] It has the same origin as the precession and may be explained similarly; just as the precession corresponds to the rotation of an inclined spinning top around a vertical axis, the nutation copies its nodding, or up-and-down, cusplike swings. As we know, the aberration and nutation made very strong evidence in favor of Copernicus; but since both could be foisted on the stars, they did not deliver the knock-out blow that parallax would have done. The Congregation of the Index did not flinch from condemning Algarotti's *Neutonianismo per le dame* for, among other indiscretions, asserting that aberration proved the annual motion of the earth.[51]

· THE OBLIQUITY AT LAST ·

With the discovery of nutation, the question of the obliquity of the ecliptic ε sharpened: is there in addition to the oscillation in ε, with an amplitude of 9.5 seconds of arc and a period of 19.2 years, a secular change identifiable by comparing solsticial measurements made over many years? To answer the question, earlier observations would have to be corrected for the nutation and aberration valid at the time they were made. As an example, Manfredi reexamined the old determinations of the latitude of San Petronio made by Cassini and Malvasia, and by Riccioli and Grimaldi. He found that the coincidence of their results to a second arose from their neglect of aberration.[52]

About ten years before Bradley made his discovery of aberration, Jacques-Eugène d'Allonville, Chevalier de Louville, upset his colleagues at the Paris Ácademy with the news that the seasons would be obliterated in 140,000 years, or a trifle more. A taciturn and precise man, who had devoted himself to astronomy since retiring from the army in 1713, Louville repeated Gassendi's pilgrimage to Marseilles to observe the summer solstice for comparison with the celebrated measurement by Pytheas. Louville detected a notable difference. Pytheas had found ε = 23°49'10", or so Louville reckoned, after correcting his predecessor's observation for refraction and parallax; whereas he measured ε = 23°28'24" by means of a quadrant of three-foot radius borrowed from the local college of Jesuits. Louville made out that the obliquity had decreased by 20' over the previous twenty centuries. Ransacking the records of the ages, he compiled the data displayed in Table 7.1.

It appeared that the obliquity had suffered a constant decrease of 1'/century from Pytheas to Louville. How long had it been falling before the Greeks first measured it? For many hundreds of thousands of years according to Louville, who credited the Egyptian priests who had told Herodotus that their records extended back to a time when the equator stood perpendicular to the ecliptic (ε = 90°).[53]

Table 7.1 Louville's obliquities

Observer	Date	$\varepsilon - 23°$	$\varepsilon - \varepsilon^{*a}$
Pytheas	360 B.C.	49'10"	21'0"
Eratosthenes	200	51'20"	23'30"
Almamon	830 A.D.	36'31"[b]	8'6"
Albategnius	969	36'31"[b]	7'50"
Arzachel	1070	34'50"[b]	6'26"
Thabit	1130	34'17"[b]	5'53"
Prophatius	1300	35'50"[b]	4'25"
Regiomontanus	1490	29'13"	0'49"
Copernicus	1540	30'3"[b]	1'39"
Tycho	1595	29'25"	1'1"
Gassendi	1636	29'12"	0'48"
Hevelius	1661	29'7"	0'43"
Richer	1672	28'54"	0'30"
Flamsteed	1691	28'32"	0'8"
Bianchini	1703	28'25"[b]	0'1"
Louville (ε^*)	1715	28'24"	0'0"

Source: Adapted from Louville, *AE,* 1719, 294.

a. (ε^*) = 23°28'24" (Louville's measurement of 1715).

b. As corrected by Louville.

That was too much for the Paris Academy. Fontenelle allowed that philosophers soused with physics might follow Louville, but not mathematicians. "Although physics liked variations, even the largest ones, in celestial motions and in angles [of intersection] of circles, astronomy is so opposed to them that they cannot be accepted without solid proof. [Indeed,] this constant [observed] uniformity is becoming a very difficult problem for physics."[54]

Louville wrote up his solid proofs for the *Acta eruditorum,* where they would reach a wider range of philosophers than the Academy afforded. He showed that Gassendi's value for ε, properly corrected to 23°29'12", fit in with the general decline he had worked out between Pytheas and himself, as did almost all the dozen other measurements, ancient and modern, he thought worth noticing. Except for Eratosthenes and Regiomontanus, all the astronomers Louville marshaled testified to a constant decrease in the obliquity. And why not a change in the obliquity? The result seemed to him to be in the nature of things as revealed by up-to-date astronomy. "We suppose, with Copernicus, that all the terrestrial circles are mobile."[55]

And there was more. Assume that the Egyptians correctly recorded that the earth's poles at one time lay in the ecliptic and that the obliquity has always declined by one minute in a century. Then it would have taken 397,150 Julian years for the earth's axis to have attained its inclination in Louville's time. But 397,150 Julian years of 365.25 days equal 402,942 Egyptian years of 360 days. Now, according to the ancient historian Diodorus Siculus, the Chaldeans possessed astronomical records going back 403,000 years before Alexander the Great. Louville: "A remarkable agreement in such a period." No doubt. "I conclude therefore that the Chaldeans not only observed the change of the obliquity of the ecliptic, but also that they knew its magnitude: which ought not to appear extraordinary in view of their very ancient and continuous observations." In 1721 Louville offered confirmation of his estimate of the secular change in ε from his own observations, made over five years, using a quadrant divided and deployed according to a method of his own. The result — 3" in five years or 60" a century — agreed perfectly with the number he had deduced with the help of Pytheas.[56] Louville had a reputation as an excellent observer. His colleagues had to take him seriously.[57]

Louville's mixture of measurement and nonsense brought back an opinion that, according to Manfredi, had been dropped even by astronomers who believed that the obliquity changes. The believers tended to restrict the excursions of the earth's axis to within a very small angle, whereas Louvillle supposed it to revolve through an entire ninety degrees. Among those who held to strict confinement, perhaps even to zero, was Maraldi, who had access to a new gnomon erected around the time that Louville left the army for astronomy. Its builder, Nicholas de Malezieu, tutor to the royal family, philologist, wit, mathematician, and courtier, had grown disgusted with the unreliable divisions on the portable quadrant with which he liked to take the altitudes of stars. Around 1713 he ordered that a fifteen-foot tower of rigid masonry be anchored in the solid rock of his estate near Paris; that a low masonry wall, similarly secured, be run for fifty feet due north of the tower; and that a ribbon of metal, suitable for receiving graduations, be carefully leveled and cemented to the top of the wall. He then invited Maraldi to join in observations of the summer solstice of 1714 and to oversee the final preparation of the instrument. "One can be confident in the ability of M. Maraldi to take all the precautions necessary." With this instrument they looked for pairs of days on either side of the summer solstice on which the sun's declination was equal. Three such pairs provided data that, when properly reduced, specified nearly the same time for the precise occurrence of the solstice. Apparently later measurements at the same instrument, "which is not at all subject to change," persuaded Maraldi and Malezieu that the obliquity did not change either, and they so informed Manfredi.[58]

But Manfredi inclined to believe in a measurable oscillation of the obliquity. Already in 1715 he had written that earlier measurements, when corrected for re-

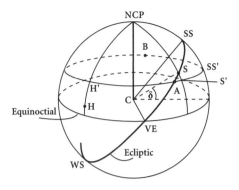

FIG. 7.12. Manfredi's method for determining the time of the summer solstice by means of a reference star H. The primed letters indicate positions on the parallel of declination δ.

fraction and Bianchini's results at S. M. degli Angeli, showed a continuing decline in ε. In 1722, making use of the accumulated observations at San Petronio, Manfredi discerned a change of 30" in 68 years, or E = −43" (E will indicate the change in ε per century). And that, he said, he would have proclaimed to the world, along with Louville, were it not that "the authority of the two gentlemen [Maraldi and Malezieu] recommended suspension of judgment."[59]

We arrive again at the classic dilemma of the empiricist working at the limits of his instruments. How to pick among conflicting data from equally competent observers? The most knowledgeable astronomers continued to entertain divergent opinions about the inconstancy of the ecliptic for three quarters of a century after Louville's results had negated the null finding of Maraldi and Malezieu. A review of the question up to the time that refined calculations based on Newtonian gravitational theory prompted consensus will demonstrate the inevitable fuzziness of exact science at the edge of knowledge.

Manfredi gave the observational program a new start by developing a method for obtaining ε indicated earlier by Flamsteed. That was in 1734, when Manfredi had the help of Celsius at San Petronio.[60] The method consisted of finding the time of the solstice by reference to the meridianal crossing of a prominent star visible by telescope in daylight. The procedure has the very great merits that the observer does not have to know the refraction at the common declination (it cancels out in calculation) or his latitude, and that he can observe far from the solstices, where the sun's declination changes so slowly that its maximum cannot be specified very closely in measurements of the type undertaken by Maraldi and Malezieu.

Figure 7.12 shows the method.[61] H represents the reference star, Arcturus for example; A and B, the sun's positions at the same declination before and after passage through the summer solstice; and δ any value of the declination of the sun S on the ecliptic WS·VE·SS. Manfredi used Arcturus as a clock: by measuring the differences between its meridianal crossing and the sun's on the days when the sun came to A and B, he could know the exact difference (Manfredi called it the "solstitial distance") in time between the star's meridianal crossing and the solstices. The difference in meridianal crossings between the star H and the sun at A measures the arc H'A; that between the crossings of H and the sun at B, the arc H'B; the difference, the arc AB; and Σ = H'·SS' = H'A + AB/2, the "solstitial distance."[62] Manfredi knew that the fixed stars moved and took into account whatever reductions were necessary: he was one of the first astronomers in Europe to accept, and extend, Bradley's discovery of aberration.[63]

The solstitial distance Σ holds a value for ε. For let the sun on the day of the summer solstice come to the meridian with an observed declination of δ. In general it will be slightly off the solstice, say by the grossly exaggerated ecliptical arc S·SS (Figure 7.12); the noon deficit v is therefore measured by the arc SS'·S'. To convert arcs parallel to the equinoctial to hours, divide by 15 (it takes the equinoctial, and any cirle parallel to it, 24 hours to make a complete revolution of 360°). In hours, therefore, v = SS'·S'/15. Since SS'·S' = H'·SS' − H'S', $v = \Sigma/15 - T$, where T is the measured time interval on the day of the solstice between the meridianal transits of the sun and the reference star H. From v and the trial theory of the sun to be perfected by the measurements one calculates how much to add to δ to obtain ε.

Both Jacques Cassini and Lemonnier judged Manfredi's method to be the only route to correct elements of the solar orbit.[64] Neither party succeeded with it. With himself at the mural quadrant and his father at the *meridiana* in the observatory, Cassini de Thury obtained in 1741 a value for ε, which, when combined with one he had made in 1738, gave E = −400"; whereas observations of 1744 and 1747, made with fixed and movable quadrants, gave E ≈ 0.[65] As for Lemonnier, who had an excellent quadrant with a micrometer eyepiece, he inferred a change in ε of +15" in five years, making E = +300, or so he said in 1743.[66] Next, reviewing the literature from Pytheas on, he decided that Louville had made his case only by ignoring contradictory evidence. *Sed contra,* Lemonnier could make his case only by ignoring that most earlier measurements indicated a decline, which, if spurious, implied that, somehow, astronomers had a perennial bias toward error in the same direction. By 1745 Lemonnier regarded the diminution of the obliquity as highly doubtful.[67]

He was then engaged in his measurements at Saint Sulpice, which he rated far more reliable than other *meridiane* owing to the suppositious immovability of the parts of the church supporting the hole and the solstice plate and to the clarity of the solstitial image cast by his innovative lens. His first observations reenforced his growing conviction that ε was very small if not zero, certainly far less than Louville's minute.[68] Further viewing at Saint Sulpice settled the matter. After following the sun for nineteen years, over one period of nutation, he found that its midsummer image came back exactly to the space it had filled in 1745.[69] The secular change in the obliquity was neither plus nor minus, but precisely nothing, as Ric-

cioli had concluded a hundred years earlier. When Lemonnier reached this null result in 1764 the consensus, in which Ximenes' measurements played an important
part, favored a value of E somewhere between –100" and –30". To explain the discrepancy between Lemonnier's results and all others, Lalande suggested that the
wall bearing the gnomon in Saint Sulpice had subsided slightly. Only a shift of one
ligne, one-twelfth of an inch, would have been required to wipe out the consequences of an E as large as –50".[70]

Lemonnier gruffly rejected this asylum, continued his observations, and managed, by 1774, to reach E ≈ –33", in close agreement with the values then put forward by Ximenes and several French astronomers.[71] Lemonnier's vacillations and
premature conclusions irritated the scrupulous Ximenes, who criticized the
Parisian protocols, the instability of the building, the small size of the gnomon,
and the picayune extent of the observations, which amounted to nothing compared with the "size, authenticity, and continuity" of meridianal measurements in
Florence.[72] Ximenes had determined the obliquity as ε = 23°28'16" in 1756, after all
corrections had been made. He had read his *meridiana* to one-twelfth of a ligne,
or about 150th of an inch. Since, around the solstice, a second of an arc amounted
to one-fourth of a ligne, he could record the sun's solstitial declination to under a
second. His determination was astonishingly close: modern astronomy gives for
the secular value of the obliquity

$$\varepsilon = 23°27'8.26" - 46.84"T, \tag{7.1}$$

where T indicates time (in centuries) from 1900. For 1756, $T = -1.44$, making ε =
23°28'15.71", which differs from Ximenes by precisely the smallest unit that he
could measure.[73]

It remained to compare this number with the precious value deducible from
the solstitial plate of 1510. Ximenes massaged this value carefully, correcting for
the error in Toscanelli's line, for the settling of the building (inferred from records
of the *fabbrica*), for the nutation, and so on, to obtain ε = 23°29'43.12", accurate,
Ximenes estimated, to within 7" of arc. That made E = –35". His first calculations
had given E = –29", a value Algarotti acclaimed as agreeing perfectly with Manfredi's estimate of –30" deduced from the records of San Petronio, "a marvelous
agreement . . . , which more than ever reserves the decision in this important astronomical dispute to Italy." Perhaps not to threaten this happy coincidence,
Ximenes adjusted his later value downward. He gave as his best estimate in 1757 a
secular change of E = –31" together with a periodic oscillation of about 15" in 18.6
years. This last was the value of the nutation he took from Manfredi.[74]

In 1775, with the help of several experienced astronomers, Ximenes made another careful observation of the summer solstice. He chose the year, the nine-

teenth after his first observation, so as to eliminate the effect of the periodic part of the change of the obliquity. This time he reviewed all the needed corrections — for refraction (there was still disagreement of a few seconds at 45°), for noon deficit, for solar parallax (fixed at 8.80" from the transit of Venus of 1761), for the penumbra, for temperature and pressure (through their effect on refraction). "In such a rivalry of measurements it will be well not to neglect such circumstances." The upshot: $\varepsilon = 23°28'9.46"$, about 2" too high. That made Ximenes' final value E = −34", agreeing well with his finding in 1757. No wonder that he insisted on his methods and results, and on the excellence of the heliometer of S. M. del Fiore, against all competitors. Boscovich unfairly complained about his confrère's faith in the solidity of his church and hence of his value for E. But the Florentine Duomo did not rattle like Saint Sulpice and Ximenes' E had its supporters in 1780.[75]

The leading opponent, $E \approx -45"$, had a high pedigree. The great mathematician Leonhard Euler had turned his attention around 1750 to calculating the consequences of the mutual gravitational pull of the planets. Whereas the precession arises predominantly from the attraction of the sun, and the nutation from that of the moon, the perturbing force of the planets, primarily Venus and Jupiter, causes the change in the obliquity. According to Euler, this last consequence of the theory of universal gravitation had not occurred to astronomers; unable to find a cause for the change in obliquity, they tended to consider it a constant. And yet he had shown that the mutual pull of the planets Saturn and Jupiter change the inclination of their axes to the ecliptic. Copernican theory in Newton's form therefore required a changing obliquity. "So, if the latitude of the fixed stars changes for the inhabitants of all the other planets, how can one assert that earth dwellers are exempt from a similar occurrence?"[76]

Euler's tedious calculation brought him to E = −47.5", which, he said, would continue in force for a few more centuries, gradually declining in absolute magnitude until, at a time he could not foresee, it would reach zero. The obliquity would then begin to increase; no transformation of the conditions of terrestrial life, as threatened by Louville, will ever occur.[77] Perhaps by attraction from Ximenes' value, Euler's number, E = −47.5", slipped to −45".[78] Lalande later calculated the same number.[79] It had some observational support. The abbé Nicolas-Louis de Lacaille, comparing his own determination of ε made between 1749 and 1757 with a corrected Chinese observation of the thirteenth century, got E = −47".[80] Perhaps guided by the prestige of theory, several other astronomers found for 47", which, according to Ximenes, was the consensus value in 1775. It was, in fact, just right. But Ximenes would not accept it; the calculations rested on uncontrollable assumptions and so exuded the sour odor of systems. In such circumstances observations must be preferred to theory, and no observations could compete with those made at Santa Maria del Fiore.[81]

He did not stand alone. Lalande, who had followed Euler to −45", decided at last in favor of Ximenes' −35".[82] Reviewing the literature at great length in 1780, he found that by combining various earlier determinations of ε with the value of 23°28'18" measured in 1750, he could obtain, for E, −35" (using Louville, 1716), −36" (Rømer, 1706), −33" (Richer, 1672), and −35" again (many Arabs). In the same volume of the *Mémoires* of the Paris Academy in which Lalande argued for E = −35", Cassini de Thury emphasized the observations that supported E = −45". There were Lacaille and the Chinese (−47"); Lacaille and San Petronio (−44"); and a family factoid (−45"), deduced from measurements made at the Paris Observatory by Jacques Cassini in 1730 and Gian Domenico in 1671. But, in Thury's judgment, not all this information, nor a "series of observations extending over a hundred years, made in the Royal Observatory," nor the convergence of measurements around the theoretical value of −45", decided the matter.[83]

His son, Jean Dominique, disagreed. Only two years earlier, in 1778, he had used what he regarded as the most reliable data ever obtained about the obliquity to arrive at E = −60". These data had been obtained between 1739 and 1778, over two nutation periods, primarily with a six-foot quadrant made by Langlois, which Cassini rated the best instrument the Observatory had ever had. Nothing could be gained, he said, from comparison with numbers obtained from lesser apparatus. Nor were all observations made with Langlois' quadrant (of which more later) equally good. Cassini admitted only those for years that saw at least three observations, and only then if the greatest discrepancy among them in seconds was less than their total number.[84] On these criteria, he recommended the value for the decline of the obliquity obtained from the summer solstices of 1755 and 1778: −14" in twenty-three years, or −60" a century. It agreed, he said, with the result from the longer spread obtained by admitting his great-grandfather's observations of 1669 and 1689, and with one of Lalande's theoretical guesses (E = −56").[85]

The result of a hundred years of measurement was a standoff. The best from the *meridiane* fell exactly as far under the preferred theoretical value as the best from the optical instruments deployed at the Paris Observatory came above it. Theory then became more insistent. Laplace entered the computations, carried them further, and deduced that the ε of 1800 exceeded that of 1700 by all of 47". On his theory, the obliquity will continue its decline until the year of grace 3560 when, *laus Deo,* it will reverse itself. Thus has mathematics confidently and reliably foretold where observation could not even retrodict. Laplace's calculation agrees with modern precision measurements.[86]

Their Competitors

In a little book on the telescope published in 1660 by one of the minor mathematicians of Bologna one comes across an ecstatic pun:

> O art more than human, incomparable, and, I might say, almost divine! And all the more because there now lives in Rome one Eustachio Divini, whose fame in this art leaves all others' behind; wherefore, on account of his excellence, I think that in the future this art, as brought by him to the highest pitch, can be called the Arte Divina.[87]

This divine art as practiced by Divini and his rival Campani made possible the discovery of many interesting features of the solar system before 1700. But it needed a hundred years of human improvement before it could produce telescopes capable of defeating lensless holes in churches as instruments of exact observation of the sun.

· THE LONG AND SHORT OF IT ·

Campani and Divini and the lesser lens makers of the seventeenth century ground their glasses to spherical surfaces. They thereby introduced two inescapable flaws, or "aberrations." Figure 7.13, which represents a plano-convex objective ABD, indicates the problem. Parallel rays from the star S fall on all points of the lens surface between I and I'. The upper ray most distant from the axis, SIP, meets the radius CP, which is perpendicular to the refracting surface PDQ, at an angle β, and bends on emergence to the angle γ to meet the axis at F. A ray closer to the axis, striking the plane surface of the lens at J, is refracted to an axial point beyond F; F_0 is the limit approached for points J closer and closer to H. The eye collects the largest bundle of best focused rays when placed half way between F and F_0. The radius of this bundle, $F_0F/2$, is inversely proportional to the square of the focal length DF_0 (see Appendix J). Hence enlarging the focal length diminishes the radius of the circle of spherical aberration. It amounted to nothing at all in the long lenses made by Campani and Divini. Typically the diameter of the lens II' was one three-hundredth of the focal distance. For a lens of diameter 5 inches and index of refraction 1.5, the radius of aberration would be a little over one ten-thousandth of an inch, and so altogether insensible.[88]

A long lens has another, and a more important, advantage over a small one. Fig-

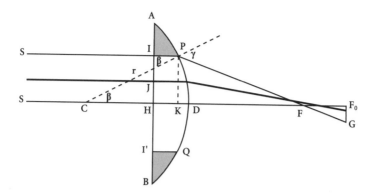

FIG. 7.13. Spherical aberration of a plano-convex lens.

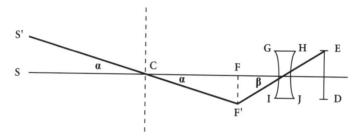

FIG. 7.14. The optics of a telescope.

ure 7.14 shows the rays from the axial star S and a neighboring star S' that pass through the center of the objective lens C, which images their rays at F and F', respectively. The eyepiece GHIJ is placed so that FF' constitutes its focal plane too; it renders the rays from F' parallel, so that they enter a properly focused eye DE at the same angle β to the axis of the instrument. The eye will therefore perceive the star S' at an angular distance β from the axial star S, whereas an unaided eye would make their separation the lesser angle α. The effective magnification of the telescope, β/α, is about equal to the ratio of the focal lengths of the objective and the eyepiece, f_o/f_e, being greater the longer the objective and the shorter the eyepiece.[89] A typical value of f_e was 3 inches; used with it, an objective of 35 feet would magnify 140 times, and one of 150 feet, 600 times.[90] That was considered the largest magnification practicable.[91]

Size does not cure the ill arising from the otherwise pleasant phenomenon of color. The geometry of light rays just presented assigned the same value of the index of refraction n to all the rays within the objective. But in fact the bending of the light at the lens' front surface creates a spectrum, the rays of which travel under different indices. In Figure 7.15, the extreme ray SP splits at P into an optical

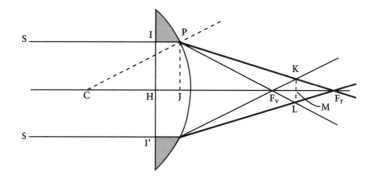

FIG. 7.15. Chromatic aberration of a plano-convex lens.

spectrum between the violet PF_v and the red PF_r. Since these rays cannot be imaged together, the best one can do is to pick the smallest circle that contains them all. KL is the diameter of that circle. It can be calculated from the pairs of similar triangles PF_vJ, LF_vM and PF_rJ, KF_rM. Set $y = KL/2$, $z = F_vM$, $V = JF_v$, $R = JF_r$. Then $y = az/V = a(R - V - z)/R$, or $y = a(R - V)/(R + V) = a(\mu - \nu)/(\mu + \nu)$, where μ and ν represent the indexes of refraction of violet and red light, respectively. With the values typical of the time, $y = a/55 = 0.036''$ for an objective two inches in aperture. In practice the indistinctness of the image is as the area of the circle of aberration to the area of the entire image, that is, as y^2/f^2, and so could be rendered nugatory by increasing the focal length of the objective.[92]

The result of the lengthening could be comical. Figure 7.16 shows the environs of the Paris Observatory, not yet landscaped, in 1705. A moderately long telescope, perhaps of 30 feet, hangs from a pole and points to the moon. Two longer lenses, each over 100 feet in focal length, are also in play, one affixed to the Observatory's roof, the other to an ugly wooden derrick, 130 feet high, dragged there in 1685 from Marly, where it had been used to raise water for the fountains of Versailles. Neither long lens is connected to its eyepiece. Astronomers using them had to jiggle and juggle their eyes and oculars to catch the rays from the objectives some hundred feet away. The maneuver was assisted by a frame for the eyepiece, ropes to position the objective, and a clockwork, invented by Cassini, to allow the objective to follow its star.[93]

The tubeless telescope, developed by Huygens, avoided the need for structures of the sort that Campani designed to test his long lenses. Figure 7.17, taken from an engraving of 1681 made for Ciampini's academy and republished by Bianchini, shows the test bed for terrestrial and celestial telescopes; rigged like a ship, it allowed the observer using the counterweight S to orient the heavy telescope almost at the touch of a finger. Unfortunately, an equally light touch from a passing breeze could knock it out of alignment.[94] Alternatives were proposed. Ciampini's Fisico-

FIG. 7.16. Aerial telescopes in use at the Paris Observatory and the Tour de Marly.
From Wolf, *Histoire* (1902), plate XI.

matematici commissioned an elaborate scaffolding 100 palms (70 ft.) long. The
Paris Academy approved a lesser contraption for a 100-foot telescope, which required the observer simultaneously to stare at the stars and work five stout pul-
leys.[95] The best mount of all, the brainchild of Gilles-François de Gottignies, S.J., a
student of Tacquet's and friend of Divini's who became professor of mathematics
at the Roman College, is shown in Figure 7.18. The 50-foot beam appears to be suspended by faith. A crowd came to see it; "everyone praised the father and his work."
Nullum problema insolubile, as Gottignies liked to say.[96]

According to Bianchini, who preferred Campani's test bed, a 50-foot telescope
was needed to view the disks and diameters of the planets, the spots on Jupiter,
and Saturn's rings, and one of 100 feet to see all the satellites of Saturn; whereas
people content with observing nothing more exotic than the eclipses of Jupiter's
moons could do with something between 15 and 25 feet.[97] These instruments did
not point to the future. As lenses continued to lengthen, the expense, weight, and
inconvenience of the great tubes became unmanageable. Explorers of the odd corners of the solar system had to learn to fumble with tubeless telescopes. Figure
7.19 shows Huygens' design. Number 613 indicates the necessary pieces; a long
pole ab carrying a platform ff bearing the housing for the objective lens ik turning
on the universal joint m; a stand x, on which the observer leans while trying to
align the ocular in the tube oq with the distant objective lens via the silk string ul;

FIG. 7.17. Campani's system of working a long telescope.
From Bianchini, *Phaenomena* (1728), plate I.

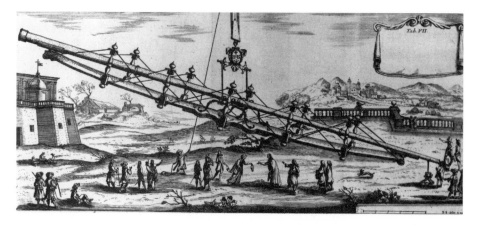

FIG. 7.18. A Jesuit telescope suspended by faith. From Bianchini, *Phaenomena* (1728), plate II.

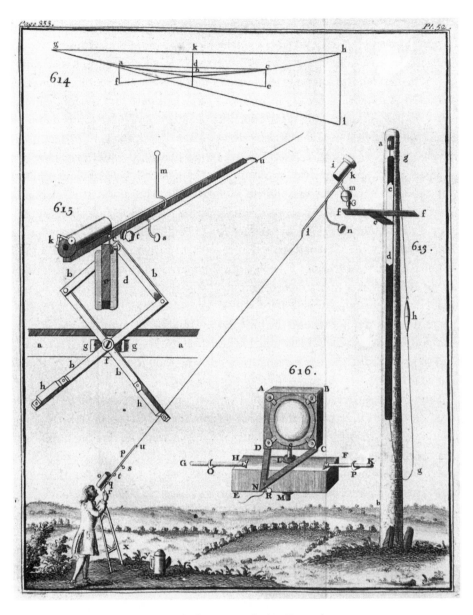

FIG. 7.19. Aerial telescope worked in Huygen's manner.
From Smith, *Opticks* (1738), 2, plate 52, p. 353.

the sticks kl and qu to help in the positioning; and the weights n and s to balance the lens housings. Number 614 presents the sagging of the string; number 615, a contrivance for raising and lowering the eyepiece (the beam aa rides on the stand x); number 616, a mounting for an objective later introduced by another Parisian astronomer.[98]

Louis XIV's minister Colbert materially assisted the shift to tubeless telescopes by providing much money for lenses and little money for mounts. It was Cassini's doing. He showed Colbert sun spots through the 17-foot Campani lens with which he had discovered the rotations of Mars and Jupiter while still in Italy. Soon after he moved into the Observatory in 1671, this same lens gave him the first glimpse man ever had of Iapetus, the second satellite of Saturn, an object with a diameter exactly one-fourth that of the largest satellite, Titan, which Huygens had found by tubeless telescope in 1655. Cassini located a third one, Rhea, in 1672, again with a Campani lens.[99] Supposing that more might be bought with better means, Colbert instructed the French Ambassador to Rome, Cardinal César d'Estrés, Bishop of Laon, to commission what he could from Campani: "besides the advantage he will have in selling them, the king will make him a substantial present." Three months later, in January 1672, Colbert extended the order to Divini and specified a focal distance of 120 palms, about 85 feet. Both Campani and Divini offered excellent lenses, albeit shorter than Colbert wanted. Should the French buy both when one might do? The Bishop advised that it would be neither royal nor prudent to stint: "Since whoever did not receive the preference would be discouraged, and since it is important to push them to do well, I believe that it would be wiser to take both." And so Colbert did, though together they cost the King more than half Cassini's annual salary.[100]

Thus encouraged, Campani supplied lenses of 80, 90, 100, and 136 feet. Unfortunately, Colbert died before they could be paid for and his successor had other priorities. Before returning them to Italy for Queen Christina, who meditated setting up an observatory, Cassini tried them out. With the two longest lenses he found two more moons of Saturn, now known as Dione and Tethys, both smaller than Iapetus, a performance that, as Delambre observed, should be credited not only to the astronomer but also "to the technician who had made the lenses and the king who had ordered them made." Cassini used the occasion to draw up instructions for viewing through long tubeless telescopes. The novice should begin with stars near the meridian, where their diurnal motion appears slowest, fix the objective so that it receives the stars' rays perpendicularly, set up a rigid support for the eyepiece, and move it about until the lenses line up. This last step, he advised, required patience and practice.[101]

Few astronomers had the patience to practice, or, perhaps, the strength and dexterity needed to manipulate the lenses. The long telescopes fell into disuse after the great Cassini's death; "astronomers were powerless to see what he had seen and had lost even the memory of some of his discoveries."[102] An anticipation of the loss may be found in Flamsteed's correspondence with Newton. When he learned about Cassini's detection of Dione and Tethys, Flamsteed looked for them with the longest lens he had, of 24 feet, but in vain. He could see only Titan. "I am as suspi-

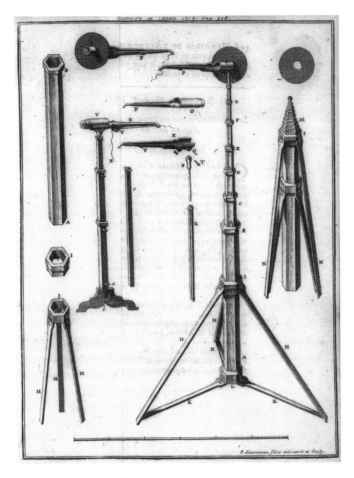

FIG. 7.20. Bianchini's portable aerial telescope.
From Réaumur, *MAS*, 1713, 306.

cious of the others as Mr. Halley or Hugenius can be. Were they placed as Cassini makes them I see not of what use they will be to us considering what long tubes are required to discover them." Flamsteed never saw them. The first man in England to do so was probably Bradley's uncle James Pound, who succeeded in 1718 using the 123-foot glass that Huygens had left to the Royal Society of London.[103]

Among the few who managed to work successfully with the big telescopes was Cassini's emulator Bianchini, who invented a mounting that reduced the difficulty of handling them. Two telescoping stands supported on tripods hold the objective and eyepiece; a string running in channels in the lens housings indicated the focal length and alignment (Figure 7.20). The seven segments of the mount supported the objective at a height of around 35 feet when fully expanded. The equipment could be carried by the observer; the objective tripod collapsed, as pictured on the right of the figure, under one arm, and everything else, boxed, under the other. Stones found at the observation site could be placed on the tripod's feet KKK to help steady it. Bianchini brought this machine to Paris in 1712 to demonstrate to the Academy, which, in an unusual show of democracy, approved it for promoting astronomical observations among private individuals unable to afford or deploy the ordinary long telescopes.[104] Jacques Cassini soon proposed a costly improvement. Making Bianchini's portable mount into a sturdy fixed post and adjusting the height of the objective by a pulley rather than by nested tubes, Cassini transformed Bianchini's "very ingenious machine" into an instrument long enough to hunt for Saturnian moons.[105]

Cassini's post, which he used in preference to the decaying Marly tower, had some advantages. He could adjust his lens both in height (via the pulley) and azimuth (by rotation of a fork carried by the frame riding on the pulley): "There can scarcely be a method simpler in construction and at the same time more convenient in practice." He could run a lens about 30 feet in focal length up the post before its shaking became apparent; to follow objects high in the sky, he attached his lens and the apparatus that moved it to the roof of the Observatory, 84 feet above the ground. In this way, with a lens of 114 feet in focal length, Cassini saw all five of Saturn's satellites at the same time, determined their periods and distances from their primary, and confirmed his father's finding that the Saturnian system obeyed Kepler's laws.[106] Cassini's representations of the moons' orbits no doubt helped Pound to see the satellites from England in 1718.

Bianchini showed what his method might accomplish in his examination of the surface of Venus. Cassini had reported seeing some spots on her face in 1666 and 1667, and guessed from their motion that Venus revolves once in 23 hours. But he did not see the spots again, nor could Huygens; and, when Bianchini sat down to spy on Venus in the 1720s, her complexion had been unblemished, as far as terrestrial astronomers could see, for sixty years. The spots reappeared for him, but, with one exception, only when he looked at Venus through lenses over 60 feet in focal length; then the surface looked much like the moon's seen with the unaided eye. With the help of the splotches Bianchini determined the period of rotation of the planet as about 24 days and defined the surfaces areas that, as reported earlier, he allotted to Portugese heros and modern astronomers.[107]

These results astonished the astronomers. A Jesuit from Florence doubted the report. "Am I forced to admit true spots visible to mortal eyes even in the brightest mirror of Divine Beauty among corporeal things, in the shining morning star, in the planet Venus?" No. Bianchini almost certainly did not see the surface of Venus, which has a thick permanent cover of carbon dioxide gas. This atmosphere is opaque to normal eyes. The surface features can be photographed in the ultraviolet; perhaps Bianchini possessed x-ray vision along with his other powers. His drawings do resemble the surface markings caught by ultra-violet cameras. But his value for the rotation period is considerably out, 24 against around 243 days. He probably tracked transient features in the cloud tops rather than permanent markings on the planet. He went to Bologna to talk to Manfredi about his visions and wrote to Maraldi for details about Cassini's sightings. He was particularly concerned to identify the spot Cassini had reported so that he could name it after its great discoverer, "for the glory of Italy." But Bianchini's Venusian land grants proved as fugitive as the fame of the heros he named them after.[108]

· THE QUADRANT ON THE WALL ·

The long telescope could not replace the *meridiana* as a precise chronicler of the movement of the sun. The length brought detail about objects seen at one sitting; but, since the telescope could not be returned accurately to the same orientation with respect to the horizon and the meridian, it was not useful for reference to the usual system of celestial coordinates. Hence it could throw no light on the obliquity of the ecliptic. A telescope competitive with a *meridiana* would have to be reliably fixed to an unmoving wall and constrained to rotate strictly in the plane of the meridian; it would need a focal length long enough to distinguish the sun's noon heights from day to day; and it would have to be associated with a scale on which its altitude could be read accurately to within a few seconds of arc. No such instrument existed in Paris as late as the mid-eighteenth century. Cassini de Thury, Jacques Cassini's son, writing in 1741 on the basis of experience at the observatory and in the field, had little good to say about the graduated sectors fitted with telescopes that he had used. "All our experience along the meridian line [from the North to the South of France through the Paris Observatory] shows over and over again the imperfection of these instruments." His son, Jean Dominique Cassini, writing in 1780 as the director of the Observatory, still esteemed *meridiane* as the best means of obtaining "a very precise determination" of the obliquity.[109]

His was a minority opinion in 1780. Two decades earlier Lalande, extrapolating from his conclusion that the settling of the walls of Saint Sulpice had subverted Lemonnier's measurements, expected that astronomers would abandon gnomons when they accepted his structural analysis. "Then they will recognize more and more that a good six-foot quadrant easily preserved from all changes and readily checked is preferable to a gnomon eighty feet high." The day of the *meridiana* was done. "This type of instrument was useful before the application of lenses to quadrants and before we knew the art of dividing instruments to two or three seconds of arc."[110] To these essential advances must be added mechanical improvements that reduced the stress and flexure in the mountings of the telescopes and the dis-covery that Newton had been wrong in thinking that lenses could not be corrected for chromatic aberration.

The incorporation of the telescope into an exact apparatus began with Picard, the first astronomer to apply lenses to instruments of angular measurement. His innovation can be dated precisely. He and his colleagues observed the summer sol-stice of 1667 using an instrument with pinnules (open sights). On 2 October, he took the altitude of the sun for the first time with a quadrant and a sextant fur-nished with lenses. Otherwise these were standard equipment for the time: large (they had radii of 9.5 ft. and 6 ft., respectively), bimetallic (they had bodies of iron and graduated limbs of copper), tarnished, and unwieldy. In 1668 Picard had a smaller instrument built to carry a telescope rather than the usual alidade with pinnules.[111]

The telescope in Picard's new sector boasted a micrometer. This essential bet-tering of the means of measurement, which had occurred to several astronomers in Britain, Italy, and France, had been made serviceable by Picard's colleague Adrien Auzout in the late 1660s. He fixed narrow threads at equal intervals in the common focal plane of the objective and eyepiece and also one or more wires mov-able from outside the telescope by an accurate screw. The number of turns of the screw needed to bring a movable wire from alignment with a fixed thread to coin-cidence with a point of interest on the image measured the angular distance of the point from the axis of the telescope.[112] With Auzout's micrometer, Picard could measure the apparent horizontal diameter of the sun with an accuracy, as deter-mined by modern analysis, of between five and ten seconds of arc. Some three sec-onds of this error arose from errors in reading, while five or more were systematic, the consequences of imperfection in the micrometer screw, of flexure of the mounting of the telescope, and of other fixed features of the instrumentation.[113]

Picard's accuracy of, say, ten seconds in obtaining the size of the sun was at least six times better than the previous standard. That follows from crediting Cassini's estimate, in connection with the rationale of building the meridian line in San Petronio, that no telescope existing in 1655 could be relied upon to deter-mine the diameter of the solar disk to within a minute of arc.[114] The geodetic expeditions begun by Picard and extended by Cassini prompted further improve-ments in the design and use of astronomical instruments. Picard used a ten-foot zenith sector with which he did not do very well, for reasons easily imagined from a picture of the instrument at work (Figure 7.21).[115] Although the scale could be read to 20 seconds of arc, the insecure mounting of the instrument and the un-comfortable attitude of the observer helped to throw off Picard's determinations of latitude by as much as 5 minutes. For surveying in the plane, Picard used a quadrant with a diameter of 38 inches (1.02 m) that allowed measurements to 15

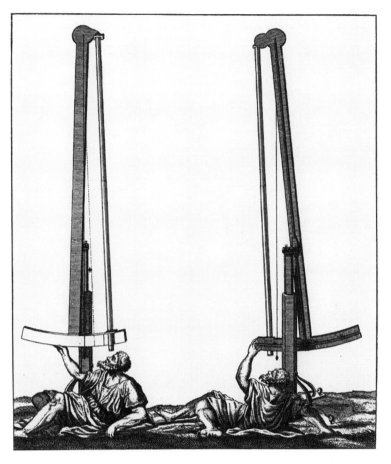

FIG. 7.21. A zenith measurement in 1670. From Picard, *Mesure* (1671).

or 20 seconds with the help of transversals. Retrospective calculations show that his angles were good to 20 seconds.[116] Even with his errors in latitude, Picard's geodetic work had an accuracy perhaps 20 to 30 times that of his predecessors.[117]

Improvements beyond Picard's came slowly. Two obvious avenues — checking graduation by dividing an entire circle rather than a quadrant or sector and making and mounting instruments more robustly — were opened by Picard's younger colleague Olaus Rømer, but few followed for many decades.[118] A five-foot quadrant, planned by Picard and mounted against a wall of the Observatory in 1683, frequently bent out of the plane of the meridian. Nevertheless, it continued in use until 1719. A six-foot quadrant of Cassini's, carrying a heavy telescope, hung from a wall in the north tower, where it suffered severely from the weather.[119] As we know, he preferred the excitement of exploring with long telescopes to the tedium of measuring with less powerful sectors.

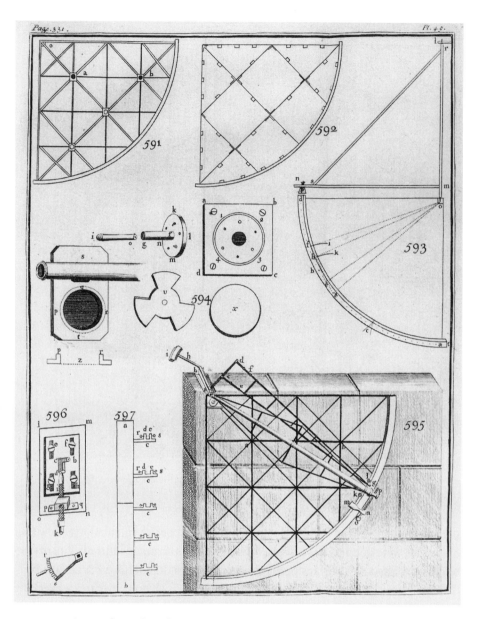

FIG. 7.22. A typical mural quadrant of English manufacture of the mid-eighteenth century. From Smith, *Opticks* (1738), 2, plate 40, p. 331.

Lacking long telescopes, the Greenwich Observatory perforce devoted itself to the more routine measurements. The best fixed graduated instrument in Europe around 1700 was Flamsteed's seven-foot mural quadrant, divided to 5 minutes on the limb (and by transversals to seconds) by Abraham Sharp, "the first person that cut accurate and delicate divisions upon astronomical instruments." Estimates of its errors range from 11 seconds to over half a minute. Sharp had done his job well — it had taken him fourteen months — but the quadrant bent under its own weight and when it warmed or cooled its brass limb and iron backing dilated or contracted at different rates. Also, the wall that carried it slowly sank under its weight. By 1715 the wall had descended so far that Flamsteed had to add 14'20" to all observations made with the instrument. With corrections and inspired adjustments, it provided the information that allowed Newton to develop, and confirm, his theory of the motions of the moon.[120]

Sharp's arc was replaced in 1725 by an eight-foot quadrant commissioned by Edmund Halley. It marked the entry of its builder, George Graham, then fifty years old, into precision astronomical instrumentation. A clockmaker by trade and training, he had succeeded to the business of the leading clockmaker in England, Thomas Tompion, by improving Tompion's escapements and marrying his niece. Some of Graham's technical improvements in clocks and watches remained standard in the industry for two centuries. The skills he developed in metalworking, especially his mastery of micrometry, made an epoch in astronomy when he applied them to forming and graduating telescope mounts. We already know two consequences of his intervention. He made the zenith sectors with which Bradley detected the aberration of starlight and the nutation of the earth.[121]

Figure 7.22 analyzes Graham's epoch-making quadrant of 1725. Number 591 shows its trellis of iron bars 2.9" wide with securing pins a and b, b being movable to allow adjustment to the horizontal. Number 592 reveals the bracing at and between joints. Number 595 represents the thing itself, carrying its reinforced telescope movable by the micrometer screw mp. The English authority on optical systems, Robert Smith, knew the best when he saw it. "A large quadrant (with a telescopic sight) fixt in the plane of the meridian to a free-stone wall . . . is by far the most accurate, expeditious and commodious instrument for the chief purposes of astronomy."[122]

Graham's rigid trellis mounting probably increased accuracy by a factor of two, to 7 or 8 seconds (on an eight-foot quadrant readable to 5 seconds); and the switch to solid brass construction, pioneered by his successors, helped to realize another factor of four, bringing the accuracy of determination of angular separations to 1 or 2 seconds of arc for the very best fixed instruments. The meticulous John Bird, who made several six-foot and eight-foot quadrants in Graham's style around the middle of the century, engraved the most delicate lines on his creations only in the morning, in the spring or in the fall, with a compass he had set the night before and allowed to come into full thermal equilibrium with the piece under construction. He would allow but one assistant to work with him, lest the combined body heats of several should expand the limb during division.[123] So far had the art and compulsion of measurement been carried by 1750.

Another of Graham's important innovations was to drive division by bisection as far as possible. The trick cannot easily be played in a quadrant graduated in degrees, but it can be carried directly to quarter divisions in one graduated into 96 primary parts. Graham would draw an arc a little longer than a quadrant with a beam compass and, picking a spot close to one end, would set one point of his compass there without changing its opening. He could then strike off an angle of two-thirds of a quadrant (60°), and, bisecting that, could add the pieces together to construct a right angle. To divide this angle into 96 parts, he had only to return to the two-thirds quadrant and bisect it six times; the result, an angle 1/64 of two-thirds of a quadrant, was the desired 1/96 of a right angle. Two more bisections of the 1/96 part created an angle smaller than 15 minutes of arc by almost exactly one minute. Graham also divided his right angle into the conventional 90 parts, subdivided both the 90-part and the 96-part scales to a twelfth of a part (that is, to 5 minutes of arc), added a vernier for interpolation, and used one scale to check the other. They never disagreed by more than 5 or 6 seconds, which made an important improvement, a factor of two, over Sharp's very best divisions.[124]

Bird extended the principle of successive bisections to the 90-degree scale by beginning with an angle of 85°20'. After 1,024 bisections — it took Bird at least 52 days to graduate a quadrant — he arrived at an angle of 5'. But how to construct the original 85°20'? Bird computed the lengths of chords that would subtend angles of 4°40', 10°20', 15°, 30°, and 42°40' at the center of the quadrant he had under construction. He made an auxiliary linear scale of great accuracy, readable to one-thousandth of an inch by vernier, with which he could set the opening of his compass to the calculated lengths of the chords, to generate his 85°20'. He made a 60° angle in Graham's manner, then added 30°, subtracted 15°, and added 10°20', all from the corresponding chords, to generate his 85°20'. He confirmed the accuracy of the work by adding 4°40' to obtain the full quadrant and by subtracting 42°40' twice to attain zero. He also inscribed divisions in 96 parts for continual cross-checking in Graham's manner.[125]

The estimates of accuracy so far presented refer to single, absolute measurements (not small-angle differences) made on quadrants or sextants of 6- to 8-foot radius. Better results could be obtained in special circumstances. Bradley's 12.5-foot zenith sector was to zenith sectors what Graham's Greenwich instrument was to mural quadrants. Firmly pivoted between two walls, the iron tube of the telescope could be displaced by a fine micrometer screw. The mechanism probably permitted accurate measurements of angular distances as close as 1.5 seconds.[126] By 1750, the precision astronomy built on the innovations of Picard's group could boast some notable discoveries and, in the best instruments of English manufacture, an exactness of measurement of close to one second of arc, an improvement by a factor of ten or twenty in some eighty years. This exactness in manufacture became accuracy in observation through protocols developed by Bradley that used zenith sectors, quadrants, and transit instruments as independent cross-checks on one another.[127]

The limit of accuracy of an absolute measurement made with the equipment of classical astronomy is just under a second of arc.[128] The reliable attainment of this limit rested on the introduction of achromatic lenses; on mechanical improvements owing to the diminution of instruments made possible by the new lenses; and, above all, on the replacement of quadrants by full circles. By graduating over 360°, the maker could check the faithfulness of diametrically opposed marks. The earliest important instruments so divided were the work of Jesse Ramsden. He proceeded tediously, by locating approximately the major points of division by compasses and then relocating them under the eyepiece of a micrometer microscope into their correct diametrically opposite positions. It took him 150 days to come full circle.

His second major machine in this style, an equatorial completed in 1793 for George Shuckburgh, had a probable error in graduation of 1.5". That would be the case, however, only if Shuckburgh read the scale through one of the six micrometer microscopes with which Ramsden enriched his telescope. By reading all six and applying appropriate corrections, Shuckburgh could have confidence in the result to 0.5". Even Bird had had to accept errors of 3". The tedium of Ramsden's method was very much reduced by Edward Troughton, who introduced a gear-driven roller to find the preliminary divisions and tabulated the errors between these divisions and those required by the microscopic inspection of diametrically opposite points. He then calculated where the definitive divisions should go and managed to put them there, he claimed, in only thirteen eight-hour working days.[129]

The works of the great English makers from Graham to Ramsden were the envy of the world. Looking on enviously, Lemonnier gave pride of place in his text on astronomical instruments of 1774 to Graham's quadrant, which brought in "six or seven perfections" its predecessors lacked.[130] Among those predecessors was an instrument designed by Louville, who had never met with a quadrant whose division did not err somewhere by 20" or more. He tried to avoid such mistakes by proceeding to 10' intervals in the usual manner and obtaining finer measurements by micrometer; everything depended upon the accuracy with which he could calibrate screw-turns with seconds of arc. He had done so, he said, almost to perfection. "If there is anything left to do in the future, I dare say that it will not amount to much, and that the two machines [he had also improved the clock] are now brought to the fullest perfection possible." He no doubt bettered measurements of small angular distances since, as Cassini de Thury later pointed out, Louville's technique allowed the observer to change his line of sight slightly without moving the telescope. A quadrant so outfitted, pointed at the upper limb of the sun at noon on the day of the summer solstice and left undisturbed for a few years, could in principle obtain a reliable value for the change in the obliquity.[131]

Langlois made a mural quadrant of six-foot radius on Louville's principles and divided it to the minute on its copper limb. But he used iron for its frame and the bars attaching it to its wall; it suffered from thermal and mechanical stresses and could be read to under a minute only by guesswork. This unhappy instrument, installed in the Paris Observatory in 1732, gave way to a six-foot version, also by Langlois; through its two heavy telescopes, which caused it to flex, the astronomers of the Observatory made the series of observations of the obliquity from which, in 1778, Jean Dominique Cassini deduced that $E = -60''$. One year earlier the Observatory put up a much smaller instrument, occupying a half rather than a quarter of a circle, and carrying two achromatic telescopes. Although less than a foot in radius, it could be read as closely, and magnified as well, as Langlois' six-foot quadrant, and gave a much better image. It was the work of the Duc de Chaulnes, who had managed to make a dividing engine without being an Englishman. The excellent little half-circle later acquired by the Observatory could be read by micrometer accurately to two seconds of arc. Chaulnes had made it in 1765; the Observatory got it in 1777; the delay in its acquisition measures the decline in the Observatory's instrumentation since the middle of the century.[132]

In 1784, the decline having become intolerable, the grandson and great-grandson of the great Cassini appealed to the state for funds to regularize the Observatory. They wanted a library, assistants, a program of observations, new instruments, and, above all, a mural quadrant, "built in the English manner." With this last provision, official French astronomy recognized the danger of not recognizing the superiority of English instruments. "It is incomprehensible to me [a touring Italian astronomer wrote home from Paris in 1786] how they can regard French observations so highly and make such a fuss about them when their instruments are so badly placed and so old." The belated modernization would not be cheap. How was it to be justified? Straightforwardly: "Astronomy is the most expensive of all the sciences." The Crown accepted the Cassinis' proposals. Not until 1788, however, could the Observatory place an order with Ramsden for two instruments made as only he could make them. It then had a still longer wait, until 1803 to be precise.[133] The English artist took his time and overcommitted his establishment.

When the French gave their commission, Ramsden had had in hand a meridian transit for the Duke of Saxe-Gotha for five years. The Duke had not sat so idly as Ramsden. When reminders and demands failed, he tried to procure an eight-foot mural quadrant that Bird, then (1786) dead a decade, had made for Pierre Jacques Onésine Bergeret, who had lent it to the Ecole Royale Militaire. When Bergeret died, his family put the quadrant up for auction. Several big bidders entered, including the Austrian government of Lombardy (in favor of the Brera Observatory), the Duke of Saxe-Gotha, and the King of France. The bidding went high. "It is the largest and best astronomical instrument we have," Lalande wrote the French minister who put up the money to acquire it. The Duke drove up the price. "The Duke sets the highest price on the possession of this instrument," wrote the Duke's agent to the same minister, pointing out the faithful service that the Duchy of Saxe-Gotha had always rendered the Kingdom of France. The French minister decided to keep the Bird in hand. The Duke's agent made the cost of the keeping very high.[134] Thus the procuring of precision astronomical instruments had become a matter of state, and the French received a reminder they did not need that their inferiority to the English in exact instrumentation exposed them to embarrassment if not to jeopardy.

It remains to say that the English quadrants settled the matter of the obliquity. Already in 1767 Bird advertised that Bradley in Greenwich and Tobias Mayer, at the observatory of Göttingen University, both using instruments he had made, got values of ε differing by only two seconds.[135] Sixty years later, Niccolò Cacciatore, a collaborator on the *meridiana* in the cathedral of Palermo and director of the city's observatory, published an analysis of the measurements of ε he had made over the previous decade with a Ramsden instrument. He confirmed the common finding that the obliquity derived from the midwinter sun exceeded that from the midsummer sun by a few seconds of arc, owing, presumably, to improper correction for refraction. By comparing like with like, however, he obtained a most satisfactory number: $E = -45.46$ seconds of arc. It fell at the upper end of the range that he obtained from reduction of measurements made by others using graduated full circles. The weighted average, between 44 and 45 seconds, agreed perfectly with theory.[136] Knowing the answer eases the forging of consensus.

The sun had set on the *meridiana* as a scientific instrument. Delambre observed in the 1820s that nothing had been heard from the old gnomons at Saint Sulpice and S. M. del Fiore and guessed, rightly, that no new ones would be built. But even he, who disliked them for their connection with Cassini the Great, had to concede that they had established the reality of the decline of the obliquity and had fixed its value to within 50 percent. To be sure, as Zach observed, the persistent problem of the penumbra and the fuzzy edges of the image ultimately prevented the *meridiane* from yielding an accurate value of E.[137] But to better what Cassini, Bianchini, and Ximenes did with their simple apparatus required 150 years of effort by the best astronomers, opticians, instrument makers, and mathematicians of Europe, and very substantial expenditures by their patrons. The simplicity and elegance of means with which the Italian meridian men posed and solved their problems deserve admiration as (to quote Giuseppe Bianchi, tutor to the sons of the Austrian Archduke of Modena, director of the Archduke's observatory, and member of many of the learned societies of Italy) "vestiges of scientific history and national greatness."[138]

In contrast, the last episode in the story of the obliquity and the Church has an unpleasantly modern ring. In 1803, the astronomers of the small observatory built at the Roman College in 1787 dedicated the first volume of their *Opuscoli* to the Pope, Pius VII. They reminded his Holiness that his predecessors Gregory XIII, Clement IX, Benedict XIV, and Pius VI (who had founded the Vatican Observatory) had been strong supporters of astronomy. The most recent and subtle developments, they said, demonstrated that astronomy still was necessary to the defense of religion. They rehearsed the tale of the Egyptian priests about the former orthogonality of the ecliptic and the equator. Now, if the obliquity had ever been 90°, it would have taken the earth's axis over 500,000 years to have arrived at its present position at the accepted rate of around −45″ per century. That would violate the "truthful narrative of Moses." Fortunately, Lalande, Laplace, and others had shown that the change in the obliquity occurs and occurred within narrow limits; the ecliptic and equator never have met at right angles and never will coincide. "A profounder astronomy alone imposed silence, and demonstrated that the ecliptic can only have an oscillatory motion within the narrow and close limits of a degree." The Pope increased the Observatory's budget.[139]

8: Time Telling

The supersession of the cathedral *meridiane* as instruments of exact astronomy by no means arrested their multiplication. Two of the most beautiful and serviceable were built after consensus had been reached about the magnitude of E and the superiority of mural quadrants with achromatic telescopes. These new *meridiane,* installed in the cathedrals of Milan and Palermo, served the civic and civil purposes of gently converting Italians from peninsular to European time. Contemporaneously, the heliometers of San Petronio and S. M. degli Angeli were adapted to time telling and many lesser instruments multiplied in and out of churches to provide convenient correctives to the pocket watches and carriage clocks that came increasingly into use as the eighteenth century aged. In the next century, some of these instruments, for example those at the Academy of Sciences in Siena and in the Piazza Vecchia in Bergamo, carried conversion further, to initiation into the grand mystery of the "equation of time."

Some Means of Conversion

· THE DUOMO OF MILAN ·

The plenipotentiary through whom the Holy Roman Empress Maria Theresa governed her province of Lombardy, Count Carlo Firmian, tried to rationalize her business. A patron and judge of the arts and sciences, Firmian encouraged astronomy, principally by supporting the transformation of the Brera, a small observatory at the former Jesuit college in Milan, into a useful institution. The Austrian authorities had taken an interest in strengthening the observatory when the Jesuits, caught up in the difficulties that prompted their suppression in 1773, seemed to be neglecting it. On the day after Christmas in 1771, Firmian's superior in Vienna, full of holiday good cheer, directed him to find out what was needed in the way of books, instruments, and assistants to carry on a full program of astronomical and meteorological investigations at the Brera. Firmian's subsequent funneling of money into the Brera earned him high praise from many, including the by then ex-Jesuit Ximenes.[1]

The generosity continued under Firmian's successor, Count Giuseppe di Wilczek, who authorized the purchase of an eight-foot mural quadrant from Ramsden and a trip to England to discuss details with the maker. The Brera astronomers justified their request by asserting the necessity of knowing their geographical position with the latest accuracy to serve as the node for a trigonometrical survey of Lombardy.[2] By then the mathematicians had mastered the steps in this minuet, which they and their patron-partners in the government had to dance for the officials of the treasury to obtain the tools of the trade. They began their survey in 1787 and completed it in 1794. The Ramsden quadrant arrived in 1788.[3]

Giovanni Angelo Cesaris, ex-S. J., formerly a student of Boscovich's, directed the Brera during its upgrading. An exact, energetic, and unimaginative calculator and observer, he was also a deeply religious man, who remained a priest after the suppression of his order relieved him of his vows. The love of his life was the Ramsden quadrant. "Again and again I am caught up in the greatest admiration of its [curves] and for hours on end I am detained most pleasantly contemplating the lines derived purely geometrically from circles, the most delightful proportions of the intervals, the almost infinite number of points and little lines, so elegant, so polished, so just, so beautifully marked, that nothing more can be [desired]."[4] Cesaris' enduring achievement was the construction of a meridian line in the metropolitan church of Milan.[5]

Table 8.1 Time of local noon in Italian hours

Date	Time	Date	Time
20 March	17:30	15 July	16:06
30 March	17:16	30 July	16:19
15 April	16:54	15 Aug.	16:37
30 April	16:34	30 Aug.	16:57
15 May	16:17	15 Sept.	17:19[a]
30 May	16:05	30 Sept.	17:40
15 June	15:57	15 Oct.	18:01
30 June	15:58	30 Oct.	18:22

Source: Gilii, *Memoria* (1805), 14. Cf. Paltrinieri et al., *Meridiane* (1995), 36–9.

a. 23 Sept. = 17:30. The table is nearly symmetric around the solstices and equinoxes, so that the differences in minutes between noon (Italian time) at, say, the vernal equinox and noon *t* days earlier and *t* days later are the same in magnitude but opposite in sign.

It was not his idea. Continuing Firmian's rationalization, Wilczek wanted to suppress the inconvenient system of Italian hours. The local custom of beginning the day with vespers, a half hour after sunset, and continuing the count through to 24 at the following vespers, had the amazing consequence, in the eyes of the Austrian regulators, of putting noon at a different time every day. On the days of the equinox, conversion between Italian hours and Austrian time was easy: vespers came at 6:30 P.M. Austrian and the following noon at 17:30 Italian. But as the days grew shorter (or longer) only a mathematician could calculate in advance when local noon would occur in Milan. Table 8.1 expresses the outcome.

The table's entries are the numbers of a mathematician. In practice, people consulted the vesper bell, rather than sunset, to fix the beginning of their day. But the bells themselves seldom told the same hour: before regulations of the type the Austrians put in force in Lombardy were adopted in Rome, it might take twenty minutes or more for all the churches in the city to ring in the same hour, for by so much did their ringers differ in time according to the season and the ease of specifying twilight.[6] The disparity in sounding the hours among church clocks throughout Europe was notorious. A Jesuit teaching manual rejects confusion of time as an excuse for tardiness: "The clocks almost never agree: you must overcome this defect by your diligence."[7] Italian schools were among the first institutions to give up Italian time. To take but one example, the trustees (*Reformatori*) of the University of Padua decreed that beginning with the fall semester of 1788 the clocks at the university and the campanile of the public schools would keep European time, that is, would ring twelve equal hours from local noon to midnight and another twelve from midnight to noon. The university's ornament in physics and mathematics, Galileo's editor Giuseppe Toaldo, regarded the change of time as an exemplar of the social utility of his science, which he also promoted by installing a lightning rod on his observatory and a *meridiana* in the municipal palace of government in Padua. "It is hoped [he said] that this example will encourage the abandonment everywhere of the barbarism of Italian hours."[8]

As Toaldo acknowledged, however, the barbarism had had its logic. Sundown can be determined, roughly, by anyone everywhere without an instrument. The extra half-hour, however, seemed to him irrational. Not so, said Giuseppe Piazzi, the maker of the *meridiana* at Palermo. The extra half-hour is the minimum time between sunset and the first appearance of stars and, consequently, gave a natural end to the day's work, especially in the fields. The problem was not in determining the rationale or the extent of the half-hour, but in fixing sunset. Some officials set city clocks according to apparent sunset; others, better instructed in astronomy, to true sunset; thereby creating, according to Piazzi, "a scandal." He recommended to people stuck on Italian time that they cleave to apparent sunset and expect to set their clocks forward or backward by one minute or so a day according to the season. Better yet, they should convert to European time. But why change, if the Italian scheme worked so well in the fields? Because, answered Piazzi, it will not do for towns and cities, where business is more complicated and sunset harder to specify.[9]

Piazzi and Toaldo suggested that good *meridiane* be set up in public spaces to control the new clocks telling European time.[10] That is exactly what Wilczek ordered when decreeing that from 1 December 1786 all public clocks in relatively urbanized Lombardy would tell ultramontane time.[11] Every city in the province was to have a *meridiana*. He had already seen to the needs of Milan. The preceding May he had commissioned the Brera astronomers, who then again included their former director Boscovich, to lay down in the cathedral "an unembellished *meridiana* to regulate exactly the time of local noon with great precision."[12] The commission was one of four. Besides making a *meridiana*, the astronomers were to calculate the expense of a trigonometrical survey of Lombardy, draw up a schedule of field irrigation according to European time, and arrange that the Brera's clock serve as the norm for all cities in Lombardy.[13] Cesaris and his younger colleague, Francesco Reggio, did the work; Boscovich consulted; while the fourth and youngest Brera astronomer, Barnaba Oriani, traveled to London to buy the Ramsden quadrant that would reward their public service.

The Duomo of Milan made a good site for a *meridiana*. Since it lies perfectly east–west, the meridian runs perpendicularly to its long axis. It is dark, central, and large. To these technical advantages can be added a political one: the Austrian authorities considered the structure a municipal building, available for civic pur-

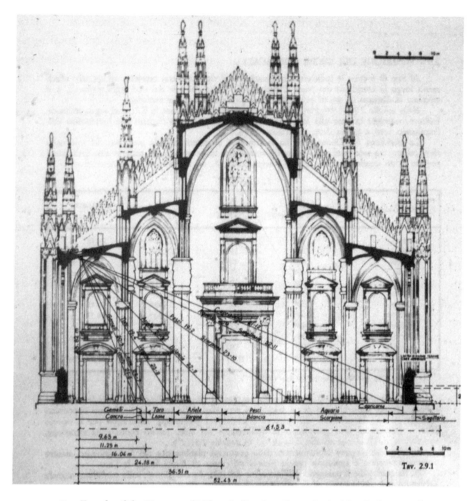

FIG. 8.1. Façade of the Duomo of Milan, indicating, from the inside, the layout of its *meridiana.* From Passano et al. (1977), 42.

poses, and not the property of the church or the *fabbricieri.* They had demanded accounts of the *fabbricieri's* expenditures, threatened to unite the Duomo's maintenance with that of the royal theater, and reserved the right to hire and fire architects.[14] They now insisted that the cathedral serve their policy of suppressing Italian time. Although the authorities insinuated that better regulation of time would make the observance of holy offices more punctual and reliable, the placement of the *meridiana* in the Duomo had no religious purpose or rationale.

Nor had the construction any scientific purpose. Nonetheless Cesaris and Reggio built it with great care, so that it might be used for astronomy in some unspecified "particular contingency." The Brera astronomers established their north by a signal from their observatory, which they could see from the hole they made in the Duomo's roof for the gnomon; that enabled them to use the orientation of the meridian, accurately established at the Brera, rather than the traditional method of dividing the arc between the sun's images at equal altitudes before and after noon. Being exact astronomers, they leveled their lines and graduated their scales for the art, as if it mattered, to within 0.10 ligne, or 1/120 of an inch. "Anyone who expects anything more accurate than this shows that he does not understand the limits that nature has set to our senses, or the perfection and defects of machines and observations." They reckoned the height at 73 feet, 8.73 lignes Paris measure, or 23.82 meters shorter than Saint Sulpice, but high enough that the midwinter image would not fit in the church. They had recourse to Sully and Lemonnier's device and ended their line in an obelisk about three meters tall.[15] Figure 8.1 represents the layout of Cesaris' *meridiana* looking west from the inside of the cathedral; the obelisk stands against the buttress on the right.

Since the nave lies strictly east-west, the line runs parallel to the façade; since the church is broad and squat, almost all the line fits across the nave and side aisles, without recourse to the transept; and, since the west end is uncluttered, the line encounters no obstacles from wall to wall. Cesaris's task had been made easy by the first architects of the cathedral, who had given it its unusual breadth and squatness after a great argument with foreign master masons. The episode provides one of the few literary indications that have survived from before the fifteenth century of the mathematical principles of medieval church design.

The Duomo was begun in 1386 by Lombard masons less experienced in such large projects than the master builders of France and Germany. Apparently the Italians started to raise the pillars before settling on the height. According to one common convention of the fourteenth century, a great church should be as high as it was broad; since the architects had chosen the breadth as 96 braccia, they would have had a nave 57.6 meters high had they built according to this convention, that is, *ad quadratum*. The plan would have made the roof of the side aisle, where Cesaris was to put his gnomon, half the height of the nave, or 48 braccia (28.8 m). That would have placed the image of the midwinter sun some 4.8 meters above where it now falls, which literally would have put the *meridiana* above the heads of the ordinary citizens who were supposed to consult it. The old masons of Milan evidently feared that their pillars might not support a church 58 meters tall. On what basis, however, could they decide how much to bring it down? Because they could not calculate from engineering principles, they, or, rather, a northern architect they consulted, suggested a geometrical one: they should build *ad triangulum*, to a height equal to the altitude of the equilateral triangle made on the 96-braccia base.[16]

Any school-aged child used to know that the height h of an equilateral triangle of side b is $(\sqrt{3}/2)b$, which, for $b = 96$, comes to 83.14 braccia. The masons of Milan

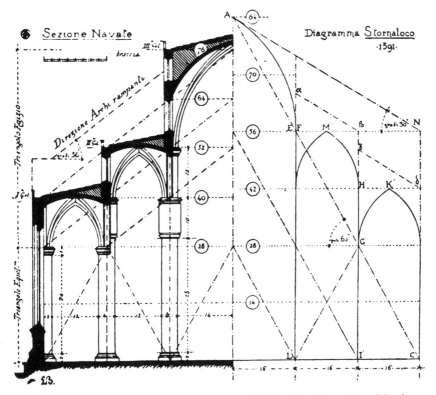

FIG. 8.2. Fiddling of the dimensions of the Duomo of Milan. The left portion of the diagram schematizes the final design. From Ackerman, *Distance points* (1991), 226.

knew that *h* was a hard number, but they did not know how to calculate it; and even if they had known how, they would have had trouble using it, since they worked ordinarily with lengths easily derivable from standard units without recourse to measurement. They called in a mathematician from Piacenza, Gabriele Stornalco, who recommended pretending that 83.14 equals 84, and dividing it into six parts of 14 braccia each. Then the heights of the outer aisles could be $3 \times 14 =$ 42, those of the inner aisles $4 \times 14 = 56$, and that of the nave the whole 84 braccia (Figure 8.2, right side). But even this elegant scheme seemed too ambitious and, after much consultation with foreign architects, who suggested strengthening the structure and building to convention, the Lombards shortened Stornalco's scheme as best they could. After 28 braccia, to which they had raised the pillars before knowing when to stop, they reduced the unit from 14 braccia to 12: the nave thus now stands at $28 + 4 \times 12 = 76$ braccia and the lower aisle roof at 40 braccia or 24 meters (Figure 8.2, left side).[17] A similar shortening occurred at San Petronio, one of whose architects visited Milan in 1390 for hints about heights. The final decision in Bologna, 44.27 meters, was 6.5 meters less than the height required *ad tri-*

angulum: which, had it been adopted and the side chapels raised in proportion, would have pushed the midwinter solar image out the door.

In 1976 astronomers at the Brera and architects of the Duomo carefully examined Cesaris' line. They found that it erred in azimuth by a maximum of 7 millimeters and in level by twice that much. Trials confirmed Cesari's claim to be able to fix noon to under a second or two. The error, if any, was swamped by the delay in transmitting the news that noon had arrived. To save citizens the trouble of attending the rendezvous of sun and rod in person, a functionary observing the line in the Duomo would notify another, waiting on the tower of the Palazzo della Ragione, who would signal to a third, stationed with a cannon on the Castello Sforzesco, who would fire his piece in announcement of noon.[18]

· THE DUOMO OF PALERMO AND ELSEWHERE ·

Down South

In the early spring of 1795, the Theatine priest, Giuseppe Piazzi, "the first great Italian astronomer after Galileo," received a commission to install a *meridiana*, required to be "very beautiful," in the cathedral of Palermo, then being refloored.[19] Piazzi had come to Palermo in 1780, as professor of mathematics at its university. Six years later the state, the Kingdom of the Two Sicilies, decided to build observatories in Naples and Palermo. Piazzi was put in charge at Palermo. There he had the advantage, as the southernmost observer in Europe, of seeing stars invisible from established centers. He took full advantage of his location and of the English instruments, including a Ramsden quadrant, that he acquired. He made over 125,000 observations in twenty years; drew up better catalogues of stars than any previous one; and, in 1801, discovered the first of the asteroids, which he named Ceres after the patron goddess of Sicily and ferdinandea after the reigning king of the Sicilies.

As he explained the big name of the little planet, Ceres ferdinandea would not have come to light at the bottom of Europe had it not been for the financial support of the rulers of Sicily. "The sciences can prosper only with the greatest difficulty without great philanthropists, and it is fitting that in turn the philanthropists receive the praise they deserve from the cultivators of science. It is not adulation, but tribute, a just and fair respect." In turn again, the King, Ferdinando III, gave Piazzi a pension for the planetoid and proposed to strike a commemorative coin. Piazzi had him put the money into a new instrument for the observatory instead.[20]

Ferdinando ran Sicily through a viceroy. In 1795 he appointed the Archbishop of Palermo, Filippo López y Rojo, to the post. A strong and rigid man, López treated Turkish corsairs, the French fleet, and his political enemies equally badly. Within four years he was neither viceroy nor archbishop. But when he was both he had desired a beautiful *meridiana* to decorate his cathedral and to serve the public.[21] Piazzi happily accepted the commission as a means of advertising astronomy and geodesy. He had his own agenda: "to promote astronomical studies, which in this country are not cultivated at all." Through astronomy Sicily would regain its position as "the blessed seat and fertile mother of practice and learning," from which, alas, it had been slipping since the time of Archimedes; through astronomy, the greatest of sciences, which ensures "the perfection of the mind, the progress of the arts and sciences, the increase and prosperity of commerce, and, in all, the splendor and richness of nations." He disclosed this reasoning, and the little trigonometrical survey with which, among other projects, he intended to implement it, to Oriani. "Excellent," Oriani replied, "both for the progress of geography and to intimate to the ignorant the immediate usefulness of mathematical sciences."[22] Piazzi's *meridiana* was certainly an advertisement, from the decorative gnomon to the winter-solstice marble.

In the construction, which followed Cesari's, Piazzi had the help of Niccolò Cacciatore, who became the senior observer as Piazzi's eyesight failed; he succeeded to the directorship in Palermo in 1817, when Piazzi left to take over the observatory in Naples. That proved a good move for Piazzi. During a premature revolt against the Bourbons in 1820 a mob stormed the vice regal palace, on the roof of which, as a symbol of aristocratic science, perched Piazzi's former nest. The mob broke the instruments, ransacked the library, and mishandled Cacciatore and his family, who lived in the observatory. It took several years, and Piazzi's help, to restore the place to its routine.[23]

Piazzi oriented the Palermo line by signal from his observatory. He made the diameter of the hole a thousandth part of the height, the canonical ratio set by Cassini and used by Cesaris. He took Cesaris' advice to level the line via a canal of water rather than by the overly sensitive modern method of an air bubble. Unfortunately, the architecture was less favorable at Palermo than at Milan: because the church is not oriented east–west, Piazzi had to run the line right to the altar (Plate 7), as at Saint Sulpice; and even then, obstacles limited the height to 46 or 47 palms (12 m), about half that at Milan. Since the latitude of the Palermo Observatory by Piazzi's precise measurement was 38°6'45.5", the line from vertex to winter solstice covered only 72.5 palms (18.7 m), shorter by 12 meters than a *meridiana* of the same height would be if situated in Milan. The work went well until suspended to await the new pavement. The intermission caused Piazzi many sleepless nights and drove him to the use of opium.[24] The pavement had not been finished in 1801, when the cathedral reopened for worship; which shows that even the most painstaking astronomers are faster than building contractors.[25]

The Palermo *meridiana* helped the chief clocks of the city tell European time. The average citizen of Sicily, however, had declined too far from Archimedes to give up intuitive time without a fight. The clocks at the Holy Office, the Church of San Antonio, and the royal palace were literally and figuratively put back to Italian time to appease the public. This retrogression resulted in the construction of two very fine lines in Sicily in the 1840s, the last handsome church *meridiane,* as part of a belated effort to get the country to run on time.[26]

Up North

The new interest of civil authorities in time conversion and the new concern for accuracy in time telling gave the established *meridiana* a new purpose. In 1776, the *fabbricieri* of San Petronio commissioned Manfredi's successor Zanotti to re-set the old *meridiana* in brass and to re-raise the hole; and, most important of all, to inscribe a scale, which still exists, recording the Italian time (in hours and minutes) of noon as marked by the true sun. Where the middle of the sun's midsummer image falls, the scale reads XV.58; where the sun enters Gemini and Leo, XVI.6; where it enters Taurus and Virgo, XVI.44; and so on. The innovation delighted the puzzled public who came to set their clocks "by observing [the sun] at noon." Thus the principal astronomer at the university fulfilled beyond expectation the ancient obligation of the mathematics professors at Bologna to regulate the city clocks.[27]

Similarly, according to an inscription on the wall of S. M. degli Angeli, Bianchini's *meridiana* served to regulate clocks in Rome until 1846, when the observatory at the Roman College started to announce noon noisily by firing a cannon.[28] Meanwhile the Vatican, used to proceeding *sub specie aeternitatis,* was winding up time. The regulator of the instruments by which it regulated its clocks in the early nineteenth century was Filippo Luigi Gilii, cameriere d'honore to Pius VII, meteorologist of the Tower of the Winds, a compulsive measurer who read his barometer twice a day and inscribed, on the floor of Saint Peter's, the lengths of the principal churches in Christendom. Observing that a *meridiana* suitable for clock setting could be made so much more easily than one for astronomical observatories that even "developing countries [*i paesi ancor meno culti*] could not lack someone able to build one," he proceeded, in 1817, to make use of the gigantic obelisk in Saint Peter's square. The coincidences of its shadow with a north–south line Gilii laid out indicated local noon, for all to see; and a few hour lines also described in the square completed the project assigned long before to Danti of making the obelisk of Sixtus V serve as the gnomon of a sundial. Previously, in 1805, Giglii had built a sundial in the Vatican, with which, and a small bell, he set the Pope's clocks so accurately that they agreed with no others in Rome.[29]

The Equation of Time

Many people who brought their clocks to *meridiane* and sundials for setting were confused by the consultation.[30] At almost every noon they would have to reset their clocks to bring mechanical time into agreement with the sun. And—a still deeper puzzle—the amount of the discrepancy changed from day to day. The fault was not only in the clock. No matter how accurate its movement, a clock made to tell equal hours throughout the year must disagree with the sun most of the time. This discrepancy was taken into account in the tables of conversion between Italian and European time. But going from sun time to European clock time without passing through Italian hours also required (and requires) a table of discrepancies for every day of the year. For the approximations acceptable in ordinary time telling, this information may be presented graphically, via a curve in the form of a figure eight called an analemma. Since this curve tells how many minutes must be added or subtracted from the clock to find the hour by the sun, it is called the "equation of time."

· AN EXCURSUS IN TIME ·

A good clock makes the intervals between successive noons equal to 24 hours. To represent its regularity in the sky astronomers invented a mean sun, which travels not in the ecliptic, as does the true sun, but along the equinoctial. In 365.25 days the mean sun, moving regularly from west to east in the equator, completes a full circuit from one vernal equinox to the next. In Figure 8.3, M designates the mean sun, ∠VE·OM its current distance from the vernal equinox; it moves toward K in

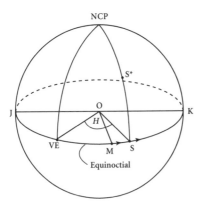

FIG. 8.3. The mean sun M, the true sun S*, and the mock sun S, the projection of the true sun onto the equinoctial. The arrows at M and S indicate the direction of the sun's annual motion; the diurnal rotation operates in the contrary sense and carries S and S* across the meridians (such as NCP·VE, NCP·J) together.

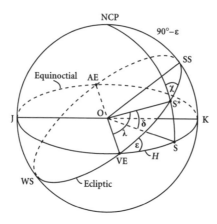

FIG. 8.4. The geometry of the equation of time; S and S* again represent the mock and the true sun, respectively.

the equinoctial JMK at the constant rate of $\omega = 360°/365.25 = 0.9856$ degrees/day. The diurnal motion sweeps M, like everything else in the sky, real or imaginary, from east to west, in the direction K·VE·J. If J·NCP·K is the meridian of observers at O, then the daily crossing of the mean sun at J represents local mean noon for them. Their clocks should read noon when this fictitious body comes to their meridian.

All points on the great circle through NCP and VE cross NCP·J along with the vernal equinox: the diurnal motion brings all the great circles through the poles successively into coincidence with every observer's meridian. Let the true sun be at S* on the great circle NCP·S*S meeting the equinoctial at S; then S and S* will cross the meridian NCP·J at the same time. The presence of the fictitious body S at J marks true local noon for the observer at O. The interval between the meridianal crossing of M and S is Q, the equation of time. By definition, $Q = \angle MOS/\Omega$, where $\Omega = 360°/1440 = 15$ minutes of arc per minute of time, the rate of the diurnal rotation.

The calculation of Q requires knowing how $\angle VE·OS$, the "right ascension" H of the sun, behaves in time. It behaves badly, for two reasons. For one, it is the projection of S* in the equinoctial; even if S* moved equably in the ecliptic, S would speed up and slow down as the direction of the annual motion changed relative to the direction of the diurnal motion. For example, around solstices (Figure 8.4) the true sun moves almost parallel to the equinoctial, so that the daily change in H is about equal to the average daily change in the sun's longitude, about one degree. At the equinoxes, in contrast, the sun's motion in the ecliptic takes place at an angle ε to the equinoctial so that H increases by only 90 percent of the daily average. The second complication in the behavior of H is that S* does not move equably in the ecliptic and, consequently, its mimic S does not move equably in the equinoctial.

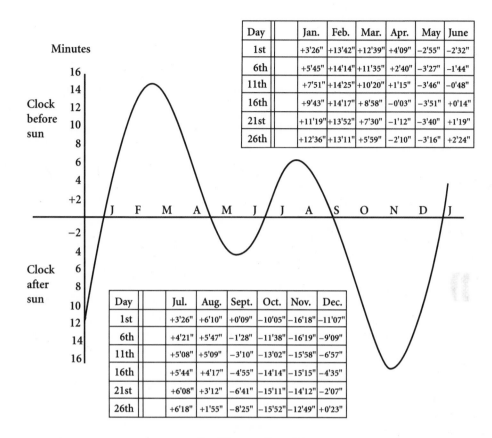

Day		Jan.	Feb.	Mar.	Apr.	May	June
1st		+3'26"	+13'42"	+12'39"	+4'09"	−2'55"	−2'32"
6th		+5'45"	+14'14"	+11'35"	+2'40"	−3'27"	−1'44"
11th		+7'51"	+14'25"	+10'20"	+1'15"	−3'46"	−0'48"
16th		+9'43"	+14'17"	+8'58"	−0'03"	−3'51"	+0'14"
21st		+11'19"	+13'52"	+7'30"	−1'12"	−3'40"	+1'19"
26th		+12'36"	+13'11"	+5'59"	−2'10"	−3'16"	+2'24"

Day		Jul.	Aug.	Sept.	Oct.	Nov.	Dec.
1st		+3'26"	+6'10"	+0'09"	−10'05"	−16'18"	−11'07"
6th		+4'21"	+5'47"	−1'28"	−11'38"	−16'19"	−9'09"
11th		+5'08"	+5'09"	−3'10"	−13'02"	−15'58"	−6'57"
16th		+5'44"	+4'17"	−4'55"	−14'14"	−15'15"	−4'35"
21st		+6'08"	+3'12"	−6'41"	−15'11"	−14'12"	−2'07"
26th		+6'18"	+1'55"	−8'25"	−15'52"	−12'49"	+0'23"

FIG. 8.5. The equation of time.

To obtain the equation of time Q thus requires a solar theory to capture the changing speed of the sun in the ecliptic and a bit of geometry to handle the inclination of the ecliptic to the equinoctial. Kepler's circular orbit with bisected eccentricity suffices for the theory. Combined with the geometrical consequences of the obliquity of the ecliptic, it indicates that the equation of time vanishes, that sun time and clock time agree, on four days during the year. This convenience occurs on 24 December, 15 April, 13 June, and 2 September (see Appendix K). A more exact calculation via Kepler's ellipse and more decimals gives 25 December, 15 April, 13 June, and 1 September. Between the four zeros the equation of time has the shape indicated in Figure 8.5. It shows that the maximum advance of the sun over the clock, some 16 minutes, occurs in November, and that the maximum delay, some 14 minutes, occurs in February.

An analemma is an obvious and useful embellishment of *meridiane* used for

clock-setting. The earliest example may be the line made for the Comte de Clermont's quarters in the Palais du Petit Luxembourg by Jean Paul Grandjean de
Fouchy, a member of the count's Société des Arts and later a perpetual secretary
of the Paris Academy of Sciences. He may have had his inspiration from his fellow-
artiste Sully, although he was barely twenty when Sully began at Saint Sulpice.
The count's analemma probably dates from the 1730s. Thereafter many private
houses were similarly outfitted and, by 1780 if not earlier, at least one public building, the church of Saint Pierre in Geneva, from which a signal emanated to alert
the punctual Swiss to the arrival of noon.[31] In that year the city of watchmakers
officially adopted mean time as legal time. Other cities followed, but slowly: London, 1792; Berlin, 1810; Paris, 1816. In 1784 Lalande urged the multiplication of
"mean-time *meridiane*" of Fouchy's type; fifty years later the improvement in
clocks and the elimination of true time from public and social life had destroyed
their usefulness.[32]

One must invert the traditional relationship between meridians and clocks,
and know in advance the current value of the equation of time, in order to arrive
at a *meridiana* when the sun does. Even that, however, will not prevent disap-
pointment. The equation indicates the difference between true sun time and local
mean time, and sufficed to convert between the sun and the clock when clocks
told local time. But — the world advances — clocks no longer tell local time. When,
after the middle of the nineteenth century, railroads made it possible to change
meridians rapidly, a passenger moving west (or east) would find his clock gradually running ahead of (or behind) local mean time. Hence the introduction of time
zones, in each of which all clocks keep the same time as that appropriate to a standard meridian.

All clocks in Italy run one hour ahead of the standard clock on the meridian
through Greenwich. But Italy extends over 10° in longitude. Hence noon in Lecce,
in the southeast, comes about 40 minutes before noon in Turin, in the northwest;
10° of longitude corresponds to 40 minutes in mean time because the mean sun
moves 360°/24 hours = 15°/hour in its apparent diurnal motion around the earth.
The difference in local time between Lecce and Turin exceeds that between Turin
and Greenwich; nonetheless, clocks in both Italian cities show the same standard
time, which is that of the fifteenth meridian east of Greenwich, which runs
through neither of them.

To obtain clock time for true noon you must know not only the equation of
time but also when, according to your clock, which tells zone time, local mean
noon occurs. For Italy that can be done for the places of interest, all of which are
west of the fifteenth meridian whose time they keep, by subtracting the longitude
east of Greenwich from 15°, multiplying by four to convert it to minutes of time,
and adding the result to 12:00. Suppose you are 9° east of Greenwich. If your watch

Table 8.2   Correction for longitude (λ)

City	λ, east of Greenwich	Correction (mins.)[a]
Bologna	11°22'	+14.5
Milano	9°11'	+23
Palermo	13°23'	+6.4
Rome	12°30'	+10

a. (60 – 4Δλ) minutes, to be added to European standard time.

keeps European standard time, it will read noon when it is 12:00 not at your location, but 6° further east. The mean sun requires an additional (6/15) hours = 4 × 6 = 24 minutes to come to your meridian. Your clock then will read 12:24. The longitude corrections for the working *meridiane* previously described are given in Table 8.2.

An easier way to secure the same information is to compare your watch, adjusted to local standard time, with a good sundial; if the watch is fast (or slow) by x minutes, you will meet the sun at the meridian x minutes after (or before) your watch shows noon. Or you can go to San Petronio, where, to ease public perplexity over time, the obliging *fabbricieri* ordered, as early as 1758, that four clocks be installed east of the meridian line, "because by equable motion alone clocks and the sun cannot be made to tell the same time." The clocks mark a conventional central European time, local mean time, old Italian time, and, what was more difficult, true solar time. The difference between the readings of the second and the fourth clock is the equation of time.

· THE ANALEMMAS OF SIENA AND BERGAMO ·

Of the many *meridiane* equipped with the figure eight of the equation of time, those of Siena and Bergamo deserve special notice. Although neither resides in a church, each would have been built in a cathedral had there been room. The Archbishop of Siena was willing to house a *meridiana*, and to rip up the beautiful pavement of his church to install it, and to pay all expenses; which, according to the eventual builder of the Siena line, Pirro-Maria Gabbrielli, would have been most appropriate, since it had many ecclesiastical uses and "ecclesiastics frequently meet in churches to give praise owed to the Most High." But owing to the thickness and position of the pillars, this brilliant reasoning could not be implemented, and no other large church could be found in Siena that did not suffer from some crippling disability.[33] As for Bergamo, neither of the great churches—the Duomo or its neighbor, S. M. Maggiore—would do, owing to their awkward orientations and cluttered interiors.[34] By default, the *meridiana* in Siena occupies the main room in the home of an academy of natural philosophers, the Accademia dei Fisiocritici, and that in Bergamo lies unhappily in the portico of the Palazzo Vecchio near the churches that could not accommodate it.

Gabbrielli chose the meeting room of the Fisiocritici against the strong and repeated advice of his former student, Ludovico Sergardi, who, as the official responsible for maintaining the fabric of Saint Peter's in Rome, knew something about building. He also knew Bianchini. They had entered the service of the popes in the same way, and at the same time, via the household of Cardinal Pietro Ottoboni (Alexander VIII). Sergardi consulted his old friend on behalf of his former teacher in 1702, when Bianchini was at work in S. M. degli Angeli. The advice, as reported to Gabbrielli: do not worry about orientation or levels; choose for height and stability; Bianchini knows everything and will forward all the necessary plans and techniques. Bianchini provided directions for overcoming difficulties encountered in installing the line in the Accademia's chamber and for finding the altitudes of the sun from the images of its limbs.[35]

The location had the merit, for Gabbrielli, of bringing together his two main contributions to erudition: the Accademia, chartered in 1691 on his initiative by the then reigning Medici cardinal; and the *meridiana*, whose installation, in the words of the Accademia, allowed Siena to boast of being the fourth city in the world enriched by so beautiful an instrument, on a level with Rome, Paris (the academicians had Cassini's uncompleted line in the Observatory in mind), and Bologna.[36] The man who made Siena the equal of Paris was a professor of medicine, an assiduous anatomist (he took three hundred cadavers apart in his youth), a vacuous natural philosopher (he enjoyed experimenting with an air pump), and a mediocre astronomer. Like Cassini and Manfredi, his interest in the stars was stimulated by a belief in astrology that further acquaintance with astronomy subverted. Like Bianchini, his concern with *meridiana* arose in connection with calendrics; he was intrigued by the suppression in the Gregorian calendar of the leap day in 1700 required in the Julian.[37] Gabbrielli thought that the rationale of this suppression should be demonstrable in Siena as well as in Rome.

Abandoning the cathedral as locus meant abandoning the church as patron. Gabbrielli found his Maecenas in a local lawyer, Girolamo Landi, and his assistants in the mathematicians of the University of Siena. It took two years to install the iron line, which ran for 24 braccia (14.4 m); Gabbrielli had trouble calculating the exact position of the hole, some 10 braccia (6.0 m) high, in finding the vertex, and in overseeing the work of placing the marbles and inscribing the signs and scales. The resultant *meridiana* resembled Bianchini's in having both a polar and

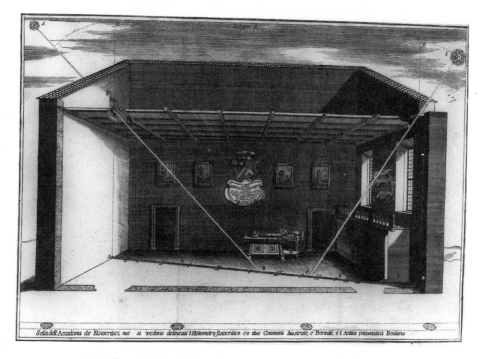

FIG. 8.6. The original meridian line at the Accademia delle Scienze, Siena.
From Ricci, Acc. fisiocritici, *Memorie*, 2 (1985), fig. 2.

a solar gnomon (Figure 8.6), in inscribing the celestial coordinates of stars at the places where they crossed the line, and in indicating the duration of daylight and twilight at the different seasons of the year.[38]

Thus equipped, the line that elevated Siena could serve various ecclesiastical, civic, and astronomical purposes. Gabbrielli specified, among the ecclesiastical, measuring the year to define movable feasts; identifying dawn, noon, and midnight, to fix the times of divine office and of feasting and fasting; and, of special importance, specifying the occurrence of twilight, when prayers can bring special indulgences. Civic purposes included conversion of sun time to Italian hours and the regulation of clocks. Astronomers too could profit from a better knowledge of the lengths of the day and the year, and an easy way to convert time; they could apply the *meridiana* also to measure the declination and apparent diameter of the sun, the motion of the polestar, the obliquity of the ecliptic, the latitude of Siena, and the right ascension of stars and planets.[39] Alas! None of this vast program came to pass. Gabbrielli died in 1705, soon after finishing his great work, and the academicians appointed to exploit it found other things to do. Bianchini observed the noon sun there in 1726 without energizing the Fisiocritici. The worry expressed by Gabbrielli's Roman correspondents about stability proved better

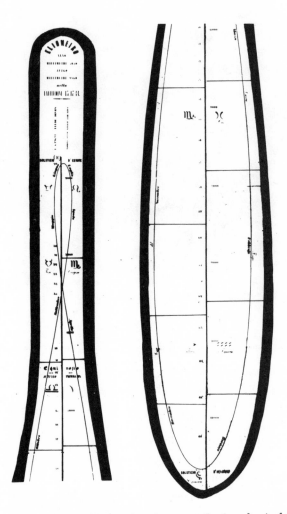

FIG. 8.7. The final (correct) analemma at the Accademia delle
Scienze, Siena. From Ricci, Acc. fisiocritici, *Memorie, 2* (1988).

founded than the Accademia's building. In forty years the instrument was unusable. A major earthquake in 1798 completed its ruin.[40]

The Fisiocritici decided to rebuild it, as a tribute to their founder. They relocated their headquarters to a suppressed convent, to which they hoped to bring the pieces of the *meridiana;* but they had no space large enough and no income great enough for the purpose. They decided to build a smaller line and to fund it in part with public monies made available for the tenth meeting of the Congresso degli Scienziati Italiani, which was to convene in Siena in 1848. A railroad engineer and Fisiocritico, Giuseppe Pianigiani, had charge of the work. He did away with star coordinates and information about twilight. Instead, he surrounded the line with an analemma, for the easy conversion of sun time to clock time (Figure 8.7).[41] When the sun touches the figure eight at the calendar date corresponding to its zodiacal position, it is mean noon. From January to mid-April, and again from mid-June to September, the sun arrives at the analemma before it meets the *meridiana* (as the sun moves from east to west, its image moves from west to east, from right to left in the figure); during the rest of the year, true noon comes before local mean noon.

Pianigiani made two errors in execution that may be more interesting and instructive than the project itself. For one, he did not take into account the political situation. The revolutions of 1848 mobilized the Fisiocritici to help repel the Austrians from Italy and canceled the Congress of Scienziati. That freed the municipality of Siena and the government of Tuscany from supporting the meeting and plunged the Fisiocritici, who had advanced the money for Pianigiani's work, deeply into debt. For a time it appeared that the *meridiana* might dissolve the Academy. But in 1862 the Scienziati convened in Siena and the public helped to pay off the debt.[42]

Pianigiani's second error had come to light the year before. On 29 January 1861, the tower clock had suddenly gone back an entire half hour. The author of this irregularity was the Piarist priest Everardo Micheli, immediate past president of the Academy and the man responsible for its *meridiana*. Micheli had been setting the tower clock by the sun in accordance with Pianigiani's analemma. On that famous January day, he noticed that the accumulated adjustments were tending in the wrong direction. He rushed to the *meridiana*. With academic horror he realized that Pianigiani had drawn the figure eight backward: he had been adding time when he should have been subtracting, and vice versa. Micheli discovered the error a dozen years after it was made. Either the *meridiana* had not been used to set clocks previously or the Sienese had adjusted to disagreeing with their neighbors about the time of day. Fortunately the Fisiocritici had time to correct their analemma, and save their reputation, before the Scienziati at last met in Siena.[43]

Bergamo's *meridiana* began life as an imitation of Milan's, as a noon mark for clocks. Because of the unevenness of the ground and the poor delimitation of the image in the open air, it was never intended to be a precision instrument. The municipality hired a former painter turned natural philosopher and physics teacher, Giovanni Albrici, to lay it out. Albrici hung a perforated plate 7.64 meters above the ground at the top of the archway leading from the palace to the cathedral. Since, originally, the east side of the portico was closed, the *meridiana* had a friendlier situation than it now has, which gives rise to an image too weak to be useful. The times of sunrise, sunset, and noon in Italian hours ran down the *meridiana*'s sides. Sometime between 1806 and 1819, two lines were added to mark the sun's position 15 minutes before and 15 minutes after true noon; they served to alert viewers that they were early or late for the show. The constant foot traffic and the removal of the portico's wall to the east degraded the instrument; by the 1850s its rod had been cracked and its inscriptions effaced. In 1857 the municipality engaged an engineer, Francesco Valsecchi, to make repairs, "bearing in mind that since the object in question serves a limited purpose, the work should be done economically."[44]

Valsecchi put down new marble, set the quarter-hour advisory lines in marble, re-incised the hours of sunrise and sunset, and added the analemma. He got it right the first time. It works in the same manner as Siena's: the sun's crossing of the appropriate branch of the analemma indicates local mean noon. Since Bergamo lies 9°40' east of Greenwich and keeps the time of the fifteenth meridian, mean noon will occur there $(15 - 9\,\frac{2}{3}) \times 4 = 22$ minutes after mean European noon. Hence to obtain the clock time for viewing true noon at Bergamo, add (or subtract) the equation of time from 12:22 (standard time) or from 13:22 (summer time). The visitor will look in vain, however, for the times of sunrise and sunset, which disappeared in 1982, when the line was last refurbished.[45]

More Light Play

Many *meridiane,* more or less ornate and accurate, appeared in churches, convents, observatories, and private houses to serve as noon marks during the eighteenth and early nineteenth centuries. The kings of France, who owned a great many clocks, had equipped all their houses with *meridiane* by 1732;[46] sixty years later the Bishop of Siebenbürgen could think of no better way to call attention to his cathedral than to install a *meridiana* in it;[47] in between, the Palazzo della Ragione at Padua, the observatories there and in Bologna, and convents in Brescia, Catania, and Naples were similarly outfitted. Two sets of these *meridiane* merit a look here. One served the up-to-date purposes of a modernizing state, the other the traditional values of a stagnant society.

By decree of 22 February 1836, the King of Belgium ordered that every principal town in his realm be furnished with an accurate *meridiana* in a cathedral, municipal building, or other public place; and that five of them — Antwerp, Bruges, Ghent, Liège, and Ostende — have, in addition, a small observatory for checking the line. The execution of this extraordinary project was entrusted to Adolphe Quetelet, then only thirty years old but known throughout Europe for his statistically based "social physics."[48] Since he was also an astronomer, he made the perfect intermediary for applying the sun to the more accurate regulation of the lives of his countrymen.

The coming of the railroads, the spread of the factory, and the need for a dependable post combined to put a premium on time; or, as Quetelet expressed it, "to make it necessary to measure with greater care a commodity whose price increases as civilization advances." In what now appears a reversal of roles, he pointed to Britain as the acme of accuracy in transportation and communication, a country in thrall to the chronometer; thus proving itself (according to the precise Piazzi) "the most cultivated, best run, richest country in Europe." To begin to raise itself to British standards, Belgium required no fewer than forty-seven meridian lines. The task would have been impossible, as Quetelet knew, if he had had to build like Cassini, Bianchini, Lemonnier, or Ximenes: "Laying out a great *meridiana,* as a work of science, would have meant much lengthy detailed work." But for clock accuracy it could be done. Quetelet began by ordering telescopes for his five small observatories from England.[49]

Quetelet began in Brussels, in the large church of Sainte Gudule, whose size and orientation admitted a line 40 meters long from vertex to winter solstice. The solar image moved at about 16 cm/min at winter solstice and 5 cm/min at summer solstice, good enough to allow specification of noon to two seconds of time; which, perhaps, was more accurate than necessary, since true local noon then varied by seven seconds across Brussels, east to west. On both sides of the *meridiana* Quetelet laid down satellite lines showing the position of the sun's image at five-minute intervals for half an hour or so on either side of noon, so that the hour could be known if clouds briefly obscured the sun around the moment of truth.[50]

Next came the chief provincial cities. Two of them, Antwerp and Termonde, had big churches appropriate for a meridian line; in the first the cathedral, in the second Notre Dame, which was decked out with the finest *meridiana* in Belgium. At Bruges and Ostende, Quetelet used the Grand'Place; the lines, picked out in white marble, received the shadow of a large sphere (at Bruges) or a small figure (at Ostende) when the sun stood in the south. At Ghent he chose the aula of the university, so perfectly situated that it seemed made for the purpose. At Malines, finding that no church would do, Quetelet had recourse to the train station. He thus completed the secularization of *meridiane,* but not the task of running Belgian railways by the sun. The deployment of the electric telegraph and the electric clock in the early 1840s rendered the increase of even highly secularized *meridiane* superfluous.[51]

The second set of intriguing *meridiane* erupted in churches near Mount Etna around 1840. The conventional church of the Benedictines of Catania, San Nicolò, cried out for a *meridiana.* The largest church in Sicily, it allowed a gnomon 23 meters high, double that of Palermo. Moreover, the monks, recruited from prominent families, could afford it. In the early 1830s the rich convent called in the obvious man, Niccolò Cacciatore, who opened the hole, leveled the floor, and hired the workmen. But the monks or the builders faltered or died and San Nicolò, unfinished in any case, remained with a taunting hole in its cupola and an unused foundation of crushed lava in its floor. Then came a happy coincidence. Just as the citizens of Catania decided to "imitate the example of the greatest cities" and complete the *meridiana,* two foreigners competent to oversee the project arrived in town.[52]

The elder of the pair, Wolfgang Sartorius, freiherr von Waltershausen, was, though a baron, the son of a professor; Goethe was his godfather and the great Gauss his tutor in mathematics. In 1834, at the age of twenty-five, Sartorius went to Sicily, where he became compulsive about volcanoes. He returned to Göttingen to work with Gauss on terrestrial magnetism.[53] There he met Christian Heinrich Friedrich Peters, a Dane from Schleswig, already a competent astronomer though not yet twenty-five, and, as Sartorius noticed, unusually healthy for a savant. He also possessed in abundance the Lutheran capacity for hard labor; like Gauss, he did not feel well unless he did four hours of calculations a day. A perfect geodesist. "His thorough mathematical and astronomical knowledge, his endurance and capacity for work, characteristics seldom united in this way, identified Peters as an excellent travel companion." He set off with Sartorious to survey Etna in 1838.[54]

Their adventures up and down the mountain recall the heroic geodetic expeditions of the eighteenth century. They battled the elements, the heat of the lowlands in summer, the snow and ice of the slopes in winter, and also death-dealing rocks, lava flows, and volcanic dust. Outside the towns, the inhabitants were as ignorant as the natives of Lapland and Peru, "surpassing all conception"; and inside the towns, monks, priests, and nuns vigorously pursued the "advancement of superstition."[55] (In a treatise on Halley's comet published in 1835, Cacciatore had taken the trouble to combat superstitions about hairy stars as if he had been writing in the seventeenth century.) Pausing in their adventures, Sartorius and Peters did the calculations necessary to finish the *meridiana* Cacciatore had started. They knocked a new gnomon in the vault so that the line would be centered directly under the cupola. It stood higher than the *meridiane* of Saint Sulpice and San Petronio. A local sculptor, Carlo Calì, made the zodiacal plaques of red and black lava. When put into service in 1842, the instrument could signal noon to within two seconds of time. That meant nothing astronomically in 1840. Still the *meridiana* of Catania added something to the art, if not the science, of church observatories: the inscription of irrelevant data, like height above sea level, constant of gravity, and altitude of Etna, as a decorative border. Of course it presented the equation of time. Time in turn dealt harshly with Calì's work. Fortunately, the local authority for the cultural heritage of Sicily, echoing the feelings that had prompted the creation of the *meridiana,* had it restored in 1966.[56]

Baron Sartorius saw nothing worthwhile in Sicily besides ruins, volcanoes, and *meridiane*. He returned to Germany in 1843 convinced, correctly, that the grip of the Bourbons and the clergy on the country could be loosened only by force.[57] Peters stayed on, as director of the geodetic survey of Sicily. The first stage of the revolution anticipated by Sartorius drove him from office in 1848; siding with the revolutionaries, he served as an engineer in their army until the fall of Palermo the following year. He fled to Paris and thence, disgusted with life, to Constantinople, to service under the sultan. In 1854 he landed in the United States, where he became director of a small college observatory in upstate New York. There the *Arbeitskraft* remarked by Sartorius earned him the discovery of forty-eight planetoids, a comet or two, the proper motion of sunspots, and the reputation of being the world's leading expert on solar physics. In 1876 he undertook to review all printed and manuscript sources and versions of Ptolemy's *Almagest* in order to purge ancient star positions from corruption in transmission. Again his capacity for work, and his knowledge of Latin, Greek, Hebrew, Arabic, Turkish, and all the languages of Western Europe, fitted him peculiarly for the task. He dropped dead one night in 1890, on the doorstep of his observatory. The Carnegie Institution of Washington later made possible the publication of his observations of sunspots and his purgation of Ptolemy.[58]

The labors of the sophisticated Europeans Sartorius and Peters in the Benedictine church in Catania confronted North with South, Cold with Warm, and modern science with other superstitions. The situation had the makings of a novel. Improbably, the novel exists: *Gli astronomi,* the story of Peters' inscription of a line in the Chiesa Matrice (the Duomo) of Acireale, a town at the base of Etna north of Catania. The principal priest of the Matrice, hoping to raise his jurisdiction into a diocese and himself into a bishop, tried to engage Peters and Sartorius to embellish his church with the sort of instrument they were finishing for the Benedictines of Catania. "[It would be] of the greatest advantage to the town, raising it in this respect to the level of great cities," for in Sicily the possession of a *meridiana* signified a "high level of civilization."[59]

When Sartorius returned to Germany, Peters went to Acireale. There he lived with the mayor and worked with the sculptor Calì on a *meridiana* similar in construction and decoration to their work in Catania, though much smaller, being only 9.1 meters high. They finished in 1844. The Acireale *meridiana* has the structural oddities of a vertex and summer-solstice plaque within a side chapel and a foundation raised slightly above the floor. Peters received 1,000 onze in gold and a box of bonbons for his work. So much is fact. The fictional Peters met with resistance and sabotage in transferring his calculations onto the floor of the church. Nonetheless he persevered, inspired, so he told himself, "by the presumption of bringing something precious and unique into a retrograde world, something noble, uncorruptible by vulgarity and coarseness, by confusion, or by approximation."[60]

The fictional Peter's mission faltered when the course of his line required the removal of a tombstone that concealed a passage to an underground grotto. Stupendo (the fictional head priest of the Chiesa Matrice) discovered evidence that nuns from nearby convents used to revel there. The would-be bishop of Acireale, fearing that the disclosure would undermine faith and morals, proposed that Peters adjust the line to miss the stone, which would be permanently sealed in place. Here astronomy, following cold geometry, conflicted with religion, consulting the public welfare. But the scandal of bending or breaking the line and of losing its prestige determined another solution. The revealing passageway was filled in, the stone replaced by a marble, and the marble cut to house the line. The maneuver made a good *meridiana* and, perhaps, an instant bishop. Acireale became a separate diocese in 1844.[61]

When Stupendo first saw the spot of light play upon the floor of his church he was ravished. "This is magnificent! Oh, the unequivocal sign of heaven!"[62] His fictional enthusiasm had many real-life counterparts. The lighting up of a special place by a flash from heaven at a preset time can make an impression even on ordinary minds. The tourists who happen to be in San Petronio when the sun plays like a searchlight across the rosy pavement tarry for longer than the five minutes they had allotted to the cathedral to watch a display of whose purpose and author they have not an inkling.

During the Renaissance, when the course of the sun regulated most people's lives, architects exploited the time-telling powers of sunlight in elaborate ways. Thus the great Borromini, in his unrealized design for a villa for his patrons, the Pamphili, proposed much light play, including lines of sight from the central windows toward midsummer and midwinter sunrise; a grand staircase whose steps indicated, by their shadows, the day of the month and the time of day; and a brilliant gesture toward the reigning Pope, Innocent X Pamphilii, the successor of Borromini's most generous patron, Urban VIII. Here is Borromini's gesture: "Over everything there would be a statue of Pope Innocent, so placed that every fifteenth of September a sunbeam would kiss the statue's foot in the hour when he was created Pope."[63]

The appeal of this anniversary magic has reached from Italy literally to the antipodes. Soon after World War I an Australian architect, Philip B. Hudson, visited S. M. degli Angeli in Rome. The *meridiana* moved him deeply. Perhaps he saw the solar image record local noon exactly where Bianchini had arranged within the somber vastness of Michelangelo's church. Perhaps he found inspiration in the combination of astronomy, architecture, and history encapsulated in the grid of ellipses marking out the diurnal course of the polestar for centuries to come, or in the application of one of the principal buildings of ancient Rome, dedicated to the pleasures of the flesh, to the regulation of the other-worldly mysteries of the Catholic Church. Both the *meridiana* and the church that housed it ensured a continuing commemoration of great events and great sacrifices. That was exactly what Hudson wanted to ensure in Australia.

The event requiring perpetual recall was that heinous slaughter of young men and old values, World War I. The sacrifices included two of Hudson's brothers and all the other Australians killed in the war to end all wars. With the help of a fellow-architect, Hudson designed a monument to his countrymen that would light up once a year, under clear skies, forever renewing the memory of their loss (Plate 8). The government astronomer did the calculations. The architects put down a Stone of Remembrance bearing the word "Love" and drilled a hole, 38 meters high, where the astronomer directed in the roof of the dark chapel containing the stone. According to his calculations, a sunbeam would illuminate "Love" at some moment between 10:52 A.M. and 11:13 A.M. on every November 11 for several thousand years in memory of the stillness that fell over the battlefields of Europe on the eleventh hour of the eleventh day of the eleventh month of 1918.[64]

The day of dedication, 11 November 1934, was overcast. Nonetheless the party proceeded to the exercise. The clouds parted as the wreath was laid. "The ray of light flashed through the sheltering dimness of the Inner Shrine and shone on the Stone of Remembrance, leaving the startled group within awe-struck."[65] This eerie and sudden appearance of a sunbeam exactly faithful to time and place distilled the essence of centuries of inspired viewing within the cathedral observatories. Their unique blend of art and history, science and sanctity, can still give visitors a glimpse of the sublime.

Appendices · Abbreviations

Works Cited · Notes

Credits · Index

Appendices

A. The Vernal Equinox at Santa Maria Novella

Figure A.1 represents the situation at Danti's armillary when the center of the sun stands precisely on the equinoctial at any time during the day. (Equinoxes that occur at night of course cannot be observed.) The sun's northern limb is 15' above the equinoctial. If the hoop's diameter is 1.5 m, $x = 0.65$ cm (x is the illuminated part at the back of the hoop). Say that the hoop has the realistic width of 1 inch = 2.54 cm. Then at the moment the sun's center comes to the equinoctial, an observer would see two dim bands of light at the top and bottom of the ring, each 0.65 cm wide. (Two because the bottom of the ring is illuminated by the sun's southern limb as the top is by the northern, and dim because only rays from the outer portions of the sun can reach the back surface of the hoop.) Between the bands lies a dark region about 2.54 − 1.30 = 1.24 cm (half an inch) wide.

As the sun rises toward the north, the light vanishes from the bottom of the hoop and occupies more and more of the top. The upper half of the back of the

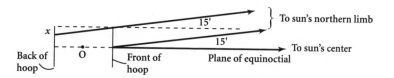

FIG. A.1 The situation at Danti's armillary with the sun at the vernal equinox at noon. The figure differs from Figure 2.25 by taking into account the finite size of the sun's disk.

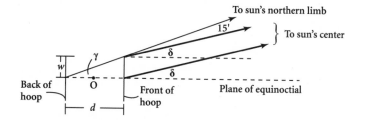

FIG. A.2 The sun with northerly declination δ at noon at Danti's armillary.

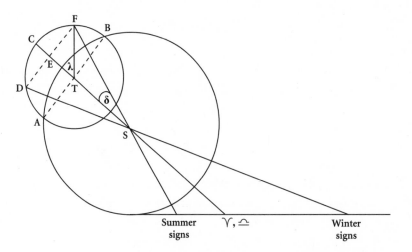

FIG. B.1 Analemma for determining the place of the zodiacal plaques at San Petronio.

hoop will be fully illuminated, though not very brightly, at a declination δ that gives $\gamma = 15' + \delta$ the magnitude required in Figure A.2. This is $\gamma =$ half-width of hoop/diameter of hoop $= a/d = 29'$, from which $\delta = 14'$. It takes the sun well over twelve hours to raise 14' of declination.

Since the time from equinox to solstice is around 90 days, during which the sun increases its declination by 23.5°, the sun would make 23.5/90 degrees/day or two-thirds of a minute of northerly declination an hour if it went at a constant rate. In fact, it moves more rapidly around the equinoxes than around the solstices. Danti offered the rule of thumb that near the equinoctial the sun moves one minute of declination in an hour of time.[1] Accordingly, illumination of half the inside of the ring would not occur until 14 hours after the sun's center arrived at the equinoctial. In order that the half-illumination be visible, the equinox would have to occur before sunrise. Precisely the same appearances would take place at the autumnal equinox, with south for north and bottom for top. Danti privileged the vernal equinox because of the timing of Easter and also, as he or his grandfather wrote in their edition of Sacrobosco, because the first point of Aries, not that of Libra, is the beginning of the ecliptic, spring being nobler than the fall as generation is nobler than corruption.[2]

B. The Analemma in the Construction of San Petronio

The vertical circles that fix the positions of the zodiacal plaques at San Petronio (Figure 3.5) are redrawn in Figure B.1, where S indicates the center of the gnomon.

The larger circle centered on S cuts the smaller centered on T so that the chord AB of the one is a diameter of the other. Let F be any point on the small circle and CTS the noon ray at an equinox. Write λ for $\angle$CTF, δ for $\angle$CSF, and r, R for the radii of the smaller and larger circles, respectively. Then, if DF is a chord parallel to AB, $EF/2 = R \sin \delta = r \sin \lambda$, or $\sin \delta = (r/R)\sin \lambda$. Now if you make $r/R = \sin \varepsilon$ and take δ to be the sun's declination, $\sin \lambda = \sin \delta/\sin \varepsilon$, from which it appears that λ is nothing other than the ecliptic longitude. For in the spherical triangle S*S·VE in Figure 8.4, $\sin \lambda/\sin 90° = \sin \delta/\sin \varepsilon$. The positions of the plaques are found by increasing λ by increments of 30°, the length of a zodiacal sign. The trick is old; Vitruvius used it in his account of sundials. And impressive. "It is incredible [says Clavius] how many and varied are the uses of analemmas in astronomy."[3]

C. The Sun's Image at San Petronio

The diameter PR of the central portion of the sun's image along the meridian line in Figure 3.17 is $h\csc^2\alpha(2\beta)$, where α is the altitude of the sun's center and $\beta = \Delta\alpha$ is half the sun's apparent size. On the equant theory, $2\beta =$ sun's diameter/sun's distance $= 2s/a(1 + e/2) = \sigma/(1 \pm e/2)$ at the absides. On the perspectival theory, the corresponding expression is $2\beta = \sigma/(1 \pm e)$. Thus the difference ΔI between the diameters of the image at winter solstice and summer solstice (which occur close to the absides) would be $(\Delta I)_k = \sigma h(6 + 4e)$ according to Kepler and $(\Delta I)_p = \sigma h(6 + 8e)$ according to Ptolemy. (The values of α at WS and SS at San Petronio are about 22° and 69°, respectively.) For a decisive confirmation of one theory over the other, Cassini had to be able to measure to within $\sigma he = 8.5$ mm, since a finding of $\Delta I = \sigma h(6 + 6e)$ would have decided nothing.

Restriction to observations around the absides could hardly be satisfactory in determining the bisection of the solar eccentricity. Cassini consequently calculated how he might decide it in principle on any day he pleased. It turns out that the relationship between the change in the apparent diameter, $\Delta\rho$, of the sun in any time Δt (which might be a day or a week) and the change in the inequality in the sun's apparent position, $\Delta(\theta - M)$, over the same period, is twice as great with a whole as with a bisected eccentricity. Here θ is the sun's position in the zodiac measured from perigee, and $M = \omega t = (360°/y)t$ is the displacement measured from the center of motion. On the first theory (Figure C.1),

$$\Delta\theta_p = \Delta M + \Delta\alpha,$$
$$\alpha = e \sin M(1 + e \cos M),$$
$$1/x = \alpha/ae \sin M = (1/a)(1 + e \cos M),$$

$$\left.\vphantom{\begin{matrix}1\\1\\1\end{matrix}}\right\} \qquad (C.1)$$

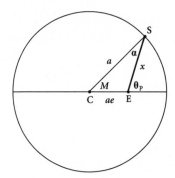

FIG. C.1. The sun's orbit according to Ptolemaic theory with undivided eccentricity.

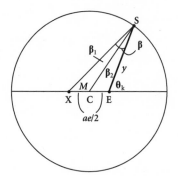
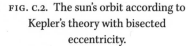

FIG. C.2. The sun's orbit according to Kepler's theory with bisected eccentricity.

the last two expressions following from the law of sines applied to ΔCES. Now $\rho = a\sigma/x$, where σ as usual indicates the sun's mean apparent diameter. Therefore, to first order in e,

$$\left. \begin{aligned} &\Delta(\theta_p - M) = e \cos M \cdot \Delta M, \\ &\Delta\rho_p = (a\sigma)\Delta(1/x) = -e\sigma \sin M \cdot \Delta M, \\ &\Delta\rho_p/\Delta(\theta_p - M) = -\sigma \tan M. \end{aligned} \right\} \tag{C.2}$$

On the second theory (Figure C.2), the law of sines applied to ΔXSC and ΔCSE gives $\beta_1 = (e/2)\sin M$, $\beta_2 = (e/2)\sin(M + \beta)$, so

$$\left. \begin{aligned} &\Delta\theta_k = \Delta M + \Delta\beta, \\ &\beta = e \sin M(1 + (e/2)\cos M), \\ &1/y = (1/a)[1 + (e/2)\cos M]. \end{aligned} \right\} \tag{C.3}$$

We now have

$$\Delta\rho_k/\Delta(\theta_k - M) = -(\sigma/2)\tan M, \tag{C.4}$$

half the earlier result.[4] Cassini's measurements confirmed that the relationship between $\Delta(\theta - M)$ and $\Delta\rho$ was that expected for the theory of the bisected eccentricity, "which much favored the opinion of Kepler."[5]

D. Ward's Cure for Ageometria

In $\triangle XQS$ (Figure 3.21), $XS = ae$, $XQ = 2a$, and therefore, $\sin \alpha = (e/2)\sin(\theta' - \alpha)$. Also, since $\theta' - \alpha = M + \alpha$, $\alpha = (\theta' - M)/2$. We have therefore $e/2 = \sin[(\theta' - M)/2]/\sin[(\theta' + M)/2]$ and

$$(1 + e/2)/(1 - e/2) = (\sin \theta'/2 \cdot \cos M/2)/(\sin M/2 \cdot \cos \theta'/2)$$

$$= (\tan \theta'/2)/\tan M/2. \tag{D.1}$$

E. The Weakness of Ward's Cure

The true anomaly is (Figure 3.20)

$$\theta = \eta - \alpha + \beta. \tag{E.1}$$

The law of sines gives, for $\triangle CPR$, $\sin \alpha = (RP/a)\sin\angle CPR$. But $RP = (a - b/a)\sin\eta$ and $\angle CPR = 90° - (\alpha - \eta)$. Hence, to second order in e, $\sin\alpha = (e^2/8)\sin\eta\cos\eta$, where η is the eccentric anomaly and $b/a = \sqrt{1 - e^2/4} \approx 1 - e^2/8$. Similarly, $\triangle CPF$ yields $\sin\beta = (ae/2PF)\sin(\eta - \alpha) = (e/2)\sin\eta/[1 - (e/2)\cos \eta]$. From the equation defining η, $\eta = M + (e/2)\sin\eta$, we have

$$\theta = M + (e/2)\sin\eta + (e/2)[1 + (e/2)\cos\eta]\sin\eta - (e^2/8)\sin\eta \cos\eta. \tag{E.2}$$

Again, from the defining equation, $\sin \eta = \sin[M + (e/2)\sin \eta)$, so that, to first order, $\sin \eta = \sin M + (e/2)\sin M \cos M$. With this substitution and replacement of η by M in the terms in e^2, the preceding equation gives, to second order,

$$\theta = M + e\sin M + (5/16)e^2\sin 2M. \tag{E.3}$$

The comparable equation for θ' can be obtained more quickly. From $\triangle XSQ$ in Figure 3.21, $\sin \alpha = (e/2)\sin(\theta' - \alpha) = (e/2)\sin(M + \alpha)$. Hence

$$\tan \alpha = (e/2)\sin M/[1 - (e/2)\cos M] \approx (e/2)\sin M[1 + (e/2)\cos M]. \tag{E.4}$$

To second order, $\tan \alpha = \alpha$. Hence, since $\theta' = M + 2\alpha$,

$$\theta' = M + e\sin M + (e^2/4)\sin 2M. \tag{E.5}$$

We end with the very small discrepancy $\theta - \theta' = (e^2/16)\sin 2M$.

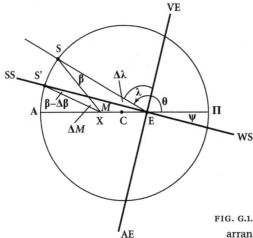

FIG. G.1. Sun's orbit with bisected eccentricity arranged for calculating the noon deficit.

F. Cassini's Cure for Ageometria

The right triangles DIG and BIG in Figure 3.23 yield

$$\tan \theta/2 = [(1 + e/2)/(1 - e/2)] \cdot \tan M/2, \tag{F.1}$$

the relationship deduced in equation D.1 between the true and mean anomalies in the approximation that the vacant focus is an equant for the motion. Since the relationship must hold for any other ecliptic point, such as C, Cassini's construction places all points of intersection of the lines BE and DA, BC and DF, and so forth, on the straight line through GH.[6]

G. The Noon Deficit

In Figure G.1, which is deceptively complicated, the earth at E sits at the center of the ecliptic, C is the center of the sun's orbit, X the equant point, ψ the direction of the line of absides with respect to the line of the solstices, θ the true anomaly, M the mean anomaly; λ, the celestial longitude, measures the angular distance of the sun from the vernal equinox, so $\lambda = \theta + \psi - 90°$. When the sun is at S', it appears precisely at the summer solstice and $\lambda = 90°$. At S, a short time Δt before it reaches S', the sun's longitude is $90° - \Delta\lambda$, where $\Delta\lambda$, the shortfall in longitude associated with the noon deficit, must be around 30'. (The sun moves along the ecliptic less than one degree a day, on average, and since the nearest noon cannot be more than 12 hours away, Δt cannot exceed 12 hours.) From ΔSXE, $\theta = \beta + M$; hence, since $\Delta\theta = \Delta\lambda$,

$$\Delta M = \omega\Delta t = \Delta\lambda - \Delta\beta. \tag{G.1}$$

From the analysis of Figure C.2 and equations C.3,

$$\beta = e\sin M + (e^2/4)\sin 2M,$$
$$\Delta\beta \approx e\Delta M\cos M. \tag{G.2}$$

Therefore, $\Delta\lambda = (1 + e\cos M)\omega\Delta t \approx (1 - e)\omega\Delta t$, since M is close to 180°.

It remains to transform $\Delta\lambda$ into the difference between the sun's declination δ at S and at the solstice S'. The basic relation,

$$\sin\delta = \sin\lambda\sin\varepsilon, \tag{G.3}$$

is deduced in Appendix B. We have for the present case,

$$\left.\begin{array}{l} \sin(\varepsilon - v) = \cos\Delta\lambda\sin\varepsilon, \quad \text{or} \\[4pt] \sin\varepsilon - v\cos\varepsilon \approx [1 - (\Delta\lambda)^2/2)]\sin\varepsilon, \\[4pt] v = [(\Delta\lambda)^2/2]\tan\varepsilon. \end{array}\right\} \tag{G.4}$$

Using the values $\omega = 2\pi/365.25 = 0.0172$ radians/day, $e = 0.033$, $\Delta t = 0.5$, and $\varepsilon = 23°30'$, we have, at last, $v = 3.5''$; which agrees precisely with Riccioli's calculation that, to find a solstice to within a day, one must be able to measure to 15 seconds of arc.[7]

H. Cassini's Representation of Atmospheric Refraction

Since $i = j + \rho$ (Figure 4.5), Snel's law requires $\sin(j + \rho) = \mu\sin j$. Since ρ is a small quantity, this equation comes to $\rho = (\mu - 1)\tan j$. But from ΔPOC, $\tan j = r\cos\alpha \cdot [r^2\sin^2\alpha + 2rt\cos^2\alpha]^{-1/2}$, if, as will be the case, $t/r \ll 1$. At $\alpha = 0$, $\rho(0) = (\mu - 1) \cdot [r/2t]^{1/2}$. Hence the stellar refractions with two parameters:

$$\rho = \rho(0)(2t/r)^{1/2}\cos\alpha[\sin^2\alpha + (2t/r)\cos^2\alpha]^{-1/2}. \tag{H.1}$$

I. Bradley's Aberration

The law of sines applied to the triangles AZS and SZΠ in Figure 7.8 (S being the sun) makes the angular diameter A'Π' of the ellipse in the plane AΠZ approximately $2(a/d)\sin\beta$, β being the star's elevation above the plane of the ecliptic as seen from the sun. The angular diameter X'Y' perpendicular to A'Π' is $2\angle SZX$, where $\tan\angle SZX \approx \angle SZX = 2a/d$. Hence for Bradley's star, for which $\beta \approx 65°$, the parallactic ellipse is close to a circle of radius a/d; for a star in the plane of the ecliptic, it would degenerate into a straight line. For the circle of aberration, the law of sines gives $\sin\gamma \approx \gamma_1 = (v/c)\sin(\beta + \gamma)$ for the recessive case and $\sin\gamma_2 \approx \gamma_2 = (v/c)\sin(\beta - \gamma)$ for the progressive one (the earth fleeing and approaching the star, respectively). Since usually $v/c \ll \beta$, we have, approximately, $\gamma_1 = \gamma_2 = (v/c)\sin\beta$, where β still represents the elevation of the star above the ecliptic plane. When the telescope moves orthogonally to the rays, the aberration γ_3 does not depend on β: from Figure 7.10e, $\tan\gamma_3 \approx \gamma_3 = v/c$.

J. Chromatic Aberration of a Thin Lens

Figure 7.13 and the law of sines give $\sin\gamma : CF = \sin(\gamma - \beta) : r$. Snel's law, $\sin\gamma = n\sin\beta$, gives in the approximation that β is very small,

$$CF_0 = nr/(n-1), \tag{J.1}$$

where the subscript indicates approximation to rays close to the axis CF. For larger values of β,

$$\sin\beta = (CF/r)(\sin\gamma\cos\beta - \cos\gamma\sin\beta). \tag{J.2}$$

Let $CF = CF_0 + x$, where $x = FF_0$ is the amount by which the focus of the extreme ray differs from that of the coxial rays; and substitute a/r for $\sin\beta$, where a is the radius of the lens' aperture. Then equation J.2 almost becomes

$$n = [n/(n-1) + x/r](n-1)(1 + na^2/2r^2). \tag{J.3}$$

From this equation, solved for x under the condition $x/r \ll 1$,

$$x = -[n^2/(n-1)]\, a^2/2r. \tag{J.4}$$

The intermediate ray, hitting the lens at J, crosses the axis between F and F_0.

From the similar triangles KFP and F_0FG, $F_0G/x = a/FK$. But FK = CF − CK ≈ $CF_0 − r = r/(n − 1)$ via equation J.1. With this substitution, the radius of the best bundle, $F_0G/2$, becomes

$$F_0G/2 = ax/2FK = n^2a^3/(2r)^2. \tag{J.5}$$

The radius $F_0G/2$ is inversely proportional to the square of the focal length $f = DF_0$ since, from equation J.1, $DF_0 = CF_0 − r = r/(n − 1)$. With this substitution, equation J.5 becomes

$$F_0G/2 = n^2a^3/[2f(n − 1)]^2. \tag{J.6}$$

K. The Equation of Time

Application of the law of sines to the spherical triangles VE·S*S and NCP·S*·SS in Figure 8.2.2 gives the "hour angle" H in terms of the sun's ecliptic longitude λ as

$$\tan H = \cos \varepsilon \tan \lambda. \tag{K.1}$$

It remains to express λ as a function of time. A good easy approximation may be obtained from Kepler's circular orbit with bisected eccentricity, redrawn in Figure K.1. The observer is at the center of the ecliptic of radius O·VE; ΠA is the line of apsides; ψ, the angle between perigee and the vernal equinox as seen from O; CS* = a, the radius of the orbit of the true sun S*; X, the equant point; OC = CX = $ae/2$, the bisected eccentricity. The law of sines applied to triangles CS*O and CS*X gives, to first order in e, $\beta = \beta_1 + \beta_2 = (e/2)(\sin \omega t + \sin \theta)$; hence, to the same approximation,

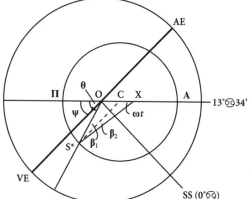

FIG. K.1 Sun's orbit with bisected eccentricity arranged for calculating the equation of time.

$$\lambda = \angle \mathrm{VE} \cdot \mathrm{OS}^* = \theta - \psi = \omega t + \beta - \psi = \omega t - \psi + e\sin \omega t. \qquad \text{(K.2)}$$

That brings the unpleasant equation

$$\tan H = \cos \varepsilon \tan[(\omega t - \psi) + e\sin \omega t], \qquad \text{(K.3)}$$

which cannot be solved algebraically.

To define Q, the equation of time, the position of the mean sun M in the equator at $t = 0$ (when the true sun is at perigee) must be specified. The convention is to place it at a distance ψ in the equator before the vernal equinox. That makes $\angle \mathrm{MO} \cdot \mathrm{VE}$ in Figure 8.3 equal to $\omega t - \psi$ and $Q = (\omega t - \psi - H)/15$. (The divisor converts angles into hours at the rate of $360°/\mathrm{day}$.) Although $H(t)$ cannot be extracted in a simple form from equation K.3, the times at which $Q = 0$, that is, at which clocks and suns agree, can be deduced without much trouble. In this case, equation K.3 becomes

$$\tan \alpha = \cos \varepsilon \tan[\alpha + e\sin(\alpha + \psi)], \qquad \text{(K.4)}$$

where $\alpha = \omega t - \psi$. Expanding the right side and assuming, what turns out to be the case, that $e\tan\alpha\sin(\alpha + \omega) \ll 1$, one may write equation K.4 as

$$\sin \alpha \cos \alpha = \cos \varepsilon(\sin \alpha \cos \alpha + e\sin \alpha \cos \psi + e\cos \alpha \sin \psi). \quad \text{(K.5)}$$

Put $x = \sin \alpha$, $\cos \varepsilon = 0.971$, $e = 0.033$, and $\psi = 76.5°$, in accordance with modern values. Then the zeros of the equation of time are the zeros of the equation

$$6.9x^4 - 4.9x^3 - 6x^2 + 4.9x - 0.86 = 0. \qquad \text{(K.6)}$$

A little trial-and-error produces the roots -0.998, 0.288, 0.430, 0.990. These indicate times 87 days before, and 25, 83, and 165 days after, the vernal equinox, which produce the dates given in the text.

Abbreviations

Archival Sources

Bianchini	F. Bianchini Papers, Biblioteca Communale, Bologna
Boscovich	R. Boscovich Papers, The Bancroft Library, University of California, Berkeley
Cassini (Bol.)	G. D. Cassini Papers, Biblioteca Centrale, Bologna
Cassini (CBC)	G. D. Cassini Papers, Biblioteca Communale dell'Archiginasio, Bologna
Cassini (COP)	Papers of the Cassinis, Observatoire de Paris, Paris
Celsius	A. Celsius Papers, Uppsala University Library, Uppsala
Gabbrielli	P. Gabbrielli Papers, Accademia dei Fisiocritici, Siena
Kircher	A. Kircher Papers, Università Gregoriana, Rome

Printed Sources

AE	*Acta eruditorum*
AHES	*Archive for history of exact sciences*
APS	American Philosophical Society
AS	Académie des sciences, Academia scientiarum, Accademia delle scienze
BJHS	*British journal for the history of science*
BU	*Biographie universelle*
CAS	Academia delle scienze, Bologna, *Commentarii*
DBI	*Dictionario biografico degli Italiani*
DSB	*Dictionary of scientific biography*
GL	*Giornale de' letterati d'Italia*
EA	*Ephemerides astronomicae*, later *Effemeridi astronomice*, Milan
HAS	Académie des sciences, Paris, *Histoire*
HSPS	*Historical studies in the physical and biological sciences*
IMSS	Istituto e museo di storia delle scienze, Florence
JHI	*Journal of the history of ideas*
Jl	*Journal*
JS	*Journal des sçavans*
MAS	Académie des sciences, Paris, *Mémoires*
PL	*Patrologia latina* (Migne)
PT	Royal Society of London, *Philosophical transactions*
SMSUB	*Studi e memorie per la storia dell'Università di Bologna*

Académie des sciences, Paris. *Recueil d'observations faites en plusieurs voyages par ordre de Sa Majesté pour perfectionner l'astronomie et la géographie*. Paris: Imp. royale, 1693.

———*Machines et inventions*. Vol. 1, *1666–1701*. Paris: G. Martin et al., 1735.

Ackerman, James S. *Distance points: Essays in theory and Renaissance art and architecture*. Cambridge, Mass.: MIT Press, 1991.

Alberigo, Giuseppe, ed. *Les conciles oecuméniques*. 2 vols. Paris: Le Cerf, 1994.

Alberti-Poja, Aldo. *La meridiana di S. Maria degli Angeli a Roma*. Rome: Palombi, 1949.

Alessandretti, Gianfranco. *La meridiana di Piazza vecchia*. Bergamo: Luchetti, 1990.

Alexander, A. F. O'D. *The planet Saturn: A history of observation, theory and discovery*. London: Faber and Faber, 1962.

Algarotti, Francesco. *Opere*. 10 vols. Cremona: L. Manini, 1778–84.

Allacci, Leo. *Apes urbanae sive de viris illustribus qui ab anno 1630 per totum 1632 Romae adfuerunt*. [1633]. Hamburg: C. Liebezeit, 1711.

Allegri, Ettore, and Alessandro Cecchi. *Palazzo vecchio e i Medici*. Florence: Studio per edizioni scelte, 1980.

Almagià, Roberto. *Carte geografiche . . . dei secoli xvi e xvii esistenti nella Biblioteca apostolica vaticana*. Vatican City: Biblioteca vaticana, 1948.

———*Le pitture murali della Galleria delle carte geografiche*. Vatican City: Biblioteca vaticana, 1952.

———*Le pitture geografice murali delle Terza Loggia e di altre sale Vaticane*. Vatican City: Biblioteca vaticana, 1955.

Amico, Rosalba d', and Renzo Grandi. *Il tramonto del Medievo a Bologna. Il cantiere di San Petronio*. Bologna: Nuova Alfa, [1987].

Angelitti, F. "Per il centenario della morte dell'astronomo Giuseppe Piazzi." Società astronomica italiana. *Memorie, 3* (1925), 369–95.

Anon. *Prognosticon concerning the frost by Monsieur Cassini, the French King's astrologer*. London: E. Witlock, 1697.

Anon. "L'Heliometro fisiocritico, overo la Meridiana senese." *GL, 6* (1711), 118–49.

Anon. "Carmina Thomae Cevae." *GL, 7* (1711), 113–36; *35* (1724), 426–7; *38* (1727), 418–20.

Anon. "Description de la méridienne de l'Hôtel de Ville de Lyon." *JS*, Jan. 1745, 154–6.

Anon. *Observations sur le bref de N. S. P. le Pape Benoit XIV au Grand-Inquisiteur d'Espagne au sujet des ouvrages du Cardinal Noris*. N.p., 20 Apr. 1749.

Anon. "Cristiano Peters." Accademia di scienze, lettere ed arti dei Zelanti, Acireale. *Atti e rendiconti, 2* (1890), xiii.

Anon. "Historical notice of the attempt made by the English government to rectify the calendar, A.D. 1584–5." *Gentleman's magazine, 36* (1851:2), 451–9.

Antonelli, Giovanni. *Di alcuni studi speciali risguardanti la metrologia, la geometrica, la geodesia e la Divina Commedia*. Florence: Calasanzia, 1871.

Arago, Dominique François. *Oeuvres complètes*. Ed. J. A. Barral. 17 vols. Paris: Gide and Baudry; Leipzig: Weigel, 1854–62.

Arckenholtz, Johann. *Mémoires concernant Christine reine de Suède*. 4 vols. Amsterdam: P. Mortier, 1751–60.

Arrighi, Gino. "Un problema geometrico meridiano in 'De gnomone meridiano bolognese' di E. Manfredi." *Archeion, 13* (1931), 320–4.

———"Attorno ad un passo del *De gnomone* di Eustachio Manfredi." *Physis, 4* (1962), 125–32.

———, ed. *Carteggi di Giovanni Attilio Arnolfini. Quaranta otto lettere inedite di Girolamo de La Lande, Ruggiero Giuseppe Boscovich e Leonardo Ximenes*. Lucca: Azienda grafica, 1965.

Auzout, Adrien. *Lettre à M. l'abbé Charles sur le "Ragguaglio di nuove osservazioni, etc., da Giuseppe Campani."* Paris: J. Cusson, 1665.

Baccini, Giuseppe. "Un opera inedita del P. Ignazio Danti." *Archivio storico per le Marche e per l'Umbria, 4* (1888), 82–112.

Backer, Aloys de, and Carlos Sommervogel. *Bibliothèque de la Compagnie de Jésus.* 11 vols. Brussels: O. Schepens; Paris: A. Picard, 1890–1932.

Badia, Iodico del. *Egnazio Danti. Cosmografo e matematico, e le sue opere in Firenze.* Florence: M. Cellini, 1881.

Baiada, Enrica, Fabrizio Bonoli, and Alessandro Braccesi. *Museo della specola.* Bologna: Bologna University Press, 1995.

Baldini, Ugo. "L'attività scientifica nel primo settecento." In G. Michele, ed. *Scienza e tecnica nella cultura e nella società del Rinascimento.* Turin: Einaudi, 1980. Pp. 467–545. (*Storia d'Italia. Annali, 3.*)

—— "Christoph Clavius and the scientific scene in Rome." In Coyne et al., *Gregorian reform* (1983), 137–69.

—— "L'astronomia del Cardinale Bellarmino." In Galluzzi, *Crisi* (1983), 293–305.

—— *Legem impone subactis. Studi su filosofia e scienza dei Jesuiti in Italia 1540–1632.* Rome: Bulzoni, 1992.

—— "La formazione scientifica di Giovanni Battista Riccioli." In Pepe, *Copernico* (1996), 123–82.

Baldini, Ugo, and Pietro Nastasi, eds. *Ruggiero Giuseppe Boscovich. Lettere ad Anton Maria Lorgna, 1765–1785.* Rome: Accademia dei XL, 1988.

Baldini, Umberto, ed. *Santa Maria Novella. La basilica, il convento, i chiostri monumentali.* Florence: Nardini, 1981.

Baleoneus, Astar. *I Baglioni.* Prato: T.p. pratensi, 1964.

Banfi, Florio. "The cosmographic loggia of the Vatican Palace." *Imago mundi, 9* (1952), 23–34.

Baroncini, Gabriele. "Alcune ipotese sulla evoluzione scientifica a Bologna nella seconda metà del '600." In *I materiali dell'Istituto delle Scienze.* Bologna: CLUEB, 1975. Pp. 78–81.

—— "La filosofia naturale nello Studio bolognese (1650–1750)." In Cremante and Tega, *Scienza e letteratura* (1984), 271–92.

—— "L'*Arithmetica realis* di Pietro Mengoli." In Mengoli, *Corrispondenza* (1986), 155–88.

Baroncini, Gabriele, and Marta Cavazza, eds. *La corrispondenza di Pietro Mengoli.* Florence: Olschki, 1986.

Barsanti, Danilo, and Leonardo Rombai. *Leonardo Ximenes: Uno scienziato nella Toscana lorense del settecento.* Florence: Medicea, 1987.

Battistella, Antonio. *Il S. Ufficio e la riforma religiosa in Bologna.* Bologna: N. Zanichelli, 1905.

Bede, the Venerable. "De computo dialogus." *PL, 90,* 647–52.

—— [Ephemeris]. *PL, 90,* 759–88.

—— "Circuli ad deprehendas cuiusque anni lunas paschales, ab anno 1 usque in annum 1595." *PL, 90,* 859–78.

—— *Opera historica.* Ed. Charles Plummer. Oxford: Oxford University Press, 1896.

—— *Opera de temporibus.* Ed. C. W. Jones. Cambridge, Mass.: Medieval Academy of America, 1943.

—— *The ecclesiastical history of the English people.* Ed. Judith McClure and Roger Collins. Oxford: Oxford University Press, 1994.

Bedini, S. A. "Christina of Sweden and the sciences." In R. G. W. Anderson et al., eds. *Making instruments count.* Aldershot: Variorum, 1993. Pp. 99–117.

Bellarmine, Robert. *Spiritual writings.* Ed. J. P. Donelly. New York: Paulist Press, 1989.

Bellinati, Claudio. *S. Gregorio Barbarigo. "Un vescovo eroico," 1625–1697.* Padua: Libreria Gregoriana, 1960.

—— "Giorgio Barbarigo, Cosimo Galileo, e il Dialogo sopra i due massimi sistemi nel seminario di Pavia." In *Galileo Galilei* (1982), 221–34.

—— "Il Dialogo con postille autografe di Galileo." In Galluzzi, *Novità* (1984), 127–8.

Bellosi, Luciano, et al. *La basilica di San Petronio in Bologna.* 2 vols. Milan: Silvana, 1983–84.

Bemporad, Azeglio. "Giuseppe Piazzi." Società astronomica italiana. *Memorie, 3* (1926), 396–413.

Ben-David, Joseph. *The scientist's role in society.* Englewood Cliffs, N.J.: Prentice-Hall, 1971.

Benedict XIV, Pope. *Bref au Grand Inquisiteur d'Espagne.* N.p. 31 July 1748.

Bertucci, Francesco-di-Paola. *Guida del Monastero dei P. P. Benedettini di Catania.* Catania: G. Musumeci-Papale, 1846.

Betti, Gian Luigi. "Il Copernicanismo nello Studio di Bologna." In Bucciantini and Torrini, *Diffusione* (1997), 67–81.

Biagioli, Mario. *Galileo courtier: The practice of science in the culture of absolutism.* Chicago: University of Chicago Press, 1993.

Bianca, Mariano, ed. *La scienza a Firenze. Itinerari scientifici a Firenze e provincia.* Florence: Alinea, 1989.

Bianchi, Giuseppe. "Discorso prelimiare." Modena, Osservatorio. *Atti, 1* (1834), xi–xxxv.

—— "Elogio dell'astronomo cavaliere G. A. Cesaris." Società italiana delle scienze. *Memorie di matematica e fisica, 22:1* (1839), cxvii–clxviii.

Bianchini, Francesco. "Nova methodus Cassiniana, observandi parallaxes & distantias planetarum a terra." *AE,* Oct. 1685, 470–8; also in Bianchini, *Observationes* (1737), 2, 7–13.

—— "Introduzione all'opera, e breve compendio della vita dell'autore." In Montanari, *Le forze d'eolo* (1694), ff. a.6–b.12.

—— *De kalendario et cyclo caesaris ac De paschali canone S. Hippoliti martyris Dissertationes duae . . . His accessit Enarratio per epistolam ad amicum De nummo et gnomone Clementino.* Rome: De Comitibus, 1703.

—— *De nummo et gnomone Clementino.* In Bianchini, *De kalendario* (1703), separately paginated.

—— *Solutio problematis paschalis.* Rome: Rev. Cam. Apost., 1703.

—— "Vita del Cardinale Enrico Noris." In Crescimbeni, *Vite* (1708), 1, 199–220.

—— "Relazione della linea meridiana orizzontale, e della elissi polare fabbricata in Roma 1702." *GL, 4* (1710), 64–86.

—— *Camera ad inscrizioni sepulcrali . . . nella via Appia.* Rome: Salvioni, 1727.

—— *Hesperi et phosphori nova phaenomena. Sive observationes circa planetam Veneris.* Rome: Salvioni, 1728.

—— *Astronomicae, ac geographicae observationes selectae.* Ed. E. Manfredi. Verona: Romanzini, 1737.

—— *La istoria universale.* [1697]. Ed. A. G. Barbazza. Rome: A. de' Rossi, 1747; 3rd ed. 5 vols. Venice: Battagia, 1825–27.

—— "Lettera . . . sopra la Meridiana." In Bianchini, *Opuscula* (1753), 2, 136–41.

—— "Descrizione della linea meridiana fatta nella chiesa di Santa Maria degli Angeli." In Bianchini, *Opuscula* (1753), 2, 123–35.

—— *Opuscula varia nunc primum in lucem edita.* Rome: Barbiellini, 1753.

—— *Carte da giuoco, in servizio dell'istoria e della cronologia.* Ed. G. B. C. Giuliari and N. Tommaseo. Bologna: G. Romagnoli, 1871.

—— *Observations Concerning the planet Venus.* Tr. Sally Beaumont and Peter Fay. Berlin: Springer, 1996.

Bianconi, Girolamo. *Guida del forrestiere per la città di Bologna.* 2 parts. Bologna: A. Nobili, 1820.

Bickerman, E. J. *Chronology of the ancient world.* 2nd ed. Ithaca: Cornell University Press, 1980.

Bienkowska, Barbara. "From negation to acceptance: The reception of the heliocentric theory in Polish schools in the 17th and 18th centuries." In Dobrzycki, *Reception* (1972), 79–116.

Bignami Odier, Jeanne, and Anna Maria Partini. "Cristina di Svezia e le scienze occulte." *Physis, 25* (1983), 251–78.

Bigourdan, Georges. "Inventaire général et sommaire des manuscrits de l'Observatoire de Paris." Observatoire de Paris. *Annales. Memoires, 21* (1895), F1–F60.

Biot, Jean-Baptiste. "Cassini, C. D." *BU, 7* (1854), 133–6.

Blaeu, Jan. *Le grand atlas.* 12 vols. Amsterdam: Blaeu, 1663.

Boinet, Amédée. *Les églises parisiennes.* 3 vols. Paris: Minuit, 1958–64.

Bonelli, Maria Luisa Righini, and Thomas B. Settle. "Egnatio Danti's great astronomical quadrant." IMSS, *Annali, 4:2* (1979), 3–13.

Bonelli, Maria Luisa Righini, and Albert van Helden. *Divini and Campani.* Florence: IMSS, 1981. (IMSS. *Annali,* 1981:1, suppl.)

Bongiovane, Silvio. *Scherzi astrologici d'alcuni avvenimenti nel mundo.* Bologna: G. Monti, 1667.

Borgato, Maria Teresa. "La prova fisica della rotazione della terra e l'esperimento di Guglielmini." In Pepe, *Copernico* (1996), 201–61.

Bortolotti, Ettore. *La storia della matematica nell'Università di Bologna.* Bologna: Zanichelli, 1947.

Boscovich, Roger. "De litteraria expeditione per pontificam ditionem." *CAS, 4* (1757), 353–96.

—— *Lettere a Giovan Stefano Conti.* Ed. Gino Arrighi. Florence: Olschki, 1980.

Boscovich, Roger, and Christopher Maire. *Voyage astronomique et géographique dans l'État de l'Eglise entrepris par l'ordre et sous les auspices du Pape Benoît XIV.* Paris: Tilliard, 1770.

Bosmans, Henri. "Le Jésuite mathématicien anversois André Tacquet (1612–1660)." *Le compas d'or, 3* (1925), 63–87. (Vereenigung der antwerpsche Bibliophilen, *Bulletijn.*)

—— "Tacquet, André." Académie royale des sciences . . . , Brussels. *Biographie nationale, 24* (1926–29), cols. 440–64.

—— "La logistique de G. F. de Gottignies." *Revue des questions scientifiques, 13* (1928), 216–44.

Boucher d'Argis, Antoine Gaspard. "Description de la ligne Méridienne traceé dans l'Eglise de St. Sulpice." In Boucher. *Variétés historiques, physiques et littéraires.* 3 vols. Paris: Nyon, 1752. Vol. 2, pp. 330–9.

Bouillet, Auguste. *Saint Sulpice.* Paris: Gaume, [1899].

Bouillier, François. *Histoire de la philosophie cartésienne.* 2 vols. Paris: Durand, 1854.

Boyer, Carl B. "Note on epicycles and the ellipse from Copernicus to La Hire." *Isis, 38* (1947), 54–6.

Braccesi, Alessandro. "Gli inizi della specola di Bologna." *Giornale di astronomia, 4* (1978), 327–50.

Braccesi, Alessandro, and Enrica Baiada. "Proseguendo sulla specola di Bologna: Degli studi del Manfredi sull'aberrazione al catalogo di stelle dello Zanotti." *Giornale di astronomia, 6* (1980), 5–29.

Braccesi, Alessandro, and G. Morgi. "È rinata la sala meridiana della Specola." *Il cielo di Bologna,* no. 5 (n.d.), 21–5.

Bradley, James. "A letter giving an account of a new-discovered motion of the fix'd stars." *PT, 35:406* (1729), 637–61.

—— "A letter . . . concerning an apparent motion observed in some of the fixed stars." *PT, 45:485* (1748), 1–43.

—— *Miscellaneous works and correspondence.* Ed. S. P. Rigaud. 2 parts. Oxford: Oxford University Press, 1832–33.

Brandmüller, Walter. "Commento." In Brandmüller and Greipl, *Copernico* (1992), 15–130.

Brandmüller, Walter, and Egon Johannes Greipl, eds. *Copernico e la chiesa. Fine della controversia (1820). Gli atti del Sant' Uffizio.* Florence: Olschki, 1992.

Brink, Sonja. "Fra Egnazio Danti, das Programm der Sala vecchia degli svizzeri im Vatikan und C. Ripas 'Iconologia.'" Kunsthistorisches Institut, Florenz. *Mitteilungen, 27* (1983), 223–54.

Brisbar, Jean de. *Le calendrier historique.* 2nd ed. Leyden: Boutesteyn, 1697.

Brizzi, Gian Paolo. *La formazione della classe dirigente nel sei-settecento.* Bologna: Il Mulino, 1976.

Broderick, James. *Robert Bellarmine, saint and scholar.* London: Burns and Oates, 1961.

Brown, James Wood. *The Dominican church of Santa Maria Novella at Florence.* Edinburgh: O. Schulze, 1902.

Brown, Lloyd Arnold. *Jean Dominique Cassini and his world map of 1696.* Ann Arbor: University of Michigan Press, 1941.

Bruin, F., and M. Bruin. "The limit of accuracy of aperture-gnomons." In Y. Maeyama and W. G. Saltzer, eds. *Prismata. Naturwissenchaftsgeschichtliche Studien.* Wiesbaden: F. Steiner, 1977. Pp. 21–42.

Bucciantini, Massimo, and Maurizio Torrini, eds. *La diffusione del Copernicanismo in Italia, 1543–1610.* Florence: Olschki, 1997.

Buchner, Edmund. *Die Sonnenuhr des Augustus.* Mainz: P. von Zabern, 1982.

Buchowiecki, Walther. *Handbuch der Kirchen Roms, 2.* Vienna: Hollinek, 1970.

Burleigh, Lord. "Report to the Lords of the Council . . . of the consultation had, and the examination of the Plain and brief discourse by John Dee . . . 25 Martii 1582." *PT, 21:* 257 (1699), 355–6.

Busacchi, Vincenzo. "Il nuovo spirito di ricerca e lo sperimentalismo nell'opera poco nota di medici e non medici nel'600 a Bologna." *SMSUB, 1* (1956), 417–33.

—— "F. M. Grimaldi (1618–1663) e la sua opera scientifica." International Congress of History of Science, *8* (1956). *Actes* (1958), 651–5.

Busignani, Alberto. *Le chiese di Firenze. Quartiere di Santa Maria Novella.* Florence: Sansoni, 1979.

Busignani, Alberto, and Raffaello Bencini. *Le chiese di Firenze. Quartiere di San Giovanni.* Florence: Le Lettere, 1993.

Cacciatore, Niccolò. "Descrizione della *meridiana* del Duomo di Palermo." *Giornale di scienze, lettere ed arti per la Sicilia, 7* (1824), 172–6.

——— *Del Real osservatorio di Palermo libri VII, VIII, e IX.* Palermo: F. Solli, 1826.

——— *Sull'origine del sistema solare discorso.* 2nd ed. Palermo: L. Dato, 1826.

——— *Riflessioni sull'imminente Ritorno della Comete di Halley.* Palermo, 1835.

Calendrelli, Giuseppe. "Dimonstrazione delle diverse formule, che possono usarsi nel calendario Giuliano e Gregoriano." *Opuscoli astronomici, 7* (1822), 1–136.

Calendrelli, Giuseppe, and Andrea Conti. [Dedication.] *Opuscoli astronomici, 1* (1803), v–xi.

Callmer, Christian. "Queen Christina's library of printed books in Rome." In Platen, *Queen Christina* (1966), 59–73.

Calmet, Agostino. "Dissertazione sovra il sistema del mondo degli antichi ebrei." In Galileo, *Opere* (1744), 1–20.

Campanacci Magnami, I. "Newton e Eustachio Manfredi." In Gino Tarozzi and Monique van Vloten, eds. *Radici, significato, retaggio dell'-opera newtoniana.* Bologna: Società italiana di fisica, 1989. Pp. 340–51.

Campani, Giuseppe. *Ragguaglio di due nuove osservazioni.* Rome: F. de Falco, 1664.

——— *Lettere . . . intorno all'ombre delle stelle Medicee nel volto di Giove, ed altri nuovi fenomeni celesti.* Rome: F. de Falco, 1665.

Camus, G., et al. "Les méridiennes de l'Eglise Saint-Sulpice à Paris." *L'astronomie, 104* (May, 1990), 195–214.

Carafa e Branciforte, Carlo Maria. *Exemplar horologiorum solarium civilium.* Mazareni: G. la Barbera, 1689.

Cardella, Lorenzo. *Memorie storiche de' cardinali della Santa Romana Chiesa.* 9 vols. Rome: Pagliaini, 1792–94.

Carini, Isidoro. "Diciotto lettere inedite di Francesco Bianchini a Giov. Ciampini." *Il muratori, 1* (1892), 145–75.

Casanovas, Juan. "The Vatican Tower of the Winds and the calendar reform." In Coyne et al., *Gregorian reform* (1983), 189–98.

Casciato, Maristella, Maria Grazia Ianniello, and Maria Vitale, eds. *Enciclopedismo in Roma barocca. Athanasius Kircher e il museo del Collegio Romano tra Wunderkammer e museo scientifico.* Venice: Marsilio, 1986.

Casini, Paolo. "Les débats du Newtonianisme en Italie, 1700–1740." *Dix-huitième siècle, 10* (1978), 85–100.

Caspar, Max. *Kepler.* Ed. C. Doris Hellmann. London: Abelard-Schuman, 1959.

Cassi Ramelli, Antonio. *Curiosità del Duomo di Milano.* Milan: Alfieri, 1965.

Cassini, Anna. *Gio. Domenico Cassini. Uno scienziato del seicento. Testi e documenti.* Perinaldo: Commune di Perinaldo, 1994.

——— "Il Cassini e la sua opera." In Perinaldo. Celebrazione cassiniana. 275° anniversario della morte dell'astronomo perinaldese Gian Domenico Cassini. Unpublished lecture, 12 Sept. 1987.

Cassini, Giovanni (Gian) Domenico. *Observationes aequinoctiales in Templo Divi Petronii habendae.* Bologna: Ducci, 1655. 1 p., folio.

——— *Novum lumen astronomicum ex novo heliometro.* [Bologna: n.p., 1656]. 4 p.

——— *Specimen observationum bononiensium.* Bologna: Ducci, 1656.

——— *Lettera astronomica . . . al sig. Abbate Ottavio Falconieri. Sopra l'ombre de pianetini Medicei in Giove.* Rome: F. de Falco, 1665. 7 p.

——— *Lettere astronomiche . . . sopra il confronto di alcune osservazioni delle comete di quest'anno MDCLXV.* Rome: F. di Falco, 1665.

——— *Lettere astronomiche . . . al Sig. Abbate Ottaviano Falconieri sopra la varietà delle macchie osservate in Giove.* Rome: F. de Falco, 1665. 12 p.

——— [*Lettere astronomiche sopra l'hipotesi solari, e la refrazione*] *Lettera prima . . . a G. Montanari. Seconda lettera astronomica ad sig. Carlo Rinaldini . . . Terza lettera astronomica al Sig. A. P.* Bologna: n.p., 1666. 33 p. Also in Roberti, *Miscellanea* (1692), 283–340.

——— "Extrait d'une lettre écrite de Rome, touchant les nouvelles découvertes faites dans Jupiter." *JS,* 22 Feb. 1666; *PT, 1:*10 (Mar. 1665/6), 171–3.

——— "Extrait d'une lettre . . . touchant la découverte . . . du mouvement de la planète Vénus à l'entour de son axe." *JS,* 12 Dec. 1667; *PT, 2:*22 (1667), 615–17.

——— *Ephemerides bononienses mediceorum syderum.* Bologna: Monolesi, 1668.

——— "Nouvelle manière géométrique & directe de trouver les apogés, les excentricitez, & les anomalies du mouvement des planètes." *JS,* 2 Sept. 1669, 548–52.

——— *Observations astronomiques faites en divers endroits.* [1672]. In AS, Paris, *Recueil* (1693), 71 p.; and in Cassini, *Ouvrages* (1736), 51–77.

——— "Three letters of Jo. Domenicus Cassinus, concerning his hypothesis of the sun's motion, and his doctrine of refractions." *PT*, 7:84 (1672), 5001–2.

——— *Découvertes de deux nouvelles planètes autour de Saturne*. Paris: Mabre-Cramoisy, 1673.

——— "A discovery of two new planets about Saturn." *PT*, 8:92 (1673), 5178–85.

——— *Les élémens de l'astronomie verifiez*. [1684]. In AS, Paris, *Recueil* (1684), 74 p.; and in Cassini, *Ouvrages* (1736), 79–178.

——— "Nouvelle découverte des deux satellites de Saturne les plus proches." *JS*, 22 Apr. 1686, 139–54.

——— *De l'origine et du progrès de l'astronomie, et de son usage dans la géographie et dans la navigation*. [1693]. In AS, Paris, *Recueil* (1693), 43 p., and in Cassini, *Ouvrages* (1736), 1–50.

——— *La meridiana del tempio di S. Petronio*. Bologna: Benacci, 1695.

——— "Planisphère céleste." [Before 1699.] In AS, *Machines* (1735), 133–42.

——— "De la Méridienne de l'Observatoire royale prolongée jusqu'aux Pyrenées." *MAS*, 1701, 171–84.

——— "Les observations de l'équinoxe du printemps de cette année 1703, comparée avec les plus anciennes." *MAS*, 1703, 41–50.

——— "Sur des mémoires touchant la correction Gregorienne, communiqués par M. Bianchini à M. Cassini." *MAS*, 1704, 142–5.

——— "Des équations des mois lunaires et des années solaires." MAS, 1704, 146–58.

——— "Du mouvement apparent des planètes à l'égard de la terre." *MAS*, 1709, 247–56.

——— "De la figure de la terre." *MAS*, 1713, 187–98.

——— *Divers ouvrages d'astronomie*. Amsterdam: P. Mortier, 1736. (AS, Paris, *Mémoires . . . contenant les ouvrages adoptez . . . avant . . . 1699*, vol. 5.)

——— "Anecdotes de la vie de J. D. Cassini, rapportées par lui-même." In J. D. Cassini, *Mémoires* (1810), 255–309.

——— "L'usage des verres sans tuyaux pratiqué dans les dernières découvertes." In Wolf, *Histoire* (1902), 164–7.

Cassini, Jacques. "Des réfractions astronomiques." *MAS*, 1714, 33–52.

——— "Nouvelles découvertes sur les mouvements des satellites de Saturne." *MAS*, 1714, 361–78.

——— *De la grandeur et de la figure de la terre*. Paris: Imp. royale, 1720.

——— "Des diverses méthodes de déterminer l'apogée et la perigée, ou l'aphelie et le perihelie des planètes." *MAS*, 1723, 143–74.

——— "De la méridienne de l'Observatoire." *MAS*, 1732, 452–70.

——— "Observations du solstice d'été." *MAS*, 1738, 404–7.

——— *Elémens d'astronomie*. Paris: Imp. royale, 1740.

——— "Observation du solstice d'hiver." *MAS*, 1742, 265–73.

——— "Mémoire sur l'obliquité de l'écliptique." *MAS*, 1778, 484–504.

Cassini, Jean Dominique. *Mémoires pour servir à l'histoire des sciences et à celle de l'Observatoire, suivis de la vie de J. D. Cassini*. Paris: Colas, 1810.

Cassini de Thury, C. F. "Sur la détermination des solstices." *MAS*, 1741, 128–48.

——— "Sur la hauteur apparente du tropique du Cancer." *MAS*, 1741, 113–22.

——— "Suite du mémoire sur les réfractions." *MAS*, 1743, 249–58.

——— "Mémoire sur les variations que l'on remarque dans les hauteurs solsticiales, tant d'été que d'hiver." *MAS*, 1748, 257–71.

——— "De la hauteur solsticiale du bord supérieur du soleil." *MAS*, 1767, 130–2.

——— "Mémoire sur l'obliquité de l'écliptique." *MAS*, 1778, 484–504.

——— "Mémoire sur l'obliquité de l'écliptique." *MAS*, 1780, 471–4.

Catteneo, E. *Il duomo nella vita civile e religiosa di Milano*. Milan: NED, 1985.

Cavazza, Marta. "La cometa del 1680–81: astrologi ed astronomi a confronto." *SMSUB*, 3 (1983), 409–66.

——— "Giandomenico Cassini e la progettazione dell'Istituto delle scienze di Bologna." In Cremante and Tega, *Scienze* (1984), 109–32.

——— "Impact du concept baconien d'histoire naturelle dans les milieux savants de Bologne." *Les études philosophiques*, no. 3 (1985), 405–14.

——— "La corrispondenza inedita tra Leibniz, Domenico Guglielmini, Gabriele Manfredi." In Cavazza, *Rapporti* (1987), 51–79.

———, ed. *Rapporti di scienziati europei con lo stu-*

dio bolognese fra '600 e '700. Bologna: Istituto per la storia dell'università, 1987. (*SMSUB, 6.*)

—— *Settecento inquieto. Alle origini dell'Istituto delle scienze di Bologna.* Bologna: Il Mulino, 1990.

Celani, Enrico. "L'epistolario di Monsignor Francesco Bianchini." *Archivio veneto, 36* (1888), 155–87.

Cellarius, Andreas Palatinus. *Harmonia microcosmica, seu Atlas universalis et novus.* Amsterdam: Jansson, 1661.

Celoria, Giovanni. *Sulle osservazioni di comete fatte da Paolo del Pozzo Toscanelli e sui lavori astronomici suoi in generale.* Milan: Hoepli, 1921.

Centre international de synthèse. *Avant, avec, après Copernic.* Paris: Blanchard, 1975.

Ceranski, Beate. *"Und sie fürchtet sich vor niemandem." Die Physikerin Laura Bassi (1711–1778).* Frankfurt: Campus, 1996.

Cesaris, Giovanni Angelo. "De linea meridiana descripta in templo maximo mediolani anno MDCCLXXXVI commentarius." *EA,* 1787, App., 123–48.

—— "De quadranti murali quem Speculae mediolanensi construxit Iesse Ramsden Londini." *EA,* 1792, App., 73–104.

Ceva, Tommaso. "Instrumentum pro sectione cuiuscumque anguli rectilinei in partes quotcunque aequales." *AE,* 1695, 290–4.

—— *Philosophia antiqua-nova.* [1704]. Milan: Bellagotta, 1718.

Chambre, Marin Cureau de la. *La lumière.* Paris: J. d'Allin, 1662.

Chapman, Allan. "The accuracy of angular measuring instruments used in astronomy between 1500 and 1850." *Jl for the history of astronomy, 14* (1983), 133–7.

—— "The design and accuracy of some observatory instruments of the seventeenth century." *Annals of science, 40* (1983), 457–71.

—— *Dividing the circle: The development of critical angular measurement in astronomy, 1500–1830.* Chichester: E. Horwood, 1990.

Charma, Antoine, and Georges Marcel. *Le père André.* 2 vols. Paris: Hachette, 1857.

Chaulnes, Michel-Ferdinand d'Albert d'Ailly, duc de. "Mémoire sur quelques moyens de perfectionner les instruments d'astronomie." *MAS,* 1765, 411–27.

Chéronnet, Louis. *Saint Sulpice.* Paris: Cerf, 1947.

Chiarini, Armando. *La meridiana della Basilica di S. Petronio in Bologna.* 3rd ed. Bologna: Grafica Emiliana, 1978.

Ciasca, Raffaela. *L'arte dei medici e speziali nella storia e nel commercio fiorentino.* Florence: Olschki, 1927.

Clavius, Christopher. *Compendium novae rationis restituendi kalendarium.* Rome: Bladi, 1577. In Clavius, *Opera* (1612), 5, 3–12, and in Schiavo, *Meridiana* (1993), after p. 32.

—— *Gnomonices libri octo.* Rome: F. Zanetti, 1581.

—— *Romani calendarii a Gregorio XIII P.M. restituti explicatio.* Rome: A. Zanetti, 1603. Clavius, *Opera* (1612), 5, [viii], 1–[623].

—— *Opera mathematica.* 5 vols. Mainz: A. Hierat, 1612.

Cognet, Louis. "Ecclesiastical life in France." In Jedin, *History, 6* (1981), 3–106, 381–428.

Cohen, I. Bernard, ed. *Puritanism and the rise of modern science: The Merton thesis.* New Brunswick, N.J.: Rutgers University Press, 1990.

Comolli, Angelo. *Bibliografia storico-critica dell'architettura civile ed arti subalterne.* 4 vols. Rome: Stamperia vaticana, 1788.

Conti, Andrea. "Formule della lettera domenicale ne' calendari giuliano, e gregoriano." *Opuscoli astronomici, 8* (1824), 61–74.

Copernicus, Nicholas. *Minor works.* Ed. Edward Rosen. Baltimore: Johns Hopkins University Press, 1985.

—— *On the revolutions.* Ed. Edward Rosen. Baltimore: Johns Hopkins University Press, 1992.

Coronelli, Vincenzo. *Epitome cosmografica.* Venice: n.p., 1693.

Cosetino, G. "L'insegnamento delle matematiche nei collegi gesuitici." *Physis, 13* (1971), 205–18.

Costabel, Pierre. "L'atomisme, face cachée de la condemnation de Galilée?" *La vie des sciences, 4* (1987), 349–65.

—— "Science positive et forme de la terre au début du xviiie siècle." In Lacombe and Costabel, *Figure* (1988), 97–113.

Costantini, Claudio. *Baliani e i Gesuiti.* Florence: Giunti Barbara, 1969.

Costanzi, Enrico. *La chiesa e le dottrine copernicane.* Siena: Biblioteca del clero, 1897.

Cotter, Charles. *A history of nautical astronomy.* London: Hollis and Carter, 1968.

Cousin, Victor. "Sur un manuscrit contenant des

lettres inédites du P. André." *JS*, 1841, 5–29, 94–122.

Coyne, George, M. Heller, and J. Zycinski, eds. *The Galileo affair: A meeting of faith and science.* Vatican City: Specola vaticana, 1985.

Coyne, George, M. A. Hoskin, and O. Pedersen, eds. *Gregorian reform of the calendar.* Vatican City: Specola vaticana, 1983.

Cremante, Renzo, and Walter Tega, eds. *Scienza e letteratura. Cultura italiana del settecento.* Bologna: Il Mulino, 1984.

Crescimbeni, Giovanni Mario, ed. *Le vite degli Arcadi illustri.* 4 vols. Rome: A. de' Rossi, 1708–28.

—— *Vita di Monsignor Gio. Maria Lancisi, camerier segreto, e medico di nostro signore Papa Clemente XI.* Rome: A. de' Rossi, 1721.

Croce, Benedetto. "Francesco Bianchini e G. B. Vico." In Croce. *Conversazioni critiche.* Series 2, vol. 2. Bari: Laterza, 1950. Pp. 101–9.

Crombie, Alistair C. *Augustine to Galileo.* 2nd ed., 2 vols. in 1. London: Heinemann, 1970.

Crosby, Alfred J. *The measure of reality: Quantification and Western society, 1250–1600.* Cambridge: Cambridge University Press, 1997.

Cruikshank, Dale P. *Venus.* Tucson: University of Arizona Press, 1983.

D'Addio, Marco. *Considerazioni sui processi di Galileo.* Rome: Herder, 1985.

Dallari, Umberto. *I rotuli dei lettori . . . dello Studio Bolognese.* 3 vols in 4. Bologna: Merlani, 1888–1909.

Daniel-Rops, Henri. *The Catholic Reformation.* 2 vols. Garden City, N.Y.: Doubleday, 1964.

Danti, Egnatio. *Trattato dell'uso et della fabbrica dell'astrolabio.* Florence: Giunti, 1569.

—— ed. *La sfera di Messer Giovanni Sacrobosco.* Florence: Giunti, 1571.

—— *Usus et tractatio gnomonis magni [Bononiensis].* Bologna: Rossi, [1576].

—— *Le scienze matematiche ridotte in tavole.* Bologna: Compagnia della stampa, 1577.

—— *Primo volume dell'uso e fabbrica dell'astrolabio.* Florence: Giunti, 1578.

—— *Trattato del radio latino.* Rome: V. Ascolti, 1583.

Daumas, Maurice. *Les instruments scientifiques aux xviie et xviiie siècles.* Paris: PUF, 1953.

David, Jean-Claude. "Grimm, Lalande et le quart de cercle de l'Ecole royale militaire." *Dix-huitième siècle, 14* (1982), 277–87.

Davis, M. Daly. "Beyond the "Primo libro" of Vincenzo Danti's 'Trattato delle perfette proporzioni.' " Kunsthistorisches Institut, Florence. *Mitteilungen, 26* (1982), 64–84.

Débaret, Suzanne. "La qualité des observations astronomiques de Picard." In Picolet, *Picard* (1987), 157–73.

Débaret, Suzanne, and Curtis Wilson. "The Galilean satellites of Jupiter from Galileo to Cassini, Römer and Bradley." In Taton and Wilson, *Planetary theory* (1989), 144–57.

Delambre, J. B. J. *Histoire de l'astronomie moderne.* 2 vols. Paris: Huzard-Coucier, 1821.

—— *Histoire de l'astronomie au xviiie siècle.* Ed. M. Mathieu. Paris: Bachelier, 1827.

—— *Grandeur et figure de la terre.* Ed. G. Bigourdan. Paris: Gauthiers-Villars, 1912.

Delisle, Joseph-Nicolas. "Sur l'observation des solstices." *MAS*, 1714, 239–46.

—— "La construction facile et exacte du gnomon pour régler une pendule au soleil par le moyen de son passage au méridien." *MAS*, 1719, 54–8.

Delitalia, Francesco. "La meridiana di San Michele in Bosco in Bologna." AS, Bologna. *Atti.* Classe di scienze fisiche. *Rendiconti, 4:1* (1977), 29–31.

Del Re, Niccolò. "Benedetto XIV e la curia romana." In Mario Cecchelli, ed. *Benedetto XIV . . . Convegno internazionale.* 3 vols. Ferrara: Centro studi "Girolamo Baruffaldi," 1981. Vol. 1, pp. 641–62.

Delumeau, Jean, and Monique Cottret. *Le catholicisme entre Luther et Voltaire.* 6th ed. Paris: PUF, 1996.

Dempsey, Charles Gates. "Some observations on the education of artists in Florence and Bologna during the later sixteenth century." *Art bulletin, 62* (1980), 552–69.

Deschales, Claude François Milliet. *Cursus seu mundus mathematicus.* 4 vols. Lyon: Anisson, 1690. (First ed., 3 vols., Lyon, 1674.)

Desiderio, Alderando. *Tavole de' cycli solari.* Rome: Chracas, 1703.

Dibner, Bern. *Moving the obelisks.* Cambridge, Mass.: MIT Press, 1970.

Dickens, Charles. *Pictures from Italy.* In Dickens, *The works.* Ed. A. Lang. 34 vols. New York, 1900. Vol. 28, pp. 309–17.

Divini, Eustachio. *Brevis annotatio in systema saturnium Christiani Eugenii.* Rome: G. Dragondelli, 1660.

——— *Pro sua annotatione in systema saturnium Christiani Hugenii adversus eiusdem assertionem.* Rome: G. Dragondelli, 1661.

Divonne, Paul de, ed. *Autour de la méridienne de la cathédrale de Nevers. Exposition.* Nevers: Regards sur la cathédrale de Nevers, 1990. 15 p.

Dobrzycki, Jerzy, ed. *The reception of Copernicus' heliocentric theory.* Dordrecht: Reidel, 1972.

——— "Astronomical aspects of the calendar reform." In Coyne et al., *Gregorian reform* (1983), 117–21.

Dohrn-van Rossum, Gerhard. *Die Geschichte der Stunde. Uhren und modernen Zeitordnung.* Munich: C. Hanser, 1992.

——— *History of the hour. Clocks and modern temporal orders.* Chicago: University Chicago Press, 1996.

Dorna, Alessandro. *Sulla rifrazione.* Turin: Loescher, 1882.

Drake, Stillman. *Discoveries and opinions of Galileo.* New York: Doubleday, 1957.

Dreyer, J. L. H. *A history of astronomy from Thales to Kepler.* [1906]. Ed. W. H. Stahl. New York: Dover, 1953.

——— *Tycho Brahe: A picture of scientific life and work in the 16th century.* [1890]. New York: Dover, 1963.

Duhamel, Jean Baptiste. *Astronomia physica.* Paris: P. Lamy, 1660.

——— *Philosophia vetus et nova.* 2 vols., rev. ed. Nuremberg: Zieger, 1682.

——— *Regiae scientiarum academiae historia.* Paris: Michallet, 1698.

Duhem, Pierre. *Le système du monde.* 10 vols. Reprint, Paris: Hermann, 1959.

——— *To save the phenomena: An essay on the idea of physical theory from Plato to Galileo.* [1908]. Chicago: University of Chicago Press, 1969.

Eschinardi, Franesco. *Microcosmi physicomathematici.* Voi. 1. Perugia: Typ. episcopalis, 1668.

——— *Ragguaglio . . . sopra alcuni pensieri sperimentabili proposti nell'Accademia fisicomatematica di Roma.* Rome: Tinassi, 1680.

——— *Cursus physicomathematicus.* Rome: G. G. Komarek Bohemi, 1689.

Euler, Leonhard. "De motu corporum coelestium a viribus quibuscumque perturbato." AS, Petersburg. *Novi commentarii, 4* (1752/3), 161–96.

——— "De la variation de la latitude des étoiles fixes et de l'obliquité de l'écliptique." AS, Berlin. *Mémoires, 10* (1754), 296–336.

——— "Recherches sur les inégalités de Jupiter et de Saturne." AS, Paris. *Recueil des pièces qui ont remporté des prix, 7:2* (1759), 83 p.

Fabri, Augustino. *Efemeridi. Premonizioni astronomiche et astrologico-mediche.* Bologna: Barbieri, 1675.

Fabri, Honoré. *Dialogi physici in quibus de motu terrae disputatur, marini aestus nova causa proponitur.* Lyon: Formy, 1665.

——— *Synopsis optica.* Lyon: Boissat and Rameus, 1667.

Fabroni, Angelo. "Elogio di Giuseppe Toaldo." In Toaldo, *Completa raccolta* (1802), *1,* vii–xxxvi.

Falco, Giorgio. "L. A. Muratori e il preilluminismo." In Fubini, *Cultura* (1964), 23–42.

Fanti, Mario. *La fabbrica di S. Petronio in Bologna.* Rome: Herder, 1980.

Fantoli, Annabile. *Galileo. Per il copernicanismo e per la chiesa.* Vatican City: Specola vaticana, 1993.

——— *Galileo: For Copernicanism and the Church.* Tr. G. V. Coyne. Vatican City: Specola vaticana, 1994.

Fantuzzi, Giovanni. *Notizie degli scrittori bolognesi.* 9 vols. Bologna: S. Tommaso d'Aquino, 1781–94.

Favaro, Antonio. "Cesare Marsili e la successione di Gio. Antonio Magini nella lettura in matematica dello Studio di Bologna." Deputazione di storia patria per le provincie di Romagna. *Atti e memorie, 22* (1904), 411–80.

——— "Oppositori di Galileo. VI. Maffeo Barberini." Istituto veneto di scienze, lettere ed arti. *Atti, 80:2* (1920/1), 1–46.

Feldhay, Rifka. *Galileo and the Church. Political inquisition or critical dialogue?* New York: Cambridge University Press, 1995.

Feldman, T. S. "Applied mathematics and the quantification of experimental physics: The example of barometric hypsometry." *HSPS, 15:2* (1985), 127–97.

Ferrato, Pietro. *Lettere di celebri scrittori dei secoli xvii° e xvii°.* Padua: L. Penada, 1873.

Ferrone, Vincenzo. *Scienza, natura, religione. Mondo newtoniano e cultura italiana nel primo settecento.* Naples: Jovene, 1982.

——— *The intellectual roots of the Italian enlightenment.* Tr. Sue Brotherton. Atlantic Highlands, N.J.: Humanities Press, 1995.

Feuillebois, Geneviève. "Les manuscrits de la bibliothèque de l'Observatoire de Paris." *Jl for the history of astronomy*, 6 (1975), 72–4.

Feyerabend, Paul. "Galileo and the tyranny of truth." In Coyne et al., *The Galileo affair* (1985), 155–64.

Fichera, Francesco. *Una città settecentesca*. Rome: Società editrice d'arte illustrata, 1925.

Finocchiaro, Maurice. *The Galileo affair. A documentary history*. Berkeley: University of California Press, 1989.

Fiorani, L. "Astrologi, superstiziosi e devoti nella società romana del seicento." *Ricerche per la storia religiosa di Roma*, 2 (1978), 96–162.

Fisch, Max H. "The Academy of the Investigators." In Edgar Ashworth Underwood, ed. *Science, medicine and history*. 2 vols. Oxford: Oxford University Press, 1953. Vol. 1, pp. 521–63.

Flamsteed, John. "Ad cl. Cassinum epistola." *PT*, 8:96 (21 July 1673), 6094–6[1]00.

———. *The Gresham lectures*. Ed. E. G. Forbes. London: Mansell, 1975.

Fondelli, Mario. "Le rappresentazione cartografiche . . . di Firenze." In Commune di Firenze. *Atlante di Firenze*. Venice: Marsilio, 1993. Pp. 35–42.

Fontenelle, Bernard le Bovier de. *Oeuvres diverses*. 3 vols. The Hague: Gosse and Nealme, 1728–29.

——— Eloges des académiciens. 2 vols. The Hague: van der Kloot, 1740.

——— "Eloge de M. Bianchini." *HAS*, 1729, 102–15.

Forbes, Eric G. "The rational basis of Kepler's laws." British Astronomical Association. *Jl*, 82 (1971), 33–7.

——— *Greenwich Observatory*, Vol. 1. *Origins and early history (1675–1835)*. London: Taylor and Francis, 1975.

Foresta Martin, Franco, Nicoletta Lanciano, and Pasquale Lanciano. *Della terra alle galassie. Un viaggio astronomico attraverso la città*. Rome: ENEA, n.d. (ENEA-MUSIS, *Quaderni*, 1.)

Francini, Marta Pieroni. "Da Clemente XI a Benedetto XIV: il caso Davia (1734–1750)." *Rivista di storia della chiesa in Italia*, 37 (1983), 438–71.

Frangenberg, Thomas. "Egnatio Danti's optics. Cinquecento Aristotelianism and the medieval tradition." *Nuncius*, 3:1 (1988), 3–38.

Fubino, Mario. *La cultura illuministica in Italia*. 2nd ed. Rome: RAI, 1964.

Gabba, Luigi. "Sulla verificazione della meridiana descritta del Duomo di Milano." Istituto lombardo di scienze e lettere. *Rendiconti, 54* (1921), 447–58.

——— "Ricordo dell'astronomo Giovanni Angelo Cesaris." Società astronomica italiana. *Memorie, 28* (1957), 149–54.

Gabbrielli, Pirro Maria. *L'heliometro fisiocritico overo la meridiana senese*. Siena: Bonetti, 1705.

Gadroys, C. *Le système du monde selon les trois hypothèses*. Paris: G. Desprez, 1675.

Galilei, Galileo. *Dialogo . . . Accresciuto di una lettera dello stesso . . . e di vari trattati di più autori*. Florence [Naples]: no printer, 1710.

——— *Opere*. 4 vols. Ed. G. Toaldo. Padua: Stamperia del seminario, 1744.

——— *Le opere*. Ed. A. Favaro. 20 vols. Florence: Barbèra, 1890–1909.

Galileo Galilei e Padova. Libertà di indagine e principio di autorità. Padua: Studia patavina, 1982. (*Rivista di scienze religiose, 29:3*.)

Gallo, Rodolfo. "Le mappe geografiche del Palazzo Ducale di Venezia." Deputazione di storia patria per le Venezie. *Archivio veneto, 32–33* (1943), 47–113.

Galluzzi, Paolo. "Galileo contro Copernico." IMSS. *Annali, 2:2* (1977), 87–148.

———, ed. *Novità celeste e crisi del sapere*. Florence: IMSS, 1983.

Gambaro, Ivana. *Astronomia e tecniche di ricerca nelle lettere di G. B. Riccioli ad A. Kircher*. Genoa: Università, Centro di studio sulla storia della tecnica, 1989. (*Quaderni, 15*.)

Garrido, Pablo Maria. *Un censor español de Molinos y de Petrucci: Luiz Pérez de Castro, O. Carm. (1636–1689)*. Rome: Institutum carmelitanum, 1988.

Garzend, Léon. *L'inquisition et l'hérésie. Distinction de l'hérésie théologique et de l'hérésie inquisitoriale*. Paris: Desclée, 1912.

Gassendi, Pierre. *Institutio astronomica*. [1647]. 2nd edn. London: C. Bee, 1653.

——— *The Mirrour of true nobility and gentility. Being the life of Nicolas Claudius Fabricius Lord of Peiresk*. Tr. W. Rand. London: H. Moseley, 1657.

——— "Proportio gnomonis ad solstitialem umbram observata Massiliae, anno M.DC.XXXVI." In Gassendi, *Opera* (1658), 4, 523–36.

——— [Vita Regiomontani,]. In Hughes, *Regiomontanus* (1967), 11–18.

———*Opera omnia.* 6 vols. Lyon: Anisson and De-venet, 1658.

Gaythorpe, S. B. "On Horrock's treatment of the evection and the equation of center." Royal Astronomical Society. *Monthly notices,* 85 (1925), 858–65.

[Gemmelaro, Carlo]. "Sul genio di Catania. Lettera di un Catanese al sig. N. N. in occasione del proseguimento de' lavori nel Molo, in luglio 1841." Accademia gioenia. Gabinetto letterario. *Giornale,* 6:4 (Apr.–May 1841), 31–54; 6:5 (July–Aug. 1841), 3–29.

———"Sulla meridiana costruita di recente nella chiesa de' Benedettini in Catania." Accademia gioenia. Gabinetto letterario. *Giornale,* 6:6 (Aug.–Sept. 1841), 23–32. (Dated 15 Feb. 1842.)

Gilii, Filippo Luigi. *Memoria sul regolamento dell'orologio italiano colla meridiana.* Rome: Caetani. 1805.

———*Risultati delle osservazioni meterologiche fatte l'anno 1807 nella specola pontificia vaticana.* Rome: Salomoni, 1808.

———*Architettura della basilica di S. Pietro in Vaticano.* Rome: Romanis, 1812.

Gingerich, Owen. "Remarks on Copernicus' observations." In Westman, *Copernican achievement* (1975), 99–107.

———"The civil reception of the Gregorian calendar." In Coyne et al., *Gregorian reform* (1983), 265–79.

Ginzel, F. K. *Handbuch der mathematischen und technischen Chronologie.* 3 vols. Leipzig: J. C. Hinrich, 1906–14.

Giordani, Gaetano. "Relazione delle feste celebrate per Cristina di Svezia in Bologna." *Almanacco statistico bolognese,* 11 (1840), 62–98.

Giuffrida d'Angelo, Francesco. *Sopra un nuovo metodo di misurare il filo della meridiana ed osservazioni sulla medesima e suoi lavori. Lettera prima . . . al signor Wolfgang Sartorius.* Catania: P. Giuntini, 1838. 6 p. (*Il caronda,* no. 3; dated 7 Oct. 1836.)

———*Brevi cenni sulla seconda meridiana nel gran Tempio dei padri cassinesi in Catania.* Messina: M. Minasi, 1841. 4 p. (*Veridico,* no. 2.)

Giuliari, G. B. Carlo. "La capitolare biblioteca de Verona. IV. Fabbrica della nuova biblioteca." *Archivio veneto,* 11 (1876), 51–74.

Giustiniani, Michele. *Gli scrittori liguri.* Rome: Tinasi, 1667.

Gomez Lopez, Susana. *Le passioni degli atomi.*

Montanari e Rosetti: una polemica tra galilleiani. Florence: Olschki, 1997.

Gotteland, Andrée. "Mesure de la méridienne de Le Monnier à St Sulpice." Société astronomique de France. *Observations et travaux,* no. 12 (1987), 23–36.

Gotteland, Andrée, Bernard Tailliez, Georges Camus, and Paul de Divonne. *La méridienne de l'Hôpital de Tonnerre.* Dannemoine: A. Matton, 1994.

Gottignies, G. F. de. *Logistica universalis.* Naples: N. de Bonis, 1687.

Granada, Miguel A. "Giovanni Maria Tolosani e la prima reazione romana di fronte al 'De revolutionibus.'" In Bucciantini and Torrini, *Diffusione* (1997), 11–35.

Grant, Edward. "In defense of the earth's centrality and immobility: Scholastic reaction to Copernicanism in the seventeenth century." APS. *Transactions,* 74:4 (1984), 69 p.

———*Planets, stars, and orbs: The medieval cosmos, 1200–1687.* Cambridge: Cambridge University Press, 1994.

Grassi, Mariano. *Sulla vita del B.ne Calì Sardi.* Acireale: n.p., s.d.

Gravagno, Gaetano. *Storia di Aci.* Acireale: Sicilgrafica, 1992.

Greaves, John. "Reflexions made on the foregoing paper [by Lord Burleigh]." *PT,* 21:257 (1699), 356–8.

Gregory XIII, Pope. *Inter gravissimas . . .* Rome, 24 Feb. 1581. In Clavius, *Opera* (1612), 5, 13–15, and Schiavo, *Meridiana* (1993), after p. 16.

Gregory, David. "De orbita cassiniana." *PT,* 24:293 (1704), 1704–6.

———*Elements of astronomy.* Tr. J. T. Desaguliers. 2 vols. London: J. Nicholson, 1715.

Gregory, James. "An account of a controversy betwixt Stephano de Angelis and Joh. Baptista Riccioli." *PT,* 3:26 (1668), 693–8.

Grillot, Solange. "L'emploi des objectifs italiens à l'Observatoire de Paris à la fin du 17ème siècle." *Nuncius,* 2:2 (1987), 145–55.

Grimaldi, Francesco Maria. *Physico-mathesis de lumine, coloribus et iride.* Bologna: V. Renati, 1665.

Grohmann, Alberto. *Perugia.* Bari: Laterza, 1981.

Grossi, Isnardo Pio. "Breve e util modo del viver christiano' di fra Benedetto Onesti O. P. Un trattatello di vita spirituale scritto in S. Maria

Novella nel 1568." *Memorie dominicani, 11*
(1980), 505–73.

Gualdo Priorato, Galeazzo. *Historia della Sacra
Real Maestà di Christina Alessandra Regina di
Svetia.* Modena: B. Soliani, 1656.

——*Scena d'huomini illustri d'Italia.* Venice: Giu-
liani, 1659.

Guarducci, Federigo. "La meridiana del tempio di
San Petronio di Bologna riveduta nel 1904." AS,
Bologna. *Memorie, 2* (1905), 285–321. Also is-
sued separately.

Guasti, Cesare. *La cupola di Santa Maria del Fiore.
Illustrata con i documenti dell'Archivio.* Florence: Barbèra and Bianchi, 1857.

Guglielmini, Domenico. "Memoria della oper-
azione fatta . . . nell'ultima ristorazione della
meridiana." In Cassini, *Meridiana* (1695),
38–75.

——"Volantis flammae a perillusti, & excellentis-
simo domino Geminiano Montanaro
epitropeia." In Guglielmini, *Opere* (1719), *1*,
1–25.

——*Opera omnia.* 2 vols. Geneva: Cramer, Pera-
chon, 1719.

Guglielmini, Giambattista. *Carteggio "de diurno
terrae motu."* Ed. M. T. Borgato and A. Fiocca.
Florence: Olschki, 1994.

Hall, Isaac H. "In memoriam." In *Dr Henry Freder-
ick Peters.* Hamilton, N.Y.: Republican Press,
1890. Pp. 4–23.

Hall, Marcia B. *Renovation and Counter-Reforma-
tion. Vasari and Duke Cosimo in Santa Maria
Novella and Santa Croce, 1565–1577.* Oxford:
Oxford University Press, 1979.

Halley, Edmund. "Methodus directa & geometrica,
cuius ope investigantur aphelia, eccentricitates
. . ." *PT, 11:128* (1676), 683–6.

——"Some remarks on the allowance to be made
in astronomical observations for the refraction
of the air." *PT, 31:368* (1721), 169–72.

Hamel, Charles. *Histoire de l'église Saint Sulpice.*
Paris: Le Cerf, 1900.

Haynes, Renée. *Philosopher king: The humanist
Pope Benedict XIV.* London: Weidenfeld and
Nicolson, 1970.

Heilbron, J. L. "Introductory essay." In Wayne Shumaker, ed. *John Dee on astronomy.* Berkeley:
University of California Press, 1978. Pp. 1–99.

——*Electricity in the 17th and 18th centuries.*
Berkeley: University of California Press, 1979.

——"Fin-de-siècle physics." In C. G. Bernhard,
Elisabeth Crawford, and Per Sörbom, eds. *Science, technology and society in the time of Alfred Nobel.* Oxford: Pergamon, 1982. Pp. 51–73.

——*Physics at the Royal Society of London during
Newton's presidency.* Los Angeles: Clark Library, 1983.

——"Science and the Church." *Science in context,
3* (1989), 9–28.

——"Fisica e astronomia nel settecento." In W.
R. Shea, ed. *Le scienze fisiche e astronomiche.*
Turin: Einaudi, 1992. Pp. 318–443. (Paolo Gal-
luzzi, ed. *Storia delle scienze, 2.*)

——"Meridiane and meridians in early modern
science." In Piers Bursill-Hall, ed. *R. J.
Boscovich. Vita e attività scientifica.* Rome: Istituto della Enciclopedia italiana, 1993. Pp.
385–406.

——*Weighing imponderables and other quantitative science around 1800.* Berkeley: University of
California Press, 1993.

——*Geometry civilized.* Oxford: Oxford University Press, 1998.

Hilfstein, Erna. *Starowolski's biographies of Copernicus.* Warsaw: Ossolineum, 1980. (*Studia
copernicana, 21.*)

Hilgers, Josef. *Der Index der verbotenen Bücher.*
Freiburg/Br.: Herder, 1904.

Hiltebrandt, Philipp. "Eine Relation des Wiener
Nuntius über seine Verhandlungen mit Leibniz
(1700)." *Quellen und Forschungen aus italieni-
schen Archiven und Bibliotheken, 10* (1907),
338–46.

Hine, William L. "Mersenne and Copernicanism."
Isis, 64 (1973), 18–32.

Hoskin, M. A. "The reception of the calendar by
other churches." In Coyne et al., *Gregorian reform* (1983), 255–64.

Houzeau, J. C., and A. Lancaster. *Bibliographie
générale de l'astronomie.* 2 vols. Brussels:
Hayez, 1887, 1882.

Howse, Derek. *Greenwich Observatory.* Vol. 3. *The
buildings and instruments.* London: Taylor and
Francis, 1975.

——*Greenwich time and the discovery of longitude.* New York: Oxford University Press, 1980.

Hughes, Barnabas. *Regiomontanus on triangles.*
Madison: University of Wisconsin Press, 1967.

Hunt, Garry E., and Patrick Moore. *The Planet
Venus.* London: Faber and Faber, 1982.

Huygens, Christiaan. *Systema saturnum*. The Hague: A. Vlacq, 1659. Reprinted in Huygens, *Oeuvres*, 15, 209–353.

——— *Brevis annotatio systemis Saturni sui*. The Hague: A. Vlacq, 1660. Reprinted in Huygens, *Oeuvres*, 15, 439–67.

——— "Projet de déterminer la méridienne et la latitude de Paris." [1666–67]. In Huygens, *Oeuvres*, 21, 25–33.

——— *Astroscopia compendaria, tubi optici molimine liberata*. The Hague: A. Leers, 1684. Reprinted in Huygens, *Oeuvres*, 21, 201–31.

——— *Oeuvres complètes*. 22 vols. The Hague: M. Nijhoff, 1888–1950.

Ianniello, Maria Grazia. "Kircher e l'ars magna lucis et umbrae." In Casciato et al., *Kircher* (1986), 223–35.

Jacono, Luigi. "Meridiana." *Enciclopedia italiana*, 22 (1949), 902–7.

Jalland, Trevor. *The life and times of St. Leo the Great*. London: SPCK, 1941.

Jedin, Hubert. "Katholische Reform und Gegenreformation." In Jedin, *Handbuch*, 4 (1975), 449–604, 650–83.

———, ed. *Handbuch der Kirchengeschichte*. Vol. 4. *Reformation, Katholische Reform und Gegenreformation*. Freiburg: Herder, 1975.

Jedin, Hubert, ed. *History of the Church*. Vol. 6. *The Church in the age of absolutism and the Enlightenment*. London: Burns and Oates, 1981.

Jedin, Hubert, and John Dolan, eds. *History of the church*. Vol. 5. *Reformation and Counter Reformation*. London: Burns and Oates, 1980.

Jemolo, Arturo Carlo. *Il giansenismo in Italia prima della Rivoluzione*. Bari: Laterza, 1928.

John Paul II, Pope. "The greatness of Galileo is known to all." In Poupard, *Galileo* (1983), 195–200. (Speech of 1979 at the Pontifical Academy of Sciences' centennial celebration of the birth of Albert Einstein.)

Johns, Christopher M. S. *Papal art and cultural politics. Rome in the age of Clement XI*. Cambridge: Cambridge University Press, 1993.

Johnson, Francis R. *Astronomical thought in Renaissance England*. Baltimore: Johns Hopkins University Press, 1937.

Jones, Charles W. "The Victorian and Dionysian Paschal tables in the West." *Speculum*, 9 (1935), 408–21.

——— "Development of the Latin ecclesiastical calendar." In Bede, *Opera de temporibus* (1943), 1–122.

Kaltenbrunner, Ferdinand. "Die Vorgeschichte der gregorianischen Kalendarreform." Akademie der Wissenschaften, Vienna. Phil.-Hist. Classe. *Sitzungsberichte, 82* (1876), 289–414.

——— "Die Polemik über die Gregorianische Kalendarreform." Akademie der Wissenschaften, Vienna. Phil.-Hist. Classe. *Sitzungsberichte, 87* (1877), 485–586.

——— "Beiträge zur Geschichte der Kalendarreform. I. Die Commission unter Gregor XIII." Akademie der Wissenschaften, Vienna. Phil.-Hist. Classe. *Sitzungsberichte, 97* (1881), 7–54.

Keill, John. *An introduction to the true astronomy*. London: Lintot, 1721.

Kelly, John N. D. *The Oxford dictionary of popes*. Oxford: Oxford University Press, 1986.

Kepler, Johannes. *Discussion avec le messager céleste*. Ed. Isabelle Pantin. Paris: Belles Lettres, 1993.

King, David, and Gerard l'E. Turner. "The astrolabe presented by Regiomontanus to Cardinal Bessarion." *Nuncius, 9* (1994), 165–206.

King, Henry C. *The history of the telescope*. Cambridge, Mass.: Sky, 1955.

Kircher, Athanasius. *Ars magna lucis et umbrae*. Rome: H. Scheus, 1646.

Kochansky, Adam Adamandus. "Considerationes et observationes physico-mathematicae circa diurnam telluris vertiginem." *AE*, Jul. 1685, 317–27.

Koestler, Arthur. *The sleepwalkers*. New York: Grosset and Dunlap, 1970.

Konvitz, J. V. *Cartography in France, 1660–1848. Science, engineering, and statecraft*. Chicago: University Chicago Press, 1987.

Koyré, Alexandre. "An experiment in measurement." APS. *Proceedings, 97* (1953), 222–37.

——— "A documentary history of the problem of fall from Kepler to Newton." APS. *Transactions, 45:4* (1955), 329–95.

Krusch, Bruno. "Studien zu christlich-mittelalterlichen Chronologie, III. Die Entstehung unserer heutigen Zeitrechnung." Akademie der Wissenschaften, Berlin. Phil.-Hist. Klasse. *Abhandlungen*, 1938:8. 87 p.

Lacaille, Nicolas-Louis de. "Recherches sur les réfractions astronomiques, et la hauteur du pole à Paris." MAS, 1755, 547–93.

Lackmann, Heinrich. *Die kirkliche Bücherzensur nach geltendem kanonischem Recht.* Cologne: Greven, 1962.

Lacombe, Henri, and Pierre Costabel, eds. *La figure de la terre du xviiie siècle à l'ère spatiale.* Paris: Gauthier-Villars, 1988.

La Hire, Philippe de. "Méthode pour se servir des grands verres de lunette sans tuyau pendant la nuit." *MAS,* 1715, 4–10.

Lais, Giuseppe. "Memorie e scritti di Mons. Filippo Luigi Gilii, direttore della Specola Vaticana." Accademia de' Lincei. Classe di scienze fisiche. *Memorie, 7* (1890), 49–62.

—— "Memorie e scritti di Mons. Filippo Luigi Gilii, direttore della Specola Vaticana e insigne naturalista del secolo xviii." Reprinted in: Rome: Specola Vaticana. *Miscellanea astronomica, 1* (1920), 3–10.

Laistner, M. L. W. *Thought and letters in Western Europe.* A.D. *500 to 900.* Ithaca, N.Y.: Cornell University Press, 1957.

Lalande, Joseph Jérôme Le François de. "Sur quelques phénomènes qui résultent de l'attraction que les planètes exercent sur la terre." *MAS,* 1758, 339–71.

—— "Sur la manière dont on peut concilier les observations faites à Saint-Sulpice, avec la dimunition connue de l'obliquité de l'écliptique." *MAS,* 1762, 267–8.

—— "Sur la dimunition de l'obliquité de l'écliptique." *MAS,* 1780, 285–314.

—— "Méridienne, ou ligne méridienne." *Encyclopédie méthodique. Mathématiques.* 3 vols. Paris: Pancoucke, 1784–89. Vol. 2, pp. 379–85.

—— *Voyage d'un Français in Italie.* [1769]. 9 vols. Paris: Desaint, 1786.

—— *Astronomie.* 3rd ed., 3 vols. Paris: Didot, 1792.

Lami, Giovanni. *Memorabilia italorum eruditione praestantium quibus vertens saeculum gloriatur.* 3 vols. Florence: Soc. ad insigne centauri, 1742–48.

Lanciano, Nicoletta, et al. "La meridiana della chiesa di S. Maria degli Angeli a Roma." *Giornale di astronomia, 6:2* (June 1980), 117–30.

Lanza, Franco. "L'Istoria universale' del Bianchini e la 'Scienza nuova.'" *Lettere italiane, 10* (1958), 339–48.

Lastri, Marco Antonio. *L'osservatore fiorentino sugli edifizi della sua patria.* 3rd ed. 8 vols. Florence: R. Ricci, 1821.

Lattis, James. *Between Copernicus and Galileo: Christoph Clavius and the collapse of Ptolemaic astronomy.* Chicago: University of Chicago Press, 1994.

Le Gentil de la Galaisière, Guillaume. "Recherches sur l'obliquité de l'écliptique." *MAS,* 1757, 180–9.

Leibniz, G. W. *Epistolae ad diversos.* Ed. Christian Kortholt. 4 vols. Leipzig: Breitkopf, 1734–42.

—— *Mathematische Schriften.* Ed. C. I. Gerhardt. 7 vols. Halle: H. W. Schmidt, 1849–63.

—— *Sämtliche Schriften und Briefe.* Series 2. *Philosophischer Briefwechsel.* Vol. 1, *1663–1685.* Darmstadt: Reichl, 1926.

—— *Sämtliche Schriften und Briefe.* Series 3. *Mathematischer, naturwissenschaftlicher und technischer Briefwechsel.* Vol. 1, *1672–76.* Berlin: Akademie Verlag, 1976.

—— *Textes inédits.* Ed. G. Grua. 2 vols. Paris: PUF, 1948.

—— *Philosophical papers and letters.* Ed. L. E. Loemker. 2nd ed. Dordrecht: Reidel, 1969.

—— "On Copernicanism and the relativity of motion." In Leibniz. *Philosophical essays.* Ed. R. Ariew and D. Garber. Indianapolis: Hacker, 1989. Pp. 91–4.

Lelewell, Joachim. *Pythéas de Marseille et la géographie de son temps.* Brussels: P. J. Voglet, 1836.

Lemesle, Gaston. *L'église Saint-Sulpice.* Paris: Blond and Gay, 1931.

Lemonnier, P. C. "Recherches sur la hauteur du pôle de Paris." *MAS,* 1738, 209–25.

—— "Sur le solstice d'été de l'anneé 1738." *MAS,* 1738, 361–78.

—— *Histoire céleste, ou Recueil de toutes les observations astronomiques faites par ordre du roy.* Paris: Brianson, 1741.

—— "Mémoire où l'on prouve qu'il y a une inégalité très sensible dans les plus grandes hauteurs du soleil au solstice d'été." *MAS,* 1743, 67–9.

—— "Construction d'un obélisque à l'extremité septentrionale de la Méridienne de l'Eglise de Saint Sulpice." *MAS,* 1743, 361–6.

—— "Dimensions de l'obélisque & du gnomon, élevés aux extrémités de la ligne méridienne de l'Eglise de S. Sulpice." *Mercure de France,* Jan. 1744, 176–85.

—— *Nouveau zodiaque.* Paris: Imp. royale, 1745.

—— *Institutions astronomiques.* Paris: Guerin, 1746.

—— "Solstices d'été." *MAS*, 1762, 263–6.

—— "Solstice d'été de 1767." *MAS*, 1767, 417–22.

—— "Diverses observations faites aux solstices." *MAS*, 1774, 252–3.

—— *Description et usage des principaux instruments d'astronomie*. Paris: AS, 1774. (AS, *Description des arts et métiers*.)

Lenoble, Robert. *Mersenne, ou la naissance du mécanisme*. [1943]. 2nd ed. Paris: Vrin, 1971.

Levallois, Jean-Jacques. "Picard géodésien." In Picolet, *Picard* (1987), 227–46.

—— "L'Académie royale des sciences et la figure de la terre." In Lacombe and Costabel, *Figure* (1988), 41–75.

Levera, Francesco. *Prodromus universae astronomiae restitutae*. Rome: Bernabò, 1663.

—— *De innerantium stellarum viribus, & excellentia*. Rome: Bernabò, 1664.

—— "Dissertatio de die quo festum paschae celebrandum est hoc anno M.DC. LXVI." In Petit, *Lettre* (1666), 23–9.

Lévy, Jacques. "Picard créateur de l'astronomie moderne." In Picolet, *Picard* (1987), 133–41.

Listing, J. B. "Zur Erinnerung an Sartorius von Walterhausen." Akademie der Wissenschaften, Göttingen. *Nachrichten*, 1876, 547–59.

Long, Roger. *Astronomy in five books*. Cambridge: The author, 1742.

Lopez, Pasquale. *Riforma cattolica e vita religiosa e culturale a Napoli dalla fine del cinqecento ai primi anni del settecento*. Naples: Mezzogiorno, 1964.

Lopiccoli, Fiorella. "Il corpuscularismo italiano nel 'Giornale de' letterati' di Roma (1668–1681)." In Magrini, *Storia* (1990), 19–93.

Lorenzoni, G. "Ricordi intorno a Giuseppe Toaldo." AS, Padua. *Atti e memorie*, 29:2 (1913), 271–316.

Louville, Jacques-Eugène d'Altonville. "Application du micomètre à la lunette du quart de cercle astronomique." *MAS*, 1714, 65–77.

—— "De mutabilitate eclipticae dissertatio." *AE*, July 1719, 281–94.

—— "Observation des hauteurs méridiennes du soleil au solstice d'été de cette année 1721." *MAS*, 1721, 167–73.

Luchinat, Cristi Acidini. *La cattedrale di S. M. del Fiore a Firenze*. Vol. 2. Florence: Cassa di Risparmio di Firenze, 1995.

Maccaferri, Davide. "Gian Domenico Cassini e la meridiana di San Petronio." *Il carobbio, 7* (1981), 243–56.

Macdonald, T. L. "Riccioli and lunar nomenclature." British Astronomical Society. *Jl, 77* (1967), 112–17.

Maffei, Paolo. *Giuseppe Settele, il suo diario e la questione Galileiana*. Foligno: Edizioni dell'Arquata, 1987.

Maffei, Scipione. "Vita di Monsignor Francesco Bianchini." In Bianchini. *La istoria universale* (1747), a.2v–a.5v.

Maffioli, C. S. "Guglielmini vs. Papin (1691–1697). Science in Bologna at the end of the XVIIth century through a debate on hydraulics." *Janus, 71* (1984), 63–105.

—— "Domenico Guglielmini . . . e la nuova cattedra d'idriometria nello studio di Bologna (1694)." In Cavazza, *Rapporti* (1987), 81–124.

Magliabecchi, Antonio. *Clarorum venetorum ad Ant. Magliabecchium nonnullosque alios epistolae*. 2 vols. Florence: Typ. ad insigne Apollonis, 1745.

—— *Clarorum belgorum . . . epistolae*. 2 vols. Florence: Typ. ad insigne Apollonis, 1745.

—— *Clarorum germanorum . . . epistolae*. Florence: Typ. ad insigne Apollonis, 1746.

Magrini, Maria Vittoria Predeval. "Introduzione." In Magrini, *Scienza* (1990), 5–18.

—— ed. *Scienza, filosofia e religione tra '600 e '700 in Italia. Ricerche sui rapporti tra cultura italiana ed europea*. Milan: F. Angeli, 1990.

Mailly, Edouard. *Essai sur la vie et les ouvrages de L. A. J. Quetelet*. Brussels: F. Hayez, 1875.

Malagola, Carlo. "Christina di Svezia in Bologna." Reale accademia araldica italiana. *Giornale araldico-genealogico-diplomatico, 8* (1800–81), 201–7, 242–7, 274–80, 319–24.

Malbois, Emile. "Oppenord et l'église Saint-Sulpice." *Gazette des beaux-arts, 75* (1933:1), 34–46.

Malezieu, Nicholas de. "Sur l'observation du solstice." *MAS*, 1714, 320–7.

Malvasia, Cornelio. *Ephemerides novissimae*. Modena: Cassiani, 1662.

Mancinelli, F., and Juan Cassanovas. *La Torre dei Venti in Vaticano*. Vatican City: Editrice Vaticana, 1980.

Mandosio, Prospero. *Biblioteca romana*. 2 vols. Rome: F. de Lezzaris, 1682–92.

Mandrino, A., G. Tagliaferri, and P. Tucci, eds. *Un*

viaggio in Europa nel 1786. Diario di Barnaba Oriani astronomo milanese. Florence: Olschki, 1994.

Manessou Mallet, Allain. Description de l'univers. 5 vols. Paris: Thierry, 1683.

Manetti, Antonio di Tuccio. The life of Brunelleschi. Ed. Howard Saalman. University Park, Pa.: Pennsylvania University Press, 1970.

Manfredi, Eustachio. "Parere sopra l'opera pasquale del Sig. Ab. Jacopo Bettazzi da Prato." [1722]. In Manfredi, Elementi (1744), 321–54.

——— De annuis inerrantium stellarum aberrationibus. Bologna: Typ. C. Pisari, 1729.

——— "De novissima meridianae lineae, quae in Divi Petronii extat, dimensione." CAS, 1 (1731), 589–98.

——— "De novissimis circa fixorum errores observationibus." [1 Oct. 1730]. CAS, 1 (1731), 599–639.

——— De gnomone meridiano bononiensi. Bologna: L. della Volpe, 1736.

——— "In Bianchini observationibus praefatio." In Bianchini, Observationes (1737), i–xiii.

——— Elementi della cronologia con diverse scritture appartenenti al calendario romano. Bologna: L. della Volpe, 1744.

——— Istituzioni astronomiche. Bologna: L. della Volpe, 1749.

Manzini, Carlo Antonio. L'occhiale all'occhio. Bologna: Benacci, 1660.

Marchese, Vincenzo. Memorie dei più insigni pittori, scultori e architetti domenicani. 2nd ed., 2 vols. Florence: F. Le Monnier, 1854.

Marini, Gaetano. Inscrizioni antiche delle ville e de' palazzi Albani. Rome: P. Giunchi, 1785.

Marzi, Demetrio. La questione della riforma del calendario nel quinto Concilio Laterano. Florence: Carnesecchi, 1896.

Masini, Antonio di Paolo. Bologna perlustrata. 3 vols. Bologna: Benacci, 1666.

Masson, Georgina. "Papal gifts and Roman entertainments in honour of Queen Christina's arrival." In Platen, Queen Christina (1966), 244–61.

——— Queen Christina. [1968]. London: Cardinal, 1974.

Matteucci, Annna Maria. "Il gotico cittadino di Antonio di Vincenzo." In D'Amico and Grandi, Tramonto (1987), 27–54.

Maugain, Gabriel. Etude sur l'évolution intellectuelle d'Italie de 1657 à 1750 environ. Paris: Hachette, 1909.

Mazzoleni, Alessandro. Vita di monsignor Francesco Bianchini. Verona: Targa, 1735.

Mazzuchelli, Giovanni Maria. Gli scrittori d'Italia. 2 vols. in 6. Brescia: G. Bossini, 1760.

McKeon, Robert M. "Les débuts de l'astronomie de précision." Physis, 13 (1971), 225–88, and 14 (1972), 221–42.

Meli, Domenico Bartolomeo. "Leibniz on the censorship of the Copernican system." Studia leibniziana, 20 (1988), 19–42.

——— "St. Peter and the rotation of the earth: The problem of free fall around 1800." In P. M. Harman and A. E. Shapiro, eds. An investigation of difficult things: Essays on Newton and the history of exact sciences. Cambridge: Cambridge University Press, 1992. Pp. 421–47.

Meliu, Angelo. S. Maria degli Angeli. Rome: Palombi, 1950.

Mendham, Joseph. The literary policy of the Church of Rome. 2nd ed. London: J. Duncan, 1830. Plus supplements, 1836, 1843.

Mengoli, Pietro. Via regia ad mathematicas . . . ornata. Bologna: Benacci, 1655.

——— Ad majorem dei gloriam. Refrattioni, e paralasse solare. Bologna: Benacci, 1670.

——— Anno. Bologna: Benacci, 1673.

——— Mese. Bologna: n.p., 1681.

——— La corrispondenza. Ed. Gabriele Baroncini and Marta Cavazza. Florence: Olschki, 1986.

Mercator, Nicholas. "Some considerations . . . concerning the geometrick and direct method of Signor Cassini." PT, 5:57 (1670), 1168–75.

Merton, Robert. "Motive forces of the new science." [1938].In Cohen, Puritanism (1990), 112–31.

Michaux, Lucien. Inventaire général des richesses de l'art de la France. L'église Saint-Sulpice. Paris: Plon, 1885.

Middleton, W. E. Knowles. "Science in Rome, 1675–1700, and the Accademia fisicomatematica of Giovanni Giustino Ciampini." BJHS, 8 (1975), 138–54.

Minois, Georges. L'Eglise et la science. Histoire d'un malentendu. 2 vols. Paris: Fayard, 1990–91.

——— Censure et culture sous l'ancien régime. Paris: Fayard, 1995.

Miotto, E., G. Tagliaferri, and P. Tucci. La strumen-

tazione nella storia dell'Osservatorio astronomico di Brera. Milan: E. S. U., 1989.

Mira, Giuseppe M. *Bibliografia siciliana.* 2 vols.
Palermo: G. B. Gaudiano, 1875.

Moller, Daniel Wilhelm. *Disputatio academica de
seculo.* Altdorf: Meyer, 1701.

Monaco, Giuseppe. "Un parere di Francesco Bianchini sui telescopi di Giuseppe Campani."
Physis, 25 (1983), 413–31.

Monchamp, Georges. *Galileé et la Belgique.* Saint-
Trond: Moreau-Schouberechts, 1892.

——*Notification de la condemnation de Galileé.*
Saint-Trond: Moreau-Schouberechts, 1893.

——"Les corespondants belges du grand Huygens. Académie des sciences, Brussels. *Bulletin,
27* (1894), 255–308. (Also issued separately,
Brussels: Hayez, 1894.)

Monconys, Balthasar de. *Voyages.* 2nd ed. 5 vols.
Paris: Delaune, 1695.

Montalto, Lina. "Un ateneo internazionale vagheg-
giato in Roma sulla fine del secolo xvii." *Studi
romani, 10* (1962), 660–73.

Montanari, Geminiano. *Cometes Bononiae obser-
vatus.* Bologna: Ferroni, 1665.

——*Lettera . . . all'ill.mo abbate Antonio Sampieri
in risposta ad alcune obiezioni.* Bologna: E.
Maria e fratelli de' Manolesi, 1667.

——*Pensieri fisico-matematici.* Bologna:
Manolesi, 1667.

——*L'astrologia convinta di falso.* Venice: F.
Nicolini, 1685.

——*Le forze d'eolo. Dialogo fisico matematico.*
Parma: A. Poletti, 1694.

Montgomery, Scott L. "Expanding the earth . . .
The case of the moon." In Montgomery. *The
scientific voice.* New York: Guilford, 1996. Pp.
196–293.

Montucla, Jean Etienne. *Histoire des mathématiques.* 2nd ed., 4 vols. Ed. J. Lalande. Paris:
Agasse, an vii–x (1799–1802).

Moreton, Jennifer. "Doubts about the calendar:
Bede and the eclipse of 664." *Isis, 89* (1998),
50–65.

Morgani, G. B. "De vita Domenici Guglielmini." In
Guglielmini, *Opere* (1719), *1*, 1–7.

Mori, Attilio. "Studi, trattive e proposte per la
costruzione di una carta geografica della
Toscana nella seconda metà del secolo xviii."
Archivio storico italiano, 35 (1905), 369–424.

Moroni, Gaetano. *Dizionario di erudizione storico-
ecclesiastica.* 103 vols. Venice: Typ. Emiliana,
1840–61.

Moyer, Gordon. "Aloisius Lilius and the 'Compendium novae rationis restituendi kalendarium' ". In Coyne et al., *Gregorian reform* (1983),
171–88.

Müller, D. Diamillo. *Biografie autografie inedite di
illustri italiani di questo secolo.* Turin: Pomba,
1853.

Müller, Kurt, and Gisela Krönert. *Leben und Werk
von G. W. Leibniz. Eine Chronik.*
Frankfurt/Main: Klostermann, 1969.

Mutus, Savinius (Levera, Francesco). *Dialogus contra duas hic transcriptas epistolas.* Rome: Bernabò, 1664.

Napoli, Carlo di. *Nuove invenzioni di tubi ottici di-
monstrate nell'Accademia fisicomatematica Romana.* Rome: G. G. Komarek Boemo, 1686. 19p.

Nardini, Bruno. "La facciata." In Baldini, *Santa
Maria Novella* (1981), 44–52.

Newton, Isaac. *Correspondence.* Ed. H. W. Turnball
et al. 7 vols. Cambridge: Cambridge University
Press, 1959–77.

Nicolini, Fausta. "Tre amici di mons. Celstini
Galiani: Benedetto XIV, il card. Davia, mons.
Leprotti." Deputazione di storia patria per le
provincie di Romagna. *Atti e memorie, 20*
(1930), 87–138.

Nikolic, Djordje. "Roger Boscovich et la géodésie
moderne." *Archives internationales d'histoire
des sciences, 14* (1961), 315–35.

Nobis, H. M. "The reaction of astronomers to the
Gregorian Calendar." In Coyne et al., *Gregorian
reform* (1987), 243–54.

Nordenmark, N. V. E. *Anders Celsius. Professor i
Uppsala 1701–1744.* Uppsala: Almqvist & Wik-
sell, 1936.

Noris, Enrico. *Annus, & epochae Syromacedonum .
. . Accesserunt nuper Dissertationes de paschali
Latinorum cyclo annorum lxxxiv ac Ravennate
annorum xcv.* Florence: Typ. Magni Ducis, 1691.

North, John. "Chronology and the age of the
world." In W. Yorgau, ed. *Cosmology, history,
and theology.* New York: Plenum, 1977. Pp.
307–33.

——"Astrology and the fortunes of churches."
Centaurus, 24 (1980), 181–211.

——"The Western calendar." In Coyne et al., *Gregorian reform* (1983), 75–113.

Odier, Jeanne Bignami, and Anna Maria Partini.

"Cristina di Svezia e le scienze occulte." *Physis*, 25 (1983), 251–78.

Oldenburg, Henry. *Correspondence*. Ed. A. Rupert Hall and Marie Boas Hall. 13 vols. Madison: University of Wisconsin Press, 1965–73 (vols. 1–9); London: Mansell, 1975–77 (vols. 10–11); London: Taylor and Francis, 1986 (vols. 12–13).

Olmsted, John. "The scientific expedition of Jean Richer to Cayenne (1672–1673)." *Isis, 34* (1942), 117–28.

—— "The 'application' of telescopes to astronomical instruments, 1667–1669." *Isis, 40* (1949), 213–25.

Oriani, Barnaba. "Obliquità dell'eclittica dedotta dalle osservazioni solstiziali." *EA*, 1830, App., 9–56.

Ortelius, Abraham. *The theater of the world*. [1606]. Ed. R. A. Skelton. Amsterdam: Theatrum orbis terrarum, 1968.

Ottaviano, G. M. *La meridiana di precisione della Specola di Padova. Teoria e storia della gnomonica*. Padova: tesi di laurea, 1989.

P., E. *Il nuovo orologio della Piazza de' Mercanti in Milano, ossia Istruzione popolare per regolare gli orologii*. Milan: Bernardoni, 1859. 7 p.

Paatz, Walter, and Elisabeth Paatz. *Die Kirchen von Florenz. Ein kunstgeschichtliches Handbuch*. 6 vols. Frankfurt/Main: Klostermann, 1940–54.

Pachtler, Georg Michael, ed. *Ratio studiorum et institutiones scholasticae societatis Jesu*. 4 vols. Berlin: A. Hofmann, 1887–94.

Pagani, Giovanni Battista. *Metodo di computare i tempi*. Palermo: G. Bayona, 1726.

Palatio, Pietro. *Novae ephemerides motuum solis*. Rome: Bernabò, 1664.

Palcani, Luigi C. "Elogio di Leonardo Ximenes." Società italiana. *Memorie di matematica e fisica*, 5 (1790), ix–xxviii.

Palmeri, Vincenzo. "Ignatio Danti." Deputazione di storia patria per l'Umbria. *Bolletino, 5* (1899), 81–125.

Paltrinieri, Giovanni, Italo Frizzoni, and Renato Peri. *Meridiane e orologi solari di Bologna e provincia*. Bologna: Artiere Edizionitalia, 1995.

Paoli, Alessandro. "La scuola Galileo . . . Corrispondenza del padre Grandi e padre Ceva." *Annali delle università toscane, 28:6* (1908), 1–44, and *29:3* (1910), 45–102.

Pascal, Alberto. "L'apparechio polisettore di Tomasso Ceva e una lettera inedita di Guido Grandi." Istituto lombardo di scienze e lettere. *Rendiconti, 48* (1915), 173–81.

Passano, Carlo Ferrari da, Carlo Monti, and Luigi Mussio. *La meridiana solare del Duomo di Milano*. Milan: Fabbrica del Duomo, 1977.

Passerin, Ettore. "Giansenisti e illuministi." In Fubini *Cultura* (1964), 209–28.

Pastor, Ludwig. *History of the Popes*. 40 vols. Vols. 1–2. London: J. Hodges, 1891. Vols. 3–40. London: Kegan Paul, 1894–1953.

Pedersen, Olaf. "The ecclesiastical calendar and the life of the church." In Coyne et al., *Gregorian reform* (1983), 17–74.

Pelikan, Jaroslav. *The Christian tradition: A history of the development of doctrine*. Vols. 4–5. Chicago: University of Chicago Press, 1984–89.

[Pell, John]. *Easter not mistimed*. London: Garthwait, 1664.

Pellegrino, Luigi. *Biografia di Antonio Ma. Jaci*. Massina: T. Capra, 1842.

Pepe, Luigi. "Gabriele Manfredi (1681–1761) et la diffusion du calcul différentiel en Italie." In Albert Heinekamp, ed. *Beiträge zur Wirkungs- und Rezeptionsgeschichte von . . . Leibniz*. Wiesbaden: Steiner, 1986. Pp. 79–87.

——, ed. *Copernico e la questione copernicana in Italia del xvi ad xix secolo*. Florence: Olschki, 1996.

Peters, Christian Heinrich Friedrich. *Heliographic positions of sunspots*. Ed. E. B. Frost. Washington, D.C.: Carnegie Institution, 1907.

Peters, Christian Heinrich Friedrich, and Edward Ball Knobel. *Ptolemy's catalogue of the stars. A revision of the Almagest*. Washington, D.C.: Carnegie Foundation, 1915.

Petit, Pierre. *Lettre . . . touchant le jour auquel on doit célébrer la Feste de Paques*. Paris: Cusson, 1666.

Philip, Alexander. *The calendar: Its history, structure, and improvement*. Cambridge: Cambridge University Press, 1921.

Piazza, Carlo Bartolomeo. "Discorso istorico, geografico, ed ecclesiastico dell'istrumento astronomico [in S. M. degli Angeli]." In Piazza. *La gerarchia cardinalizia . . . a Clemente XI*. Rome: Bernabò, 1703. Pp. 625–30.

Piazzi, Giuseppe. *Discorso . . . nell'aprirsi la prima volta la cattedra di astronomia nell'Accademia de' R. studii di Palermo*. Palermo: R. stamperia, 1790. (Also in Piazzi, *Sulle vicende* (1990), 47–70.)

—— *Della specola astronomica de' regi studi di Palermo*. Palermo: R. stamperia, 1792. (The preface, "Discorso preliminare sulle vicende dell'astronomia in Sicilia," is reprinted in Piazzi, *Sulle vicende* (1990), 71–82.)

—— *Risultati delle osservazioni della nuova stella*. Palermo: R. stamperia, 1801.

—— *Della scoperta del nuovo pianeta cerere ferdinandea*. Palermo: R. stamperia, 1802.

—— "Dell'obliquità dell'eclittica." Società italiana della scienza. *Memorie di matematica e di fisica*, 11 (1804), 426–45.

—— *Del reale osservatorio di Palermo libro sesto*. Palermo: R. stamperia, [1806].

—— "Sull'orologio italiano e europeo, riflessioni." [1800 or 1801]. *Giornale di scienze, lettere ed arti per la Sicilia*, 7 (1824), 137–72.

—— *Discorso*. Palermo: L. Dato, 1824.

—— *Sulle vicende dell'astronomia in Sicilia*. Ed. Giorgia Foderà Serio. Palermo: Sellerio, 1990.

Piazzi, Giuseppe, and Barnaba Oriani. *Corrispondenza astronomica*. Milan: Hoepli, 1874.

Picard, Jean. "Mesure de la terre." [1671]. *MAS, 1666–1699*, 7 (1729), 133–90.

Picolet, Guy, ed. *Jean Picard et les débuts de l'astronomie de précision au xviie siècle*. Paris: CNRS, 1787.

Piper, Ferdinand. *Karls des Grossen Kalendarium und Ostertafel*. Berlin: Decker, 1858.

Pizzoni, D. Pietro. "Il volo attribuito a Gio. Battista Danti." Deputazione di storia patria per l'Umbria. *Bolletino*, 42 (1945), 209–25.

Platen, Magnus von, ed. *Queen Christina of Sweden*. Stockholm: Norstedt, 1966.

Poleni, Giovanni. "Considerazioni . . . se la terra girasse per l'orbe annuo." *GL*, 8 (1711), 199–215.

Porena, Filippo. "La geografia in Roma e il mappemundo vaticano." Società geografica italiana. *Bolletino*, 25 (1888), 221–38, 311–39, 427–53.

Portoghesi, Paolo. *Borromini nella cultura europea*. Rome: Officina Edizioni, 1966.

Poupard, Paul. "Introduction: Galileo Galilei: 350 years of subsequent history." In Poupard, *Galileo* (1983), xii–xxiii.

—— *Galileo Galilei: Toward a solution of 350 years of debate*. Pittsburgh: Duquesne University Press. 1983.

Predieri, Paolo. "Della vita e della corrispondenza scientifica e letteraria di Cesare Marsigli con Galileo Galilei e Padre Bonaventura Cavalieri." AS, Bologna. *Memorie*, 3 (1851), 113–43.

Prestinenza, Luigi. "Meridiana restaurata è quella dei Benedettini di . . . Catania." *L'astronomia*, no. 164, Apr. 1996, 15–16.

Proverbio, Edoardo. "Copernicus and the determination of the length of the tropical year." In Coyne et at., *Gregorian reform* (1983), 129–34.

—— "Historic and critical comment on the 'Riposta' of R. J. Boscovich to . . . Prince Kaunitz." *Nuncius*, 2:2 (1987), 171–225.

—— "Francesco Giuntini e l'utilizzo delle tavole copernicane in Italia nel xvi secolo." In Bucciantini and Torrini, *Diffusione* (1997), 37–55.

Quetelet, Adolphe. "Aperçu historique sur les principales méridiennes connues et sur la méridienne de Ste. Gudule en particulier." Brussels. Observatoire. *Annuaire*, 1837, 216–24.

—— *Histoire des sciences mathématiques et physiques chez les belges*. Brussels: M. Hayez, 1864.

Rabanus Maurus. "Liber de computo." *PL*, 107, 669–728.

Ragghianti, Carlo L. *Filippo Brunelleschi. Un uomo, un universo*. Florence: Vallechi, 1977.

Raggi, Giuliana, ed. *Il cielo come laboratorio*. Bologna: Università di Bologna, Assessorato Coordinamento Politiche Scholastiche, 1991.

Raule, Angelo. "Intorno alla meridiana della Basilica di S. Petronio." *Strenna storica bolognese*, 10 (1960), 215–29.

Réaumur, René. "Description d'une machine portative, propre à soutenir des verres de très grands foyers." *MAS*, 1713, 299–306.

Redondi, Pietro. *Galileo heretic*. Tr. R. Rosenthal. Princeton: Princeton University Press, 1987.

Regiomontanus. [*Kalendarium*]. Venice: Ratdolt, 1476.

Reichard, Elias Caspar. *Indicum librorum prohibitorum*. Braunschweig: Keitelian, 1746.

Renaldo, John J. "Bacon's empiricism, Boyle's science, and the Jesuit response in Italy." *JHI*, 37 (1976), 689–95.

—— *Daniello Bartoli: A letterato of the seicento*. Naples: Istituto italiano per gli studi storicî, 1979.

Repsold, J. A. *Zur Geschichte der astronomischen Messwerkzeuge, 1450–1830*. 2 vols. Leipzig: Engelmann, 1908.

Restiglian, Marco. "Nota su Giuseppe Toaldo e

l'edizione toaldina del Dialogo di Galileo." In *Galileo Galilei* (1982), 235–9.

Rétat, Pierre. "*Mémoires pour l'histoire des sciences et des beaux arts*. Signification d'un titre et d'une entreprise journalistique." *Dix-huitième siècle*, no. 8 (1976), 167–87.

Reusch, Franz Heinrich. *Der Prozess Galileis und die Jesuiten*. Bonn: Weber, 1879.

—— *Der Index der verbotenen Bücher*. 3 vols in 2. Bonn: Weber, 1883–85.

Reyher, Samuel. *Mathesis regia*. Kiel: Reumann, 1693.

Reynaud, Théophile. *Erotemata de malis ac bonis libris, deque iusta aut iniusta, eorundem confixione*. Lyon: Huguetau and Ravaud, 1753.

Rheita, Anton Maria Schyrlaeus de. *Novem stellae circum Jovem visae, circa Saturnum sex, circum Martem nonnullae*. Louvain: Typ. A. Bouvetii, 1643.

Riccardi, Pietro. "Nuovi materiali per la storia della facoltà matematica nell'antica università di Bologna." *Bulletino di bibliografia e di storia delle scienze matematiche e fisiche, 12* (1879), 299–312.

—— *Biblioteca matematica italiana*. 2 parts in 3 vols. Modena: Società tipografica modenese, 1893.

Ricci, Carlo. "Gli eliometri senesi." Accademia dei fisiocritici, Siena. *Memorie, 2* (1985), 319–55.

Ricci, Giovanni. *Bologna*. Bari: Laterza, 1980.

Riccioli, Giovanni Battista. *Almagestum novum Astronomiam veterem novamque complectens*. 1 vol. in 2 parts. Bologna: Benacci, 1651. (The version with imprint Frankfurt, 1653, appears to be identical with the edition of 1651; Riccardi, *Bibl.* (1893), 1:2, col. 372.)

—— *Geographiae et hydrographiae reformatae libri duodecim*. Bologna: Benacci, 1661.

—— *Dialogus contra duas hic transcriptas epistolas in Prodromum F. Leverae*. Rome: Bernabò, 1664.

—— *Astronomiae reformatae tomi duo*. 2 vols. Bologna: Benacci, 1665.

—— *Apologia . . . pro argumento physico-mathematico contra systema copernicanum adiecto contra illud novo argumento ex reflexo motu gravium decidentium*. Venice: Salerni and Cagnolini, 1669.

Richa, Giuseppe. *Notizie istoriche delle chiese fiorentine*. 10 vols. Florence: Viviani, 1754–62.

Richer, Jean. "Observations astronomiques et physiques faites en l'isle de Caïenne." AS. *Mémoires . . . contenant les ouvrages adoptez par cette Académie avant son renouvellement en 1699*. Vol. 5. Amsterdam: Mortier, 1736. 94 p.

Rigassio, Gian Carlo. *Le ore e le ombre: Meridiane e orologi solari*. Milan: Mursia, 1988.

Rigaud, S. P. "Memoirs of Dr James Bradley." In Bradley, *Miscellaneous works* (1832), i–cviii.

Righini, Guglielmo. "La tradizione astronomica fiorentina e l'Osservatorio di Arcetri." *Physis, 4* (1962), 133–50.

—— "Il grande astrolabio del Museo di storia delle scienze di Firenze." IMSS. *Annali, 2:2* (1977), 45–66.

Righini-Bonelli, Maria Luisa, and Albert van Helden. *Divini and Campani: A forgotten chapter in the history of the Accademia del Cimento*. Florence: IMSS, 1981. (IMSS, *Annali*, 1981: suppl.)

Righini-Bonelli, Maria Luisa, and Thomas B. Settle. "Egnatio Danti's great astronomical quadrant." IMSS. *Annali, 4:2* (1979), 3–13.

Rivosechi, Valerio. "Il simbolismo della luce." In Casciato et al., *Kircher* (1986), 217–22.

Roberti, Guadentius. *Miscellenea italica physicomathematica*. Bologna: Typ. Pisariana, 1692.

Robinet, André. "Copernic dans l'oeuvre de Malebranche." In Centre, *Copernic* (1975), 271–5.

—— "G. W. Leibniz et la république des lettres de Bologne." In Cavazza, *Rapporti* (1987), 3–49.

—— *G. W. Leibniz. Iter italicum (Mars 1689–Mars 1690). La dynamique de la République des Lettres*. Florence: Olschki, 1988.

Roche, John. "Harriot's 'Regiment of the sun' and its background in sixteenth-century navigation." *BJHS, 14* (1981), 245–61.

Romano, Giuliano, and Giorgio M. Ottaviano. "La meridiana di Toaldo." Accademia patavina di scienze, lettere ed arti. Classe di scienze matematiche e naturali. *Atti e memorie, 102:2* (1989–90), 129–49.

Rømer, Olaus. "Planisphère pour les étoiles, et pour les planetes." [Before 1699.] In AS, *Machines* (1735), 81–3.

Ronchi, Vasco. "Padre Grimaldi e il suo tempo." *Physis, 5* (1963), 349–52.

Ronsisvalle, Vanni. *Gli astronomi*. Palermo: Sellerio, 1989.

Rosen, Edward. "Was Copernicus' *Revolutions* approved by the Pope?" *JHI, 36* (1975), 531–42.

Rotta, Salvatore. "Scienza e 'pubblica felicità' in Geminiano Montanari." *Miscellanea seicento, 2* (1971), 64–210.

——— *L'illuminismo a Genova: Lettere di P. P. Celesia a F. Galiani*. Florence: Nuova Italia, [1974].

Ruggiero, Constantino. *Memorie istoriche della biblioteca ottoboniana*. Rome: Tip. Vaticana, 1825.

Rusnock, Andrea, ed. *The correspondence of James Jurin (1648–1759)*. Amsterdam: Rodopi, 1966.

Russell, J. L. "Kepler's laws of planetary motion: 1609–1666." *BJHS, 2* (1964), 1–24.

——— "The Copernican system in Britain." In Dobrzycki, *Reception* (1972), 181–239.

Russell, W. B. *We will remember them: The story of the Shrine of Remembrance*. 3rd ed. Victoria: The Shrine, 1991.

Salveraglio, Filippo. "Il Duomo di Milano. Saggio bibliografico." *Archivio storico lombardo, 3* (1886), 894–943.

Salzman, Michèle Renée. *On Roman time*. Berkeley: University of California Press, 1990.

Sanblad, Henrik. "The reception of the Copernican system in Sweden." In Dobrzycki, *Reception* (1972), 241–70.

Santa Maria Novella, un convento nella città. Pistoia: Memorie dominicane, 1980.

Sartorius von Waltershausen, Wolfgang freiherr. *Gauss zum Gedächtnis*. Leipzig: Hirzel, 1856.

——— *Der Aetna*. Ed. Arnold von Lesaulx. 2 vols. Leipzig: Engelmann, 1880.

Savelli, Roberto. *Grimaldi e la rifrazione*. Bologna: n.p., 1951.

Savorgnan di Brazzà, Francesco, ed. *L'opera del genio italiano all'estero*. Ser. 11, vol. 1. *Gli scienziati italiani in Francia*. Rome: Libreria dello Stato, 1941.

Schiavo, Armando. "Santa Maria degli Angeli alle Terme." Centro di studi per la storia dell'architettura. *Bolletino, 8* (1954), 15–42.

——— *La meridiana di S. Maria degli Angeli*. Rome: Istituto poligrafico e zecca dello stato, 1993.

Schmid, Josef. "Zur Geschichte der Gregorianishen Kalendarreform." Görres-Gesellschaft. *Historisches Jahrbuch, 3* (1882), 388–415, 543–95; 5 (1884), 52–87.

Schmidt, Everhard. "Die Galleria geografica des Vatikans." *Geographische Zeitschrift, 17* (1911), 503–17.

Schneider, Burkhardt. "The Papacy." In Jedin, *History, 6* (1981), 107–34, 558–82.

Schreiber, Johannes. "Die Mondnomenklatur Ricciolis und die Grimaldische Mondkarte." *Stimmen aus Maria-Laach, 54* (1898), 252–72.

——— "Die Jesuiten des 17. und 18. Jahrhunderts und ihr Verhältnis zur Astronomie." *Natur und Offenbarung, 49* (1903), 129–43, 208–21.

Schütz, Michael. "Cassini's meridian in Bologna." *Sterne und Weltraum, 6* (1989), 362–6.

Sebastien, père. "Machine pour diriger un tuyau de lunette de cent pieds." [Before 1699.] In AS, *Machines* (1735), 93–5.

Serena, Sebastiano. *S. Gregorio Barbarigo e la vita spirituale e culturale nel suo seminario di Padova*. 2 vols. Padua: Antenore, 1968.

Settesoldi, Enzo. "Paolo Toscanelli padre dello gnomone nella cattedrale di Firenze." *Prospettiva, 16* (Jan. 1979), 44.

Settle, Thomas B. "Dating the Toscanelli meridian in Santa Maria del Fiore." IMSS. *Annali, 3:2* (1978), 69–70.

Shrine of Remembrance, Melbourne: The national war memorial of Victoria. [Melbourne: The Shrine, 1989.]

Shuckburgh, G. E. "Account of the equatorial instrument." *PT, 83* (1793), 67–128.

Siebenhüner, Herbert. "S. Maria degli Angeli in Rom." *Münchner Jahrbuch der bildenden Kunst, 6* (1955), 179–206.

Silvestre, Peter. "A letter . . . concerning the state of learning, and several particulars observed by him lately in Italy." *PT, 22:265* (1700), 627–34.

Simoncelli, Paolo. "Inquisizione romana e Riforma in Italia." *Rivista storica italiana, 100* (1988), 5–125.

——— *Storia di una censura. "Vita di Galileo" e Concilio Vaticano II*. Milan: FrancoAngeli, 1992.

Smart, W. M. *Textbook of spherical astronomy*. Cambridge: Cambridge University Press, 1960.

Smeaton, John. "Observations on the graduations of astronomical instruments." *PT, 76* (1786), 1–47.

Smith, Robert. *A complete system of opticks*. 2 vols. Cambridge: Crownford, 1738.

Soprani, Raffaele. *Li scrittori della Liguria*. Genoa: Calenzani, 1667.

Sorbelli, Albano. *Inventari dei manoscritti delle biblioteche d'Italia*. Vol. 69. *Bologna*. Florence: Olschki, 1939.

Sortais, Gaston. *Le cartésianisme chez les Jésuites français au xviie et xviiie siècle*. Paris: G. Beauchesne, 1929.

Spoletino, Vincenzo L. "Vita di Monsig. Gio. Giustino Ciampini." In Crescimbeni, *Vite* (1708), *2*, 195–253.

Stasiewski, Bernhard. "Ecclesiatical learning." In Jedin, *History, 6* (1981), 524–46.

Stein, John W. "La sala della meridiana nella Torre dei Venti in Vaticano." *Illustrazione vaticana, 9:10* (1938), 403–10.

——— "The meridian room in the Vatican 'Tower of the Winds' and the meridian line of Pope Clement XI in the Church of Santa Maria degli Angeli." Specola Vaticana. *Miscellanea astronomica, 3* (1950), 31–67.

Stenton, Frank. *Anglo-Saxon England*. 3rd ed. Oxford: Oxford University Press, 1971.

Stephan, Ruth. "A note on Christina and her academies." In Platen, *Queen Christina* (1966), 365–73.

Stephenson, Bruce. *Kepler's physical astronomy*. Princeton: Princeton University Press, 1994.

Stimson, Dorothy. *The gradual acceptance of the Copernican theory of the universe*. New York: Baker and Taylor, 1917.

Stolpe, Sven Johan. *Drottning Kristina*. 2 vols. Stockholm: Bonniers, 1960–61.

Streete, Thomas. *Astronomia carolina. A new theory of the celestial motions*. London: Lloyd, 1661; 2nd ed., London: Smith and Briscoe, 1710.

Sully, Henri. "Description de la ligne méridienne de l'Eglise de S. Sulpice." *Mercure de France*, July 1728, 1591–1607.

Suter, Rufus. "Leonardo Ximenes and the gnomon at the Cathedral of Florence." *Isis, 55* (1964), 79–82.

Sutton, R. M. "Of time and the sun." *Physics today, 9:6* (1965), 15–19.

Swerdlow, Noel H. "Science and humanism in the Renaissance." In Paul Horwich, ed. *World changes: Thomas Kuhn and the nature of science*. Cambridge, Mass.: MIT Press, 1993. Pp. 131–68.

Tabarroni, Giorgio. "Un'opera ritrovata dell'astronomo Gian Domenico Cassini nel terzo centenario del suo trasferimento da Bologna a Parigi." *Strenna storica bolonese, 18* (1968), 313–24.

——— "G. D. Cassini et la deuxième loi de Kepler."

Congrès international d'histoire des sciences, 13. *Atti* (1971), *6*, 200–3.

——— "La meridiana." In Luciano Bellosi et al. *La basilica di San Petronio*. 2 vols. Milan: Silvana, 1983–84. Vol. 2, pp. 331–6.

Tacquet, André. *Astronomiae libri octo*. In Tacquet. *Opera mathematica*. Louvain: I. Mersium, 1669.

Tagliaferri, G., and P. Tucci. "P. S. de Laplace e il grado di meridiano d'Italia." *Giornale di fisica, 34* (1993), 257–77.

Tarmot, Joseph de. *Epistola . . . in qua Gregorianum calendarium ab erroribus per D. Cassinum objectis vindicatur*. Anglipoli: n.p., 1703.

Taton, René. "Picard et la *Mesure de la terre*." In Picolet, *Picard* (1987), 207–26.

Taton, René, and Curtis Wilson, eds. *Planetary astronomy from the Renaissance to the rise of astrophysics*. Part A. *Tycho Brahe to Newton*. Cambridge: Cambridge University Press, 1989.

Tenard, Louis. "Un guide jésuite de savoir-faire." *Dix-huitième siècle, 8* (1976), 93–106.

Tenca, L. "Le relazioni epistolari tra Giov. Domenico Cassini e Vincenzo Viviani." AS, Bologna. Classe di scienze fisiche. *Rendiconti, 243* (1955).

Thirion, Maurice. "Influence de Gassendi sur les premiers textes français traitant de Copernic." In Centre, *Copernic* (1975), 257–60.

Thomassy, Raymond. "Les papes géographiques et la cartographie du Vatican." *Nouvelles annales des voyages, 136* (1852:4), 57–96; *137* (1853:1), 151–72; *138* (1853:2), 7–47; *139* (1853:3), 266–96.

Thoren, Victor E. "Kepler's second law in England." *BJHS, 7* (1974), 242–6.

——— *The Lord of Uraniborg: A biography of Tycho Brahe*. Cambridge: Cambridge University Press, 1990.

Thorndike, Lynn. *A history of magic and experimental science*. 8 vols. New York: Columbia University Press, 1923–58.

——— *The Sphere of Sacrobosco and its commentators*. Chicago: University of Chicago Press, 1949.

Toaldo, Giuseppe. *Completa raccolta di opuscoli, osservazioni, e notizie diverse*. 4 vols. Venice: F. Areola, 1802.

——— *La meridiana del Salone di Padova*. Ed. Gaetano Sorgato. Padua: Il seminario, 1838.

Torrini, Maurizio. "Giuseppe Ferroni, gesuita e galilleano." *Physis, 15* (1973), 411–23.

———— *Dopo Galileo. Una polemica scientifica (1684–1711)*. Florence: Olschki, 1979.

Turner, Gerard L'E. "The Florentine workshop of Giovan Battista Giusti, 1562–c.1575." *Nuncius, 10* (1995), 131–71.

Turner, Gerard L'E., and Elly Dekker. "An astrolabe attributed to Gerard Mercator, c. 1570." *Annals of science, 50* (1993), 403–43.

Uzielli, Gustavo. "L'epistolario colombo-toscanelliano e i Danti." Società geografica italiana. *Bolletino, 2* (1889), 836–66.

———— *Paolo del Pozzo Toscanelli . . . Lo gnomone di Santa Maria del Fiore*. Florence: Stabilimento tipografico fiorentino, 1892.

———— *La vita e i tempi di Paolo del Pozzo Toscanelli*. Rome: Ministero della pubblica istruzione, 1894.

Van Helden, Albert. *Measuring the universe*. Chicago: University of Chicago Press, 1985.

Vasari, Giorgio. *The lives of the painters, sculptors and architects*. Ed. W. Gaunt. 4 vols. London: Everyman, 1980.

Vaselli, Crescenzo. "Vita di Pirro Maria Gabbrielli." In Crescimbeni, *Vite* (1708), *2*, 29–46.

Vasoli, Cesare. *L'enciclopedismo del seicento*. Naples: Bibliopolis, 1978.

———— "Sperimentalismo e tradizione negli 'schemi' enciclopedici di uno scienziato gesuita del seicento." *Critica storica, 17* (1980), 100–27.

Vermiglioli, Giovanni Battista. "Elogio di Ignazio Danti." *Opuscoli letterari di Bologna, 3* (1820), 1–22.

Vernazza, Guido. "La crisi barocca nei programmi didattici dello studio Bolognese." *SMSUB, 2* (1961), 95–177.

Villa, Mario. "La scuola matematica bolognese." *SMSUB, 1* (1956), 479–85.

Vitruvius. *The ten books of architecture*. Tr. M. H. Morgan. Cambridge, Mass.: Harvard University Press, 1914.

[Vollgraff, J. A.]. "Biographie de Christiaan Huygens. In Huygens, *Oeuvres* (1888), *22*, 383–778.

Walker, Ez. "A simple way for determining the exact time of noon; also a way to obtain a meridian line on a small scale." *Philosophical magazine, 25* (1806), 172–4.

Wallis, John. "Concerning the alteration (suggested) of the Julian account for the Gregorian." *PT, 21:257* (1699), 343–54.

Ward, Seth. *In Ismaelis Bullialdi Astronomiae philolaicae fundamenta, Inquisitio brevis*. Oxford: Lichfield, 1653.

———— *De cometis, praelectio. Cui subjuncta est Inquisitio brevis*. Oxford: Lichfield, 1653.

———— *Idea trigonometricae demonstratae . . . Item Praelectio de cometis, et Inquisitio brevis*. Oxford: Lichfield, 1654.

———— *Astronomia geometrica*. 3 parts. London: Flesher, 1656.

Wazbinski, Zygmunt. *L'Accademia medicea del disegno a Firenze nel cinquecento*. 2 vols. Florence: Olschki, 1987.

Weld, C. R. *A history of the Royal Society*. 2 vols. London, Parker, 1848.

Westfall, Richard S. "Bellarmino, Galileo, and the clash of two world views." *Jl for the history of astronomy, 20* (1989), 1–23; also in Westfall, *Essays* (1989), 1–30.

———— "Galileo and the Jesuits." In Westfall, *Essays* (1989), 31–57.

———— *Essays on the trial of Galileo*. Vatican City: Vatican Observatory, 1989.

Westman, Robert S. "Three responses to the Copernican theory: Johannes Praetorius, Tycho Brahe, Michael Mästlin." In Westman, *Achievement* (1975), 285–345.

————, ed. *The Copernican achievement*. Berkeley: University of California Press, 1975.

———— "The reception of Galileo's 'Dialogue.' A partial world census of extant copies." In Galluzzi, *Crisi* (1984), 329–71.

———— "The Copernicans and the churches." In D. C. Lindberg and R. L. Numbers, eds. *God and nature*. Berkeley: University of California Press, 1986. Pp. 73–117.

Westphal, A. "Die geodätischen und astronomischen Instrumente zur Zeit des Beginns exakter Gradmessungen." *Zeitschrift für Instrumentenkunde, 4* (1884), 152–6, 189–202.

———— "Basisapparate und Basismessungen." *Zeitschrift für Instrumentenkunde, 5* (1885), 257–74, 333–45, 373–85, 420–32.

Whitaker, Ewen A. "Selenography in the seventeenth century." In Taton and Wilson, *Planetary astronomy* (1989), 119–43.

White, Andrew Dickson. *A history of the warfare of science with theology*. [1896]. 2 vols. New York: Dover, 1960.

Whiteside, Derek T. "Newton's early thoughts on planetary motion." *BJHS, 2* (1964), 117–37.

Wijk, Walter Emilvan. *Le nombre d'or. Etude de chronologie technique.* The Hague: Nijhoff, 1936.

Wilson, Curtis. "Kepler's determination of the elliptical path." *Isis, 59* (1968), 5–25.

—— "From Kepler's laws, so-called, to universal gravitation: empirical factors." *AHES, 6* (1970), 89–170.

—— "Predictive astronomy in the century after Kepler." In Taton and Wilson, *Planetary astronomy* (1989), 161–206.

Winter, Eduard. "Die Katholische Orden und die Wissenschaftspolitik im 18. Jahrhundert." In Erik Amburger et al., eds. *Wissenschaftspolitik im Mittel- und Osteuropa.* Berlin: Camen, 1976. Pp. 85–96.

Wolf, Charles Joseph E. *Histoire de l'Observatoire de Paris de sa fondation à 1793.* Paris: Gauthier-Villars, 1902.

Wrightsman, Bruce. "Andreas Osiander's contribution to the Copernican achievement." In Westman, *Achievement* (1975), 213–43.

Ximenes, Leonardo. *Notizie de' tempi de' principali fenomeni del cielo.* Florence: Viviani, 1752.

—— *Del vecchio e nuovo gnomone fiorentino.* Florence: Stamperia imp., 1757.

—— "Riflessioni intorno all'obliquità dell'eclittica." Accademia de' fisiocritici, Siena. *Atti, 5* (1774), 35–54.

—— *Dissertazione intorno alle osservazioni solstiziali del 1775 allo gnomone della metropolitana fiorentina.* Livorno: Falorni, 1776.

Zaccagnini, Guido. *Storia dello studio di Bologna durante il Rinascimento.* Rome: Olschki, 1930.

Zach, Franz, Xaver von. "Beweis, dass die Oesterreichische Gradmessung des Jesuiten Liesganig sehr fehlerhaft ... sey." *Monatliche Correspondenz zur Beförderung der Erd- und Himmelskunde, 8* (1803), 507–27, and *9* (1804), 32–8, 120–30.

—— *Mémoire ... sur le degré du méridien en Piémont par le pére Beccaria.* Turin: F. Galletti, 1811.

—— "Lettre à M. le Baron de Lindenau." *Correspondance astronomique, géographique, hydrographique et statistique, 1* (1819), 1–15.

—— "Gnomons et méridiennes filaires." *Correspondance astronomique, géographique, hydrographique et statistique, 3* (1819), 265–79.

—— *Briefe ... an P. Martin Alois David.* Ed. Otto

Seydl. Prague: Gesellschaft der Wissenschaften, 1938.

Zan, Mauro de. "La messa all'Indice del 'Newtonianismo per le dame' di Francesco Algarotti." In Cremante and Tega, *Scienza e letteratura* (1984), 133–47.

Zanella, Gabriele. *Bibliografia per la storia dell'Università di Bologna.* Bologna: Istituto per la storia dell' Università, 1985. *(SMSUB, 5.)*

Zanker, Paul. *The power of images in the Age of Augustus.* Tr. Alan Shapiro. Ann Arbor: University of Michigan Press, 1988.

Zanotti, Eustachio. *La meridiana del tempio di San Petronio rinnovata.* Bologna: Istituto delle scienze, 1779.

—— "Manfredi Eustachio." In Fantuzzi, *Notizie, 5* (1786), 183–93.

Zarlino, Gioseffo. *Resolutioni de alcuni dubii sopra le correttioni dell'anno di Giulio Cesare.* Venice: Polo, 1583. 35 p. In Schiavo, *Meridiana* (1993), after p.48.

Zemplén, Jolan. "The reception of Copernicanism in Hungary." In Dobrzycki, *Reception* (1972), 311–56.

[Zeno, Apostolo]. "L'heliometro fisicocritico." *GL, 6* (1711), 119–49.

Zervas, Diane Finiello. *The parte guelfa: Brunelleschi and Donatello.* Locust Valley, N.Y.: J. J. Sugustin, 1987.

Ziggelaar, August. *Le physicien Ignace Gaston Pardies, S. J.* Odense: Odense University Press, 1971.

—— "The Papal bull of 1582 promulgating a reform of the calendar." In Coyne et al., *Gregorian reform* (1983), 201–39.

Zippel, Giuseppe. "Cosmografi al servizio dei papi nel quattrocento." *Società geografica italiana. Bolletino, 11* (1910), 843–52.

Zucchini, Guido. "Intorno alla Meridiana di S. Petronio." *Coelum, 6* (1936), 25–9.

Notes

Introduction

1. Cf. Costanzi, *Chiesa* (1897), 18–24, who ruins a similar thesis by using it apologetically (e.g., pp. 19, 27, 345–7).

2. Reasons for this complicated recipe and explanations of its terms are given in Chapter 1.

3. *GL, 6* (1711), 118–19.

4. Lalande, *Encycl. meth., Math.* (1784), 2, 381–2, repeated in Lalande, *Voyage* (1786), 2, 251. Cf. Long, *Astronomy* (1742), 1, 61, 115–17, which uses the coin minted in 1695 in celebration of the line in San Petronio as an emblem of astronomy.

5. Vaselli, in Crescimbeni, *Vite* (1708), 2, 39.

6. Crosby, *Measure* (1997), 18–19, 49–74, 226–40.

7. *DSB, 12*, 60–3 (J. F. Daly); Thorndike, *Sphere* (1949); Chapter 2 below. There were at least twenty incunabula and two hundred sixteenth-century editions of Sacrobosco's *Sphere*. Lattis, *Clavius* (1994), 41.

8. Dreyer, *History* (1953), 272–5; Duhem, *Système* (1959), 2, 259–66; *DSB, 1*, 122 (P. D. Thomas).

9. Gingerich, in Westman, *Achievement* (1975), 105–6; Dreyer, *History* (1953), 289.

10. Gassendi, in Hughes, *Regiomontanus* (1967), 11.

11. Gassendi, in ibid., 14–17.

12. Hughes, *Regiomontanus* (1967), 25.

13. Hilfstein, *Starowolski's biographies* (1980), 10 (text of 1625).

14. Rosen, in Copernicus, *Minor works* (1992), 3, 5–17.

15. Copernicus, *Revolutions* [1543], ed. Rosen (1992), 4–5.

16. Cf. Dreyer, *History* (1953), 344–5.

17. See Figures 6.3 and 6.4.

18. Copernicus, *Revolutions* (1992), 3, 4.

19. Copernicus, *Revolutions* (1992), 5–6. Copernicus' expectation that his calendrical parameters were better than the received ones was not realized.

20. Thoren, *Lord* (1990), xx; Wrightsman, in Westman, *Achievement* (1975), 220, 229–30, 233.

21. Dreyer, *Tycho* (1963), 18–19; Thoren, *Lord* (1990), 17.

22. Thoren, *Lord* (1990), 107–12, 122, 133, 142–3, 188–201, 337–40; Dreyer, *Tycho* (1963), 88–109, 117–27.

23. Caspar, *Kepler* (1959), 34–9, 50–2. The charming family portraits are by Kepler; Koestler, *Sleepwalkers* (1970), 229, 236.

24. Koestler, *Sleepwalkers* (1970), 247–53; Caspar, *Kepler* (1959), 62–71, quotes on 63, 71.

25. Caspar, *Kepler* (1959), 85–90, 253–9; Koestler, *Sleepwalkers* (1970), 296–312.

26. Dreyer, *Tycho* (1963), 167–85; Westman, in Westman, *Achievement* (1975), 292–302, 307–29.

27. Thoren, *Lord* (1990), 432–40, 451–3; Caspar, *Kepler* (1959), 116–22; Koestler, *Sleepwalkers* (1970), 313–33. This velocity might be constant about a point outside the center of the circle; see Figure 3.13.

28. Caspar, *Kepler* (1959), 123–40, 321–8.

29. Pastor, *Hist.* (1891), 12, 552–3; Kelly, *Popes* (1986), 261; Daniel-Rops, *Catholic reformation* (1964), 1, 106–14; Jedin, in Jedin, *Handb., 4* (1975), 482–3, 503–4, 507, 520, 527, and in Jedin and Dolan, *Hist., 5* (1980), 456–8, 481–7, 501–6.

30. Feldhay, *Galileo* (1995), 84–7, 92, 95–107, 138–9, 144–5; Pagani, *Metodo* (1726), 1; Alberigo, *Conciles* (1994), 2, 1526–7; Delumeau and Cottret, *Cath.* (1996), 89–90, 102–3; Jedin, in Jedin, *Handb., 4* (1975), 493–4, 513–14, and in Jedin and Dolan, *Hist., 5* (1980), 491–3; Heilbron, *Electricity* (1979), 101–4, 108–9.

31. Finocchiaro, *Affair* (1989), 12–13; Jedin, in Jedin, *Handb., 4* (1975), 490, 573–4; Pelikan, *Christ. trad., 4* (1985), 276–7.

32. Delumeau and Cottret, *Cath.* (1996), 74; Jedin, in Jedin, *Handb., 4* (1975), 490, 533, and in Jedin and Dolan, *Hist., 5* (1980), 468, 510, 546–7.

33. Pelikan, *Christ. trad.*, *4* (1985), 298–302; Delumeau and Cottret, *Cath.* (1966), 80–1.

34. Delumeau and Cottret, *Cath.* (1966), 74–7; Pelikan, *Christ. trad.*, *4* (1985), 279–89; Jedin, in Jedin, *Handb.*, *4* (1975), 491–3, 564–6, and in Jedin and Dolan, *Hist.*, *5* (1980), 469–71.

35. Pastor, *Hist.* (1891), *24*, 311–59, esp. 358, and *25*, 229–51, esp. 241, 248, 251; Kelly, *Popes* (1986), 276, 278; Jedin, in Jedin, *Handb.*, *4* (1975), 568–73, and in Jedin and Dolan, *Hist.*, *5* (1980), 541–5; Chapter 6 below.

36. Delumeau and Cottret, *Cath.* (1966), 207–9.

37. Westman, in Lindberg and Numbers, *God and nature* (1986), 87; Granada, in Bucciantini and Torrini, *Diffusione* (1997), 15–20.

38. Broderick, *Bellarmine* (1961), esp. 7–13, 25–9, 41–8, 405–23; Westfall, *Essays* (1989), 7–13, 23; Jedin, in Jedin, *Handb.*, *4* (1975), 567 (quote), and in Jedin and Dolan, *Hist.*, *5* (1980), 540; Delumeau and Cottret, *Cath.* (1966), 101.

39. Baldini, in Galluzzi, *Crisi* (1983), 301–2, and *Legem* (1992), 286–303; Lattis, *Clavius* (1994), 83, 95–7, 214–15; Bellarmine, *Writings* (1989), 120 (quote), 122 (quote).

40. Lattis, *Clavius* (1994), 14, 31; Baldini, in Galluzzi, *Crisi* (1983), 293–4.

41. Lattis, *Clavius* (1994), 66, 70, 90–1, 102, 110–11, 120–3, 126–129, 132–43.

42. Cf. Duhem, *Phenomena* (1969), 42–105.

43. Galileo, in Drake, *Discoveries* (1957), 21–58; Drake, in ibid., 60–1, 65, 75; Lattis, *Clavius* (1994), 184–90; Pantin, in Kepler, *Discussion* (1993), xvi–xxvi.

44. Galileo, in Drake, *Discoveries* (1957), 181–5, 194, 199–201, 212–15; also in Finocchiaro, *Affair* (1989), 87–118.

45. Drake, *Discoveries* (1957), 165.

46. Finocchiaro, *Affair* (1989), 70, 74–5, 77, 83, 85 (quote); Galileo, in Drake, *Discoveries* (1957), 161–71.

47. Finocchiaro, *Affair* (1989), 81–2, 83 ("eccentrics and epicycles . . . undoubtedly exist in the heavens").

48. Westfall, *Essays* (1989), 11, 17–19; Drake, *Discoveries* (1957), 163–4; Broderick, *Bellarmine* (1961), 343–78.

49. Finocchiaro, *Affair* (1989), 146. Cf. Feldhay, *Galileo* (1995), 94–6.

50. Finocchiaro, *Affair* (1989), 147–9.

51. Ibid., 150, letter of 6 Mar. 1616.

52. Westman, in Galluzzi, *Crisi* (1983), 339, and in Lindberg and Numbers, *God and nature* (1986), 103.

53. Quoted in Hilgers, *Index* (1904), 541–2.

54. Westfall, *Essays* (1989), 41–52; Fantoli, *Galileo* (1993), 217–41.

55. Monchamp, *Notification* (1893), 11–12 (quote), 17.

56. Reusch, *Index* (1883), *1*, 23, and *2*, 30, 38–41, 395, 538; Mendham, *Lit. policy* (1830), 174–85.

57. White, *Hist.* (1896), *1*, 152.

58. Minois, *Eglise* (1990), *1*, 380–1, 385, and *2*, 16 (quote), 17–18.

59. Ibid., 18, 23 (quote), 27, 30, 39; cf. Minois, *Censure* (1995), 121–8.

60. Minois, *Eglise* (1990), *2*, 48.

61. Westfall, *Essays* (1989), 1.

62. Merton, in Cohen, *Puritanism* (1990), 112–31.

63. Ben-David, *Scientist's role* (1971), 45–87; Heilbron, *Sci. cont.*, *3* (1989), 9–28.

64. Bianchini, *Phaenomena* (1728), 53–4; Cruikshank, *Venus* (1983), chapt. 1; Chapter 5 below.

65. Pagani, *Metodo* (1726), "Avvertimenti." The ordinary includes elementary geometry; readers requiring refreshment may find what they need in Heilbron, *Geometry civilized* (1998).

66. Dickens, *Pictures from Italy* (1900), 386; Ciasca, *Arte* (1927), 287.

1. The Science of Easter

1. Pedersen, in Coyne et al., *Gregorian reform* (1983), 18–19.

2. Ibid., 19, 29.

3. Bickerman, *Chronology* (1980), 47–8.

4. Mark 13:24–30.

5. Philip, *Calendar* (1921), 62.

6. Clavius, *Opera* (1612), *5*, 55–62; Pedersen, in Coyne et al., *Greg. ref.* (1983), 24, 28–30, 41–2; Jones, in Bede, *Opera* (1943), 10–11, 20.

7. Pedersen, in Coyne et al., *Greg. ref.* (1983), 30–2, 42–6; Jones, in Bede, *Opera* (1943), 14–15, 21–2, 30–2.

8. Jones, in Bede, *Opera* (1943), 16; Pedersen, in Coyne et al., *Greg. ref.* (1983), 39–40.

9. Pedersen, in Coyne et al., *Greg. ref.* (1983),

47–8; Jones, in Bede, *Opera* (1943), 62–4; Jalland, *Life* (1941), 54–5, 350–8, 412–13.

10. Jones, *Speculum, 9* (1934), 409–13, 410n (quote); Noris, *Annus* (1691), 61; Krusch, Akad. Wiss., Berlin, *Abh.*, 1938:8, 11–15: "The calculator scrupulosus [Victorius] was a very limited mind and also not very honest."

11. Dionysius Exiguus, *PL, 67,* 485–88, 513; Jones, in Bede, *Opera* (1943), 70–1.

12. Dionysius, *PL, 67,* 487, 506; Pedersen, in Coyne et al., *Greg. ref.* (1983), 46–54.

13. Krusch, Akad. Wiss., Berl., *Abh.*, 1938:8, 59–60.

14. Dionysius, *PL, 67,* 497–508.

15. Wijk, *Nombre* (1936), 14, 17–18; Krusch, Akad. Wiss., Berl., *Abh.*, 1938:8, 63–4.

16. Stenton, *Anglo-Saxon England* (1971), 123–4, 132–6, 143–5; Bede, *Eccl. hist.* (1994), 104 ("befogged" from a letter from Pope John IV to the Irish bishops, ca. 650), 156–9 (quote), 269; Moreton, *Isis, 89* (1998), 53–64.

17. Quoted by Jones, in Bede, *Opera* (1943), 104.

18. Quoted by Plummer, *Opera* (1896), 348.

19. Bede, *Eccl. hist.* (1994), 181; Bede, *Opera* (1943), 295–303; Jones, in ibid., 130–1; Pedersen, in Coyne et al., *Greg. ref.* (1983), 56–7.

20. Bede, *Eccl. hist.* (1994), 281 (quote), 283 (quote), 417n; Bede, *Historia,* ed. Plummer (1896), 341.

21. Bede, *Opera* (1943), 175, 182–97, 213–91; Jones, in ibid., 135–8.

22. [Pseudo-]Bede, "Circuli," in *PL, 90,* 859–78, a volume full of spurious computistic works ascribed to Bede.

23. [Pseudo-]Bede, "Ephemeris," in *PL, 90,* 759–88. Cf. Rabanus Maurus, *PL, 107,* 717–18.

24. [Pseudo-]Bede, "De computo dialogus," *PL, 90,* 647–8.

25. Cf. Laistner, *Thought* (1957), 218–19.

26. Jones, in Bede, *Opera* (1943), 70, 117–20.

27. Piper, *Kalendarium* (1858), 13, 107–9; Stenton, *Anglo-Saxon England* (1989), 89–90, 188–90.

28. Kaltenbrunner, Akad. Wiss., Vienna, *Sb., 82* (1876), 293–7.

29. Ibid., 297–9.

30. Ibid., 300–3; North, in Coyne et al., *Greg. ref.* (1983), 79–81.

31. North, in Coyne et al., *Greg. ref.* (1983), 83–4, 89; Kaltenbrunner, Acad. Wiss., Vienna, *Sb., 82* (1876), 310–14.

32. Ibid., 315–23.

33. Ibid., 326–50; North, in Coyne et al., *Greg. ref.* (1983), 87–90.

34. Kaltenbrunner, Akad. Wiss., Vienna, *Sb., 82* (1876), 360–70 (quote).

35. Ibid., 375–97; North, in Coyne et al., *Greg. ref.* (1983), 98–100; Marzi, *Questione* (1896).

36. Kaltenbrunner, Akad. Wiss., Vienna, *Sb., 82* (1876), 401–2, 406; Ginzel, *Handbuch* (1906), *3,* 259.

37. Kaltenbrunner, Akad. Wiss., Vienna, *Sb., 82* (1876), 408–9; Schmid, Görres-Ges., *Hist. Jahrb., 5* (1884), 53–6; North, in Coyne et al., *Greg. ref.* (1983), 101–2; Gregory XIII, *Inter gravissimas,* in Clavius, *Opera* (1612), *5,* 13.

38. Moyer, in Coyne et al., *Greg. ref.* (1983), 127–8; Ziggelaar, ibid., 202–9; Kaltenbrunner, Akad. Wiss., Vienna, *Sb., 97* (1881), 12–21; Schmid, Görres-Ges., *Hist. Jahrb., 3* (1882), 389–90, 406 (on Sirleto).

39. Schmid, Görres-Ges., *Hist. Jahrb., 3* (1882), 406; Cardella, *Memorie* (1792), 97–102.

40. Danti's work is described in Chapter 2.

41. Clavius' *Explanatio* of the new calendar gives tables for two possibilities: the one adopted and one omitting leap-days for forty years; Clavius, *Opera* (1612), *5,* 10–12.

42. Schmid, Görres-Ges., *Hist. Jahrb., 3* (1882), 393–405; Moyer, in Coyne et al., *Greg. ref.* (1983), 182–4; Clavius, *Opera* (1612), *5,* 69–75 (year), 85–9, 147–75 (month).

43. Clavius, *Opera* (1612), *5,* 95–100; Tarmot, *Epistola* (1703), 17, 71–5; Manfredi, *Elementi* (1744), 189, 194, 198, 201–4; Ginzel, *Handbuch* (1906), *3,* 260–1.

44. Tarmot, *Epistola* (1703), 7–12; Manfredi, *Elementi* (1744), 207 (quote).

45. Clavius, *Opera* (1612), *5,* 95–100.

46. Ziggelaar, in Coyne et al., *Greg. ref.* (1983), 220–4; Clavius, *Opera* (1612), *5,* 16–23, 112–31; Philip, *Calendar* (1921), 69–73.

47. Ibid., 74–82; Clavius, *Opera* (1612), *5,* 258–304 (thirty-line table for converting golden numbers and epacts). Cf. Ginzel, *Handbuch* (1906), *3,* 265, 411–23.

48. Kaltenbrunner, Akad. Wiss., Vienna, *Sb., 97* (1881), 31–9, quote on 38; Ziggelaar, in Coyne et al., *Greg. ref.* (1983), 211–17.

49. Kaltenbrunner, Akad. Wiss., Vienna, *Sb., 97* (1881), 27–47; Schmid, Görres-Ges., *Hist.*

Jahrb., 5 (1884), 58–71; Clavius, *Opera* (1612), *5*, 75–81 (Copernicus).

50. Kaltenbrunner, Akad. Wiss., Vienna, *Sb., 87* (1877), 503–19; Ziggelaar, in Coyne et al., *Greg. ref.* (1983), 226–30; Nobis, ibid., 244–5, 248–9.

51. Clavius, *Opera* (1612), *1, 5*, and *5*, f.p., 66–9; Kaltenbrunner, Akad. Wiss., Vienna, *Sb., 87* (1877), 535, 545–50; Ginzel, *Handbuch* (1906), *3*, 266–71.

52. Quoted by Burleigh, *PT, 21:* 257 (1699), 355–6. Cf. Greaves, *PT, 21:* 257 (1699), 356–8 (omit leap days for forty years).

53. Dee, in Heilbron, in Shumaker, *John Dee* (1978), 15; Anon., *Gent. mag., 36* (1851:2), 453–4, 457 (quote). The bishops cited 2 Cor. 6:15 against "Antichrist and various godly wryters" on the iminence of the end of the world.

54. Tycho and Kepler, in Kaltenbrunner, Akad. Wiss., Vienna, *Sb., 87* (1877), 573–82 (quote). Cf. Nobis, in Coyne et al., *Greg. ref.* (1983), 250–1.

2. A Sosigenes and His Caesars

1. Hall, *Renovation* (1979), 4–11, 84; "wanton destruction" was the opinion of Brown, *S. M. Novella* (1902), 71.

2. Allegri and Cecchi, *Palazzo Vecchio* (1980), 189–90, 324, 330; Hall, *Renovation* (1979), 111; Brown, *S. M. Novella* (1902), 71; Palmeri, Dep. stor. pat. Umbria, *Boll., 5* (1899), 82–3, quoting Domenico Mellini (1566).

3. Allegri and Cecchi, *Palazzo Vecchio* (1980), 305–8, 310.

4. Gallo, *Arch. ven., 32/3* (1943), 51–62; Almagià, *Pitture murali* (1952), 13–16.

5. E. Danti, in Danti, *Sacrobosco* (1571), "Proemio," f. *3v.

6. Baleoneus, *I Baglioni* (1964), 176; Pizzoni, Dep. stor. pat. Umbria, *Boll., 42* (1945), 209; *DBI, 32,* 667 (F. P. Fiore).

7. Dante de Rinaldi, in Danti, *Sacrobosco* (1571), f. *rv; Palmeri, Dep. stor. pat. Umbria, *Boll., 5* (1899), 81

8. Danti, *Tavole* (1577), dedication (quote), "Prefazione."

9. Vasari, *Lives* (1980), *1,* 46; *4,* 185, 191.

10. Danti, *Scienze* (1577), tav. 44; Davis, Kunsthist. Inst., Florence, *Mitt., 26* (1982), 64–5, 68, 72; Wazbinski, *Acc.* (1987), *1,* 183–4, 187, 195.

11. Frangenberg, *Nuncius, 3:1* (1988), 12, 17, 26–7, 36; Danti, *Tavole* (1577), and *Trattato* (1569), f. 2r (quote).

12. Allegri and Cecchi, *Palazzo Vecchio* (1980), 303–4, 309, 312; *DSB, 3* (1971), 558 (M. L. Righini-Bonelli); Palmeri, Dep. stor. pat. Umbria, *Boll., 5* (1899), 83, 85–9; Grossi, *Mem. dom., 11* (1980), 514–15.

13. Wazbinski, *Acc.* (1987), *1,* 282–3; Danti may also have taught at the Florentine Accademia del Disegno, where Galileo was to learn his mathematics.

14. Almagià, *Pitture murali* (1952), 5n8; Marchese, *Memorie* (1854), *2,* 287–9.

15. Danti, *Primo volume* (1578), 323.

16. Righini-Bonelli and Settle, IMSS, *Ann., 4:2* (1979), 4–6; Ximenes, *Gnomone* (1757), li–lii; Hall, *Renovation* (1979), 22–3.

17. Danti, *Sacrobosco* (1571), 13–14.

18. Ibid., 6–7.

19. Ibid., 9.

20. Ibid., 22.

21. Roche, *BJHS, 14* (1981), 248–50.

22. Joshua 10:12.

23. Danti, *Sacrobosco* (1571), 8.

24. Danti, *Primo volume* (1578), 319 (quote), 323.

25. Busignani, *Chiese* (1979), 40; Fondelli, in *Atlante* (1993), 40–1.

26. Danti, *Primo volume* (1578), 320.

27. Ximenes, *Gnomone* (1757), li–lii. Danti, *Primo volume* (1578), 284, took $\phi = 46°20'$; it is closer to $43°47'$. An error of a minute of arc in ϕ amounts to an error of an hour in the timing of the equinox; Malezieu, *MAS,* 1714, 321, and Appendix A below.

28. Danti, *Primo volume* (1578), 284–5, 287–8; Lami, *Nov. lett., 19* (1758), col. 82. See also Chapter 8 below.

29. Danti, *Primo volume* (1578), 286; Ximenes, *Gnomone* (1757), xlvi–xlix; Righini-Bonelli and Settle, IMSS, Florence, *Ann., 4:2* (1979), 4–6

30. Malvasia, *Ephem.* (1662), 183–4; Riccioli, *Alm. nov.* (1651), *1:1,* 139.

31. Thus Danti, *Primo volume* (1578), 323. Cf. Turner, *Nuncius, 10* (1995), 132–4, 138; Righini, IMSS, *Ann., 2:2* (1977), 59–63; *DBI, 32,* 660 (E. P. Fiore).

32. King and Turner, *Nuncius, 9* (1964), 167–72, 180–1; Turner and Dekker, *Ann. sci., 50* (1993),

419–20, 437–8; Uzielli, Soc. geogr. ital., *Boll.*, 2 (1889), 856–7; *DBI*, 2, 249 (A. Stella).

33. Danti, *Trattato* (1569), dedication.

34. Palmeri, Dep. stor. patr. Umbria, *Boll.*, 5 (1899), 90–6; Allegri and Cerchi, *Palazzo vecchio* (1980), 312. Cf. Righini-Bonelli and Settle, IMSS, *Ann.*, 4:2 (1979), 7.

35. Danti, *Primo volume* (1578), 320.

36. Ricci, *Bologna* (1980), 74.

37. Cf. Righini-Bonelli and Settle, IMSS, *Ann.*, 4:2 (1979), 11–12.

38. Lalande, *Voyage* (1786), 5, 92; Lastri, *Osserv.* (1821), 3, 11n; Danti, *Primo volume* (1578), 325 (quote).

39. These values come from IC = HIctn α, AB = HI[ctn(α – 15') – ctn(α + 15')], where α = 90° – ϕ ± ε at the solstices and 90° – ϕ at the equinoxes.

40. Calculated from 100 braccia of 0.586 m each from the main door to the old rood screen; Hall, *Renovation* (1979), 201.

41. Danti, *Primo volume* (1578), 325.

42. The discovery twenty years ago of a document authorizing the puncture ended a long debate over the reliability of Danti's report that Toscanelli had installed the *meridiana* in S. M. del Fiore around 1470. Ximenes, *Gnomone* (1757), xx–xxi; Settesoldi, *Prospettiva*, 16 (Jan. 1979), 44; Settle, IMSS, *Ann.*, 3:2 (1978), 69–70.

43. Manetti, *Life* (1970), 55; Zervas, *Parte guelfa* (1987), 147–51, 161n.

44. Ragghianti, *Brunelleschi* (1977), 178, 187–8; Celoria, *Toscanelli* (1921), 63–5; Righini, *Physis*, 4 (1962), 135–6.

45. Ximenes, *Gnomone* (1757), xlii.

46. Celoria, *Toscanelli* (1921), 66; Ximenes, *Gnomone* (1757), 174; Righini, *Physis*, 4 (1962), 136. The instrument had been repaired (the hole reset) in 1511; Guasti, *Cupola* (1857), 184.

47. Maccaferri, *Il carobbio*, 7 (1981), 256; Paltrinieri, *Meridiane* (1995), 336.

48. Matteucci, in Amico and Grandi, *Tramonto* (1987), 49–50.

49. Dante, *Primo volume* (1578), 325.

50. Fanti, *Fabbrica* (1980), 173–5.

51. Ibid., 175–8.

52. Riccioli, *Alm. nov.* (1651), 1:1, 131–2.

53. Ibid. (1651), 1:1, 136; Riccioli to Kircher, 1646, in Gambaro, *Astr.* (1989), 73. We have (Fig.

2.29), with HI = h, IC = x, tan α = h/x, tan(α – $\Delta\alpha$) = $h/(x + \Delta x) \approx$ tan $\alpha(1 – \Delta x/x)$, where $\Delta\alpha$ is half the sun's angular diameter and Δx is BC; α = 90° – ϕ and ϕ = 44°30' according to Riccioli.

54. Danti, *Usus* (1576); cf. Guarducci, AS, Bologna, *Mem.*, 2 (1905), 285, and Casanovas, in Coyne et al., *Greg. ref.* (1983), 194.

55. Manfredi, *De gnomone* (1736), 2; Delambre, *Hist.* (1821), 2, 688–9, 721; Cassini, *Meridiana* (1695), 6–7. Giustiniani, *Scrittori* (1667), 30–1, was wrong to call Danti's line useless.

56. Gassendi to Wendelin, 14 Kal. Jun. (18 May) 1636, in Gassendi, *Opera* (1658), 4, 524; Gassendi, *Peiresk* (1657), 143–4.

57. Gassendi to Wendelin, Id. Jul. (15 July) 1636, in Gassendi, *Opera* (1658), 4, 524.

58. Ibid., 525–30.

59. Ibid., 536–7 (letter of Id. Jun. 1643). Pytheas obtained ϕ = 43°3.5' for Marseilles, not far off the modern value of 43°11'; Lelewell, *Pythéas* (1836), 47, 56.

60. Danti, *Primo volume* (1578), "A' lettori" and 324; Palmeri, Dep. stor. pat. Umb., *Boll.* 5 (1899), 97; *DBI*, 32, 660 (E. P. Fiore); Cardella, *Memorie* (1792), 5, 102–3.

61. Zaccagnini, *Storia* (1930), 258–9; Dallari, *Rotuli* (1888), 2, 195, 198, 201; Baccini, *Arch. stor. Marche Umb.*, 4 (1888), 87, 102–6, 110–11.

62. Zaccagnini, *Storia* (1930), 260–61; the letter was dated 13 Apr. 1577.

63. Danti, *Trattato* (1583), "A' lettori," and 107–8; Dante to Priore degli Innocenti, Florence, 23 Nov. 157[8], in Ferrate, *Lettere* (1878), 7–8; Grohmann, *Perugia* (1981), 7–8, 82, 105, 108 (work of Giulio Danti).

64. Danti to Priore degli Innocenti, 23 Nov. 157[8], in Ferrate, *Lettere* (1878), 7–8. The map was painted over in 1798; Pastor, *Hist.* (1891), 20 (1923), 619–21.

65. Palmeri, Dep. stor. pat. Umbria, *Boll.*, 5 (1899), 99–100; Brink, Kunsthist. Mus., Florence, *Mitt.*, 27 (1983), 247n17.

66. Zaccagnini, *Storia* (1930), 261; *DBI*, 32, 660 (E. P. Fiore); Almagià, *Pitture murali* (1952), 6.

67. Zippel, Soc. geog. ital., *Boll.*, 11 (1910), 843–52.

68. Almagià, *Pitture murali* (1952), 11–12; Banfi, *Imago mundi*, 9 (1952), 31–2, 34.

69. Thomassy, *Nouv. ann. voy.*, 136 (1852), 59, 62, 82, 90, 96; 137 (1853), 155–6; 138 (1853), 37–9.

70. Almagià, *Pitture geografiche* (1955), 28
(quote), reproduction after p. 28; Banfi,
Imago mundi, 9 (1952), 23–4, 28–9, 33–4.

71. Thomassy, *Nouv. ann. voy., 138* (1853), 7–28;
Almagià, *Pitture mundi* (1952), 2, and *Pitture
geografiche* (1955), 29; Ortelius, *Theater of the
world* (1606), ed. Skelton, containing Danti's
"Perugia" on folio 82; Ricci, *Bologna* (1980),
8–9, 14–15.

72. Schmidt, *Geog. Zs., 17* (1911), 506–12, 517.

73. Almagià, *Pitture murali* (1952), 1, 3–4, and
plates of the Belvedere maps; Almagià, *Carte
geografiche* (1948), 14–15, and Tav. III.

74. Almagià, *Pitture murali* (1952), 3, 8–10; cf.
Manfredi to Marsili, 1711, on Magini's maps,
in Baiada et al., *Museo* (1995), 59.

75. It later did useful work; Gilii, *Risultati* (1808),
iv–v, 26–40.

76. Danti, *Primo volume* (1578), ad fin.; Brink,
Kunsthist. Inst., Florence, *Mitt., 27* (1983),
244–5.

77. Stein, Spec. Vat., *Misc. astr., 3* (1950), 34–5.

78. Repeated by, e.g., Lalande, *Voyage* (1786), *4*,
66.

79. Pastor, *Hist* (1891), *19*, 287n2. Vermiglioli, *Op.
lett. Bol., 3* (1820), 17; Schmid, *Hist. Jahrb., 3*
(1882), 415; Casanovas, in Coyne et al., *Greg.
ref.* (1983), 191–2.

80. Palmeri, Dep. stor. pat. Umbria, *Boll., 5* (1899),
106–13; Dibner, *Moving the obelisks* (1970), 21,
32, 40.

81. Vermiglioli, *Op. lett. Bol., 3* (1820), 17; Al-
magià, *Pitture murali* (1952), 7; Marchese,
Memorie (1854), 2, 296; *DBI, 32*, 661 (E. P.
Fiore).

82. Danti to Duca d'Urbino, 25 Sept. 1576, in Almagià, *Pitture geografiche* (1955), 5.

83. Resp., Danti to Priore degli Innocenti, 23 Nov.
157[8], in Ferrate, *Lettere* (1873), 8, and Danti
to the Dominican General [?], 20 Mar. 1577, in
Almagià, *Pitture murali* (1952), 5.

84. Vermiglioli, *Op. lett. Bol., 3* (1820), 5–6.

3. Bononia Docet

1. Fontenelle, *Oeuvres* (1728), *3*, 161.

2. Ibid., 166.

3. Delambre, *Hist. astr. mod.* (1821), 2, 603. Delambre, the chief architect of the metric system, resented the ascendancy of the Cassinis.

4. Biot, *Biog. univ., 7 (*1854), 133.

5. Cf. Algarotti to E. Zanini, in Algarotti, *Opere,
10* (1784), 368, mentioning Toscanelli, Cassini,
and Cassini's successor at Bologna, Eustachio
Manfredi; Delambre, *Hist.* (1821), 2, 686–7.

6. Malvasia, *Ephem.* (1662), title page and dedication; Montanari, *Cometes* (1665), 6, 44.

7. Malvasia, *Discorsi astrologici* (Bologna, 1647
ff.); Fantuzzi, *Notizie* (1781), *5*, 359–61.

8. G. D. Cassini, in J. D. Cassini, *Mémoires*
(1810), 256–64; Fontenelle, *Oeuvres* (1728), *3*,
155–6; Grillot, *Nuncius, 2:2* (1987), 146–7; G. D.
Cassini, *Specimen* (1656), 19.

9. Riccioli, *Alm. nov.* (1651), *1:1*, x, xxvii (three
years), xli (Copernicus), 43–4 (house systems); Delambre, *Hist.* (1821), 2, 687 (quote)

10. Fantuzzi, *Notizie* (1781), *5*, 159.

11. J. D. Cassini, *Mémoires* (1810), 261, 264; Giustiniani, *Scrittori* (1667), 359; *DBI, 21*, 484 (Ferrari). Cavalieri included Copernicus' system
among several other astronmical "hypotheses" in his courses between 1642 and 1645;
Riccardi, *Bull. bibl. stor. sci., 12* (1879), 303–4;
Vernazza, *SMSUB, 2* (1961), 139.

12. Dallari, *Rotuli* (1888), *2*, 470.

13. Betti, in Bucciantini and Torrini, *Diffusione*
(1997), 67; Rotta, *Misc. seic., 2* (1971), 119.

14. G. D. Cassini, *Meridiana* (1695), title page;
Favaro, Dep. stor. patr. Romagna, *Atti mem.,
22* (1904), 413; Bortolotti, *Storia* (1947),
139–41.

15. Bianchini, in Montanari, *Forze* (1694), f.
a.9–10; Fantuzzi, *Notizie* (1781), 159; Bianchi,
Obs., *Atti, 1* (1834), xii–xiii; Rotta, *Misc. seic., 2*
(1971), 67, 72–6, 143n22, 145n34. The itinerant
Galilean was Paul del Buono.

16. Rotta, *Misc. seic., 2* (1971), 73–4, 81, 146n36;
Maffioli, *Janus, 71* (1984), 64; Cavazza, *Et.
phil.,* no. 3 (1985), 407–11.

17. Rotta, *Misc. seic., 2* (1971), 119 (quote), 120,
124–8; Montanari, *Astr.* (1685), 121–3;
Thorndike, *Hist., 8* (1958), 342–4.

18. Baiada, *Museo* (1995), 38; Cavazza, *SMSUB, 3*
(1983), 441–7; Bongiovane, *Scherzi astrologici*
(1667), 7 (quote), 55; Montanari, *Astr.* (1685),
ix (second quote), xii–xv, 131–41.

19. Fantuzzi, *Notizie* (1781), *6*, 9–11; Vernazza,
SMSUB, 2 (1961), 140–1; Mengoli to Magli-
abecchi, 5 Oct. 1675 and July 1681 (quote), in
Mengoli, *Corrisp.* (1986), 80, 118, and
Baroncini, in ibid., 158–79.

20. Mengoli to Magliabecchi, 20 Feb. and 12 Oct. 1677 (quote), 20 June 1680, 1 June 1675 (quote), and 10 Nov. 1676 (quote), in Mengoli, *Corresp.* (1986), 99–100, 103, 109, 66, 93; Baroncini, in ibid., 157; Mengoli, *Mese* (1681), 156; Zanella, *Bibl.* (1985), 197.

21. Cf. Rotta, *Misc. seic.*, 2 (1971), 70.

22. Dallari, *Rotuli* (1888), 2, 473–4, 478, 482, 487, 491, 495, 501; 3, 6, 17, 26, 31, 35–6, 40–1; J. Collins to Newton, 16 July 1670, in Newton, *Corresp.* (1959), 1, 33–4 (on Mengoli).

23. Baldini, in Pepe, *Copernico* (1996), 125, 128, 131, 135–7 (quote).

24. Ibid., 130, 147–8, 153, 156–7, 173–4; Riccioli to ?, 14 Nov. 1643, in Gambaro, *Astr.* (1989), 65, and Riccioli, *Alm. nov.* (1651), 1:1, 204.

25. The censors' questions, dated 24 Nov. 1646, are in Gambaro, *Astr.* (1989), 40; Riccioli's answers, as related to Kircher, in letters of [Dec. 1646] and 22 Dec. 1646, ibid., 72–3, 75, 78–9, 81.

26. Riccioli to Kircher, 22 Dec. 1646, in ibid., 79.

27. Ibid., 81.

28. Ibid., 176–8; Riccioli, *Alm. nov.* (1651), 1:1, f. A.2, xviii, xxvii (a notice of himself among brief biographies of important astronomers).

29. Riccioli, in Grimaldi, *Physico-mathesis* (1665), unnumbered leaf after index; Fantuzzi, *Notizie* (1781), 4, 305.

30. Riccioli, *Alm. nov.* (1651), 1:1, xiii (quote), xv.

31. Forbes, in Flamsteed, *Lectures* (1975), 6–7, and index, s.v. "Riccioli"; Riccioli, *Alm. nov.* (1651), 1:1, f. A2.

32. Riccioli, *Alm. nov.* (1651), 1:1, 204 1/2 (yes, p. 204.5)–205; Schreiber, *Stimmen aus Maria-Laach,* 54 (1898), 255–6, 260–3, 265; Macdonald, Brit. Astr. Soc., *Jl.,* 77 (1967), 112; Whitaker, in Taton and Wilson, *Planetary astronomy* (1989), 134–8.

33. Montgomery, *Scientific voice* (1996), 242, 250, 264–5, 272–4.

34. Riccioli, *Alm. nov.* (1651), 1:1, 161. Cf. *DSB, 5,* 542 (B. Eastwood).

35. Quoted from Gambaro, *Astr.* (1989), 15.

36. Riccioli, *Astr. ref.* (1665), 1, 5, and *Geog. ref.* (1661), 295; Ricci, *Bologna* (1980), 117, 119. Cf. Flamsteed, *Lectures* (1975), 242.

37. Fanti, *Fabbrica* (1980), 187–9, 191–9, 203–4, 207.

38. G. D. Cassini, in J. D. Cassini, *Mémoires* (1810), 268, and *Meridiana* (1695), 7–8

39. Maccaferri, *Il carrobio, 7* (1981), 246–7; Tabarroni, in Bellosi, *Basilica* (1983), 2, 331; Manfredi, *De gnomone* (1736), 3.

40. Guarducci, AS, Bologna, *Mem.,* 2 (1905), 285; Zucchini, *Coelum, 6* (1936), 26–7; Fanti, *Fabbrica* (1980), 209, 251.

41. Riccioli, *Astr. ref.* (1665), 1, 6; Tabarroni, in Bellosi, *Basilica* (1983), 2, 335n10; Dallari, *Rotuli* (1888), 2, 465–6, 474; Bortolotti, *Storia* (1947), 143; Vernazza, *SMSUB, 2* (1961), 137–8. The *taccuino,* or almanac, remained a charge of the astronomy professor at Bologna far into the eighteenth century; Fantuzzi, *Notizie* (1781), 5, 194.

42. G. D. Cassini, *Meridiana* (1695), 9–10; Guglielmini, in ibid., 41–2.

43. G. D. Cassini, *Meridiana* (1695), 14; Guglielmini, in ibid., 40.

44. Quoted by Guarducci, AS, Bologna, *Mem.,* 2 (1905), 286.

45. Fontenelle, *Oeuvres* (1728), 3, 159; Gualdo Priorato, *Historia* (1656), 125.

46. G. D. Cassini, *Meridiana* (1695), 10–11; Manfredi, *De gnomone* (1736), 4.

47. G. D. Cassini, *Meridiana* (1695), 11–12; Riccioli, in Mutus, *Dialogus* (1664), 2.

48. G. D. Cassini, *Specimen* (1656), dedication.

49. Gualdo Priorato, *Historia* (1656), 67–70; Arckenholtz, *Mém.* (1751), 3, x, 495–6; Giordani, *Alm. stat. bol., 11* (1840), 67, 71–3, 76–9, 89–90; Malagola, *Gior. arald. gen. dipl., 8* (1880–81), 208, 245–7; Callmer, in Platen, *Queen Christina* (1966), 60; Stolpe, *Drott. Kristina* (1960), 2, 30–1.

50. Mengoli, *Via regia* (1655), 5, 43; Baroncini, in *Materiali* (1975), 80; Christina, "Sentiments," in Arckenholtz, *Mém.* (1751), 4, 52.

51. G. D. Cassini, in J. D. Cassini, *Mémoires* (1810), 270, and *Meridiana* (1695), 14; Gualdo Priorato, "Lomellino," in *Scena* (1659).

52. Odier and Partini, *Physis, 25* (1983), 257–60.

53. Cardella, *Mem.* (1792), 7, 90; Stolpe, *Drott. Kristina* (1960), 2, 32–3; Masson, in Platen, *Queen Christina* (1966), 244; Bedini, in Anderson et al., *Instr.* (1993), 99–100.

54. G. D. Cassini, *Meridiana* (1695), 21; R. Soprani, *Li scrittori della Liguria* (Genoa, 1667), quoted in A. Cassini, *Cel. Cassiniana* (1987), 4.

55. Blaeu, *Grand atlas* (1663), *9*, 124–5; Monconys, *Voyages* (1695), *4*, 423.

56. Guglielmini, in G. D. Cassini, *Meridiana* (1695), 38; Fontenelle, *Oeuvres* (1728), *3*, 134.

57. G. D. Cassini, in J. D. Cassini, *Mémoires* (1810), 271–2; Fontenelle, *Oeuvres* (1728), *3*, 161.

58. Giustiniani, *Scrittori* (1667), 365; *DBI*, *21*, 484–5 (Ferrari); Cardella, *Mem.* (1792), *7*, 124 (quote); correspondence in Cassini Papers (Paris), A.D. 2.34.

59. Fontenelle, *Oeuvres* (1728), *3*, 161, 164; *DBI*, *21*, 484–5 (Ferrari); Monconys, *Voyages* (1695), *4*, 504–5, describes a delay at Urbano.

60. G. D. Cassini, *Lettere astr.* (1665), 14, and in J. D. Cassini, *Mémoires* (1810), 279–82; Arckenholtz, *Mém.* (1751), *2*, 146; Bedini, in Anderson et al., *Instr.* (1993), 108–9.

61. G. D. Cassini, in J. D. Cassini, *Mémoires* (1810), 282, and G. D. Cassini, *Origine* (1693), 51–2.

62. G. D. Cassini, in J. D. Cassini, *Mémoires* (1810), 282–4.

63. Righini-Bonelli and van Helden, *Divini and Campani* (1981), 11–13, 22; Campani, *Ragguaglio* (1664), 8–9, 17, 21.

64. Oldenburg to Boyle, 29 Sept. and 5 Nov. 1664, and 5 Dec. 1665, in Oldenburg, *Corresp.* (1965), *2*, 240, 293, 629; Righini-Bonelli and van Helden, *Divini and Campani* (1981), 24–5, 28, 32–8.

65. Righini-Bonelli and van Helden, *Divini and Campani* (1981), 21, 25, 28n.

66. Campani, *Ragguaglio* (1664), 38–9.

67. Cassini, *Lettere . . . Falconieri* (1665), 1–8; Campani, *Lettera* (1665), 3 (quote).

68. *JS*, 22 Feb 1666, 295–6; Huygens to Moray, 24 Dec. 1665, in Huygens, *Oeuvres* (1888), *5*, 550; Huygens to his brother, 12 Oct 1668, ibid., *6*, 267 (quote). Further to Cassini's discoveries: Giustiniani, *Scrittori* (1667), 366–70; Delambre, *Hist.* (1821), *2*, 709–10; Arago, *Oeuvres* (1855), *3*, 316–17; *DSB*, *3*, 100–1 (R. Taton); *DBI*, *21*, 485 (Ferrari).

69. G. D. Cassini, *Ephem. bonon.* (1668); Débaret and Wilson, in Taton and Wilson, *Planetary astronomy* (1989), 149–50; Heilbron, *Imponderables* (1993), 187, 221.

70. Delambre, *Hist.* (1821), *2*, 745; G. D. Cassini, *Origine* (1693), 49.

71. G. D. Cassini, *Ephem. bonon.* (1668),

"Proemium," f. †2, and *Origine* (1693), 4–6 (quote).

72. Riccioli, *Alm. nov.* (1651), *1:1*, 45–6, 114–15.

73. From $\triangle$SPC in Figure 3.8, $\chi = (r/s)\sin(90° + \alpha)$ $= (r/s)\cos \alpha$.

74. Riccioli, *Alm. nov.* (1651), *1:1*, 113–14; Van Helden, *Measuring* (1985), chap. 1–2.

75. Arago, *Oeuvres* (1855), *3*, 317.

76. *DBI*, *21*, 485 (Ferrari); Fontenelle, *Oeuvres* (1728), *3*, 165–6; *DSB*, *3*, 102 (R. Taton).

77. G. D. Cassini, in J. D. Cassini, *Mémoires* (1810), 286, 293–4; cf. Delambre, *Hist.* (1821), *2*, 692, and Wolf, *Hist.* (1902), 21–6.

78. J. Cassini, *Elémens* (1740), vi.

79. Wolf, *Hist.* (1902), viii, x–xi, 195, 205, 208–10; Wolf effectively answers Cassini's detractors Delambre and Arago (ibid., 198–204, 210–12). The eventual installation of the *meridiana* is described in Chapter 5 below.

80. Biagioli, *Galileo* (1993), 104–5, 139–50.

81. Rheita, *Novem stellae* (1643), 6; Huygens to his brother, 28 July 1673, in Huygens, *Oeuvres* (1888), *7*, 348 (Cassini's industry); G. D. Cassini, *Découvertes* (1673), "Au Roy" (play on xiv); Fontenelle, *Oeuvres* (1728), *3*, 168; Savorgnan, *Opera* (1941), 46–62.

82. Cf. Huygens, *Oeuvres* (1888), *21*, 25.

83. G. D. Cassini, *Specimen* (1656), 3; Riccioli, *Astr. ref.* (1651), *1*, 5, 15–16, 161; Mengoli, *Refrattioni* (1670), 15–17.

84. G. D. Cassini, *Specimen* (1656), 11, 16–17; Riccioli, *Alm. nov.* (1651), *1:1*, 139 ($365^d5^h48^m40^s$).

85. The relations between mathematicians and the censorship are examined in Chapter 6.

86. *Whitaker's almanack* (London, 1993, 1994); cf. Riccioli, *Alm. nov.* (1651), *1:1*, 143–4, 152–4.

87. A hint to trigonometers: apply the law of sines to $\triangle$JES and $\triangle$IES; that gives $\tan \psi = \cos[(\angle GSH + \angle HSI)/2]/\cos[(\angle GSH + \angle JSG)/2]$ and $e = \cos[(\angle GSH + \angle HSI)/2]/\sin \psi$.

88. From Malvasia's timings of the equinoxes and solstices of 1663 and 1664 (VE at 20 Mar. 4^h35^m and 19 Mar. 10^h29^m, SS at 21 June 5^h17^m and 20 June 11^h15^m, AE at 22 Sept. 20^h28^m and 22 Sept. 20^h34^m, and WS at 21 Dec. 10^h46^m and 20 Dec. 16^h42^m, all times P.M.), one deduces by Ptolemy's method $\psi = 8 \textcircled{\odot} 34$ (8°34' of Cancer), $e = 0.035$; Malvasia gave $\psi = 8 \text{♋} 18$, $e = 0.017$. Malvasia, *Ephem.* (1662), 60, 91, 185.

89. Wilson, *Isis*, *59* (1968), 6–7, and *AHES*, *6* (1970), 96–7, 99; Stephenson, *Kepler's phys. astr.* (1994), 49–61.

90. Flamsteed, *Lectures* (1975), 195 (text of 16 Nov. 1681); G. D. Cassini, *Meridiana* (1695), 24–6.

91. E.g., Keill, *Introduction* (1721), 278.

92. Riccioli, *Alm. nov.* (1651), *1:1*, 119, 151–2, 570 (quote).

93. Gassendi, *Inst.* (1653), 77–9, 96.

94. Ptolemy had given 0.042 and Copernicus 0.032; Riccioli, *Alm. nov.* (1651), *1:1*, 156.

95. Riccioli, *Alm. nov.* (1651), *1:1*, 152–6; Flamsteed, *Lectures* (1975), 197 (text of 16 Nov. 1681).

96. Gassendi, *Peiresk* (1657), 131–2.

97. G. D. Cassini, *Meridiana* (1695), 10.

98. Ibid., 10, making $2p \approx 1$ inch Paris measure. The factor of 100 comes from csc $2\alpha_{ss} = 0.01$ and $2p/\text{PR} = 2p/\sigma\text{hcsc } 2\alpha$.

99. Since $h \gg \text{QU} = h\text{csc } \alpha$, $\text{OQ} \approx \text{OU}$; strictly speaking, the center of the image on the floor is not the image of the center of the sun, but the discrepancy usually can be neglected.

100. Cf. Tabarroni, in Bellosi, *Basilica* (1983), *2*, 333.

101. Guglielmini, in G. D. Cassini, *Meridiana* (1695), 51; Manfredi, *De gnomone* (1736), 7–8. Cf. Lemmonier, *MAS*, 1762, 266, and Delambre, *Hist.* (1821), *2*, 723.

102. G. D. Cassini, *Meridiana* (1695), 10–11, 13 (quote). Cf. Riccioli, *Alm. nov.* (1651), *1:1*, 117, reporting his own and Kepler's concern about determining the place of the image precisely.

103. Riccioli, *Astr. ref.* (1665), *1:1*, 38; Guglielmini, in G. D. Cassini, *Meridiana* (1695), 51–2; Manfredi, *De gnomone* (1736), 10–11.

104. Manfredi, *De gnomone* (1736), 12–13.

105. Grimaldi, *Physico-math.* (1665), 5–6, 220–3. Grimaldi may have been alerted to diffraction phenomena by the incongruously large values of σ (between 38' and 56') given by Christopher Scheiner, a Jesuit student of sun spots; Wilson, *AHES*, *6* (1970), 105.

106. Riccioli, *Astr. ref.* (1665), *1:1*, 37; Baiada et al., *Museo* (1995), 36 (first quote); Mengoli, *Reffrationi* (1670), 28 (second quote); Masini, *Bologna perlustrata* (1666), *3*, 254.

107. G. D. Cassini, *Specimen* (1656), 29, 30.

108. G. D. Cassini, *Novum lumen* (1656), 1–3; cf. Tabarroni, *Strenna*, *18* (1968), 318.

109. Riccioli, *Alm. nov.* (1651), *1:1*, 101–3. These systems appear in the chapter openers in this book; see infra, p. 355, for identifications.

110. Riccioli, *Alm. nov.* (1651), *1:1*, 101–103; Costantini, *Baliani* (1969), 3–8, 39, 51–2, 76.

111. Cf. Russell, *BJHS*, *2* (1964), 9.

112. G. D. Cassini, *Specimen* (1656), 25–6; Gregory, *Elements* (1715), *1*, 392–400; the unfriendly quotation is from Arago, *Oeuvres*, *3* (1855), 317–18.

113. *HAS*, 1703, 68; David Gregory, "Adnota . . . cum Neutono, 4 May 1694," in Newton, *Corresp.* (1959), *3*, 314, 318, 322n16; cf. ibid., *4*, 355, Delambre, *Hist.* (1821), *2*, 724, 735, and Arago, *Oeuvres*, *3* (1855), 317–18.

114. Riccioli, *Alm. nov.* (1651), *1:1*, xvii, xvi (quote); Levera, *Prodromus* (1663), 23. Cf. C. A. Paggi to Levera, 1 July 1664, in Palatino, *Ephem.* (1664), f. †4.

115. Allacci, *Apes* (1633), 107; Mandosio, *Bibl.* (1682), *2*, 313; Levera, *Prodromus* (1663), dedication (quote), and *De inner. stell. viribus* (1664), f. ⁺2; Christina, letter of 23 Mar. 1686, in Arckenholtz, *Mém.* (1751), *4*, 53–4 (quote); Bedini, in Anderson et al., *Instr.* (1993), 112. Thorndike, *History* (1923), *8*, 321–3, discusses Levera's astrology; otherwise he seems to have escaped scholarly attention although he belonged to a group of astrologer-astronomers including Andreas Argoli and Pietro Palatio, both known calculators of ephemerides.

116. Kepler, as quoted by Riccioli, *Alm. nov.* (1651), *1:1*, 149.

117. The convention in the seventeenth century was to count time from aphelion; to ease comparison with more familiar treatments, the modern convention of beginning at perihelion is followed.

118. With $\Delta\text{FPQ} = (a/2)(\cos \eta - e/2)b\sin \eta$ and $\text{Q}\Pi\text{P} = (ab/2)(\eta - \sin \eta \cos \eta)$, the equality of areas gives $\eta - (e/2)\sin \eta = 2\pi t/y$, which is Kepler's equation.

119. Ward, *Astr. geom.* (1656), "Praefatio," ††3.

120. Russell, *BJHS*, *2* (1964), 12–16.

121. Ward, *Inquisitio brevis* (1653), "Praefatio," and 26; Riccioli, *Alm. nov.* (1651), *1:1*, 149–50; Gre-

gory, *Elements* (1715), *1*, 387–8; Boyer, *Isis*, *38* (1947), 54–5; Wilson, *AHES*, *6* (1970), 107–16, 120.

122. Ward, *Inquisitio brevis* (1653), 28. Cf. Ward, *Astr. geom.* (1656), 6–9, 24.

123. Ward, *Astr. geom.* (1656), "Dedicatio," f. †1rv.

124. Ibid., "Prefatio."

125. Ibid., 1, 3, 7–9, 18; Thoren, *BJHS*, *7* (1974), 247–8; Wilson, *AHES*, *6* (1970), 121.

126. Gregory, *Elements* (1715), *1*, 386–88; Wilson, *AHES*, *6* (1970), 118, 124–5; Thoren, *BJHS*, *7* (1974), 248–9, 256. Cf. Whiteside, *BJHS*, *2* (1964), 121–8, and Russell, *BJHS*, *2* (1964), 18–19.

127. Cassini, *JS*, *2* (1669), 549; Gassendi to Cassini, xiv kal. Feb. (19 Jan) 1654, in Gassendi, *Opera* (1658), *6*, 323.

128. Fontenelle, *Oeuvres* (1728), *3*, 157, following the quasi-official history of the early Paris Academy, Duhamel, *Historia* (1698), *1*, 55.

129. Cassini, *JS*, *2* (1669), 549.

130. Ibid., 550.

131. Wilson, *AHES*, *6* (1970), 124–6, 129–34, and in Taton and Wilson, *Planetary astronomy* (1989), 181–3; Russell, *BJHS*, *2* (1964), 18–19.

132. Mercator, *PT*, *5:57* (1670), 1171–4; Gregory, *PT*, *24:293* (1704), 1704–6; J. Cassini, *MAS*, 1723, 158–63; Manfredi, *De gnomone* (1736), 88–93; T. Narducci to G. Grandi, 30 July 1738, in Arrighi, *Physis*, *4* (1962), 131; Delambre, *Hist.* (1821), *1*, 762 (quote), 765–78.

133. Delambre, *Hist.* (1821), *2*, 761 (quote), 770, 777; Newton to John Collins, 6 Feb. 1669/70, in Newton, *Corresp.* (1959), *1*, 23; Whiteside, *BJHS*, *2* (1964), 122–8. Cf. Gregory to Collins, 5 Sept. 1670 (ibid., 41), asking for news of Mercator's debate with Cassini: "somethynge to ye same purpose . . . hath been lying by mee these several years."

4. Normal Science

1. Fabri, *Efemeridi* (1675), 6–14, 20; Fantuzzi, *Notizie* (1781), *6*, 17; Mengoli, *Anno* (1673), 4.

2. Morgani, in Guglielmini, *Opere* (1719), *1*, 2–10; Montucla, *Hist.* (1802), *3*, 714–15; Fontenelle, *Oeuvres* (1728), *3*, 135, 138; Maffioli, in Cavazza, *Rapporti* (1987), 81–3, and in *Janus*, *71* (1984), 67–9; Robinet, in Cavazza, *Rapporti* (1987), 10–14, 43–4.

3. Morgani, in Guglielmini, *Opere* (1719), *1*, 15, 17, 26–37; Fantuzzi, *Notizie* (1781), *4*, 320; Fontenelle, *Oeuvres* (1728), *3*, 134, 144.

4. Montanari, *Astrologia* (1685), 133; Fantuzzi, *Notizie* (1781), *5*, 9–10, 17–18, 183–6; Cavazza, *Settecento* (1990), 173–5, 216–19; Cassini to Manfredi, 1 May 1699, Cassini Papers (Bologna), 4705; Baiada et al., *Museo* (1995), 43–4.

5. Baiada et al., *Museo* (1995), 59–60; Fantuzzi, *Notizie* (1781), *5*, 184, 187–8, 193–4; Pepe, in Heinekamp, *Beiträge* (1986), 83–5; Robinet, in Cavazza, *Rapporti* (1987), 45–6 (on Gabriele Manfredi).

6. Fantuzzi, *Notizie* (1781), *5*, 190 (quote).

7. Manfredi, *De gnomone* (1736), 17–20, 99–397.

8. *AE*, Oct. 1740, 602–7, on 603.

9. Riccioli, *Astr. ref.* (1665), *1*, 29–36, and *passim*; Wilson, *AHES*, *6* (1970), 104–5; Chapter 6 below. Cf. Mengoli, *Refrattioni* (1670), 19–26.

10. Levera, *Prodromus* (1663), 124, citing *Alm. nov.* (1651), *1:1*, 116–17.

11. Riccioli, in Levera, *Prodromus* (1663), 2–3; Pierre Petit to Huygens, 2 Sept. 1662, in Huygens, *Oeuvres* (1888), *4*, 236; Flamsteed, *Lectures* (1975), 242 (11 May 1682).

12. Levera, *Prodromus* (1663), 124–9; Mutus, *Dial.* (1664), 14–15. Riccioli, *Astr. ref.* (1665), *1*, 7, reports the observation of the VE in 1656.

13. Riccioli, in Mutus, *Dial.* (1664), 2; G. D. Cassini, in ibid., 8, and in Cassini Papers (Paris), 22: 9–13, 46–7, 198–269, 271–6.

14. Manfredi, *De gnomone* (1736), 1.

15. Riccioli, *Alm. nov.* (1651), *1:1*, 107.

16. Van Helden, *Measuring* (1985), 4–19, 46, 50. The inventor of the eclipse method got $k = 2.5$.

17. With $r/s = 3' = 0.00087$ rad, $m = 64r$ according to Equation 4.2 with $k = 2.8$.

18. From $(R/r)^3 = (\sigma s/2r)^3 = s/r$, or $\chi^2(0) = \sigma^3/8$. With $\sigma = 30' = 0.0087$ radians, $\chi(0) = 1'$ and $s = 3438r$.

19. Van Helden, *Measuring* (1985), 78–9, 84.

20. Riccioli, *Alm. nov.* (1651), *1:1*, 108, 110, 112; Flamsteed, *Lectures* (1975), 78, 95–7, 311, 376–7, 413; Van Helden, *Measuring* (1985), 113, 117.

21. Streete, *Astr. carol.* (1661), 12 (1710), 34; Flamsteed to Cassini, 7/17 July 1673, in *PT*, *8:96* (21 July 1673), 6[1]00 ($\chi(0) < 10''$), and *Lectures* (1975), 101, 103 (text of 4 May 1681).

22. Streete, *Astr. carol.* (1661), 12 (1710), 34; Huygens, *Oeuvres* (1888), *21*, 31 (text of 1667); G. D. Cassini, *Origine* (1693), 38 (quote); Van Helden, *Measuring* (1985), 129.

23. G. D. Cassini, in Malvasia, *Ephem.* (1662), 156.

24. Riccioli, *Alm. nov.* (1651), *1:1*, 114; Streete, *Astr. carol.* (1661), 119 (1710), 137, reports and uses these values.

25. G. D. Cassini, *Elémens* (1684), 82–3.

26. Riccioli, *Alm. nov.* (1651), *1:1*, 115.

27. G. D. Cassini, *Meridiana* (1695), 15–16.

28. Cassini, *Origine* (1693), 36–7. The theory used an elliptical orbit with equant, Kepler's χ and Cassini's ρ; cf. Manfredi, *De gnomone* (1736), 67–8.

29. Malvasia to Cassini, 1 June 1662, in Malvasia, *Ephem.* (1662), 154.

30. G. D. Cassini, in ibid., 173, *Elémens* (1684), 84–6 (quote), and *Meridiana* (1695), 16; Malvasia, *Ephem.* (1662), "Proemium," f. †iv; Manfredi, *De gnomone* (1736), 43.

31. Cf. Delambre, *Hist.* (1821), *2*, 724.

32. Malvasia to Cassini, 27 July 1662, in Malvasia, *Ephem.* (1662), 174.

33. Delambre, *Hist.* (1821), *2*, 801.

34. Dorna, *Sulla rifr.* (1882), 5.

35. Flamsteed to Newton, 27 Nov. 6 and 10 Dec. 1694, in Newton, *Corresp.* (1959), *4*, 51, 54, 57 (quote); cf. Forbes, in Flamsteed, *Lectures* (1975), 67.

36. Cassini to Malvasia, 4 Idus Junii (10 June) 1662, in Malvasia, *Ephem.* (1662), 155; the decrease from the usual value, *e* = 0.018, amounted to almost 7' of longitude around the equinoxes.

37. Cassini, in Malvasia, *Ephem.* (1662), 156, 185.

38. Riccioli, *Astr. ref.* (1665), *1*, 56, 61; G. D. Cassini, *Lettere* (1666), 10–15. Riccioli compromised a bit on *e*, however, to 0.0174.

39. G. D. Cassini, *Lettere* (1666), 1–2 (quote), 5–9, reviewed in *PT*, 7:84 (1672), 5001–2; Giustiniani, *Scrittori* (1667), 370.

40. G. D. Cassini, *Lettere* (1666), 16–17, and *Meridiana* (1695), 16–17 (quote); Manfredi, *De gnomone* (1736), 43–4, 50.

41. Mengoli, *Refrattioni* (1670), 8, 11–12, 19–23, 62–3, and *Anno* (1673), 261; G. D. Cassini, *Lettere* (1666), 24–32.

42. Manfredi, *De gnomone* (1736), 40 (quote).

43. Cassini, *Elémens* (1684), 84.

44. Richer, in *Mémoires*, *5* (1736), 4; cf. Olmsted, *Isis*, *34* (1942), 118, 122–6.

45. Richer, in *Mémoires*, *5* (1736), 5–6; G. D. Cassini, *Elémens* (1684), 87–90. Cf. Van Helden, *Measuring* (1985), 131–2.

46. G. D. Cassini, *Elémens* (1684), 91, 105–6, 123, 125 (quote), and *Origine* (1693), 39.

47. Roche, *BJHS*, *14* (1981), 257.

48. Riccioli, *Alm. nov.* (1651) *1:1*, 114. Cf. Wilson, *Isis*, *59* (1968), 24.

49. Halley, *PT*, *31* (1721), 169; cf. J. Cassini, *MAS*, 1714, 38, 51.

50. From Figure 4.7, $r_{\text{mars}} = (r\sin \Delta\phi/2)\sin \chi/2$.

51. G. D. Cassini, *Observations* (1672), 53–4, 58; Delambre, *Hist.* (1821), *2*, 738–41; Van Helden, *Measuring* (1985), 137–42. Cf. Olmstead, *Isis*, *34* (1942), 125: $\chi_{\text{mars}} = 25''$, implying $\chi(0) = 9.5''$, was "the most dramatic result of the expedition."

52. Delambre, *Hist.* (1821), *2*, 728–9 (quotes), 803.

53. Lemonnier, *Hist.* (1741), ii, vii–xii.

54. Lacaille, *MAS*, 1755, 547–9; Halley, *PT*, *31* (1721), 172; Wilson, in Taton and Wilson, *Planetary astronomy* (1989), 189.

55. Manfredi, *De gnomone* (1736), 51.

56. Riccioli, *Astr. ref.* (1665), *1:1*, 26.

57. Flamsteed, *Lectures* (1975), 385 (7 Nov. 1683), 419 (16 Apr. 1684, quote).

58. Riccioli, *Alm. nov.* (1651), *1:1*, 164. Riccioli obtained ε from α(ss) – (90° – φ) in order to avoid refraction; most of the others he cites, including Cassini, used 2ε = α(ss) – α(ws). Cf. Manfredi, *De gnomone* (1736), 49–50.

59. Levera, *Prodromus* (1663), 199–200; Montanari, *Ephemeris lansbergiana* (1665), an update of Malvasia's tables, in Rotta, *Misc. seic.*, *2* (1971), 80; Mengoli, *Refrattioni* (1670), 63–4; Flamsteed, *Lectures* (1975), 225–6, 234–5, 237–9, 241–3, 246 (lectures of 1682).

60. G. D. Cassini, in Malvasia, *Ephem.* (1662), 175, 185; Malvasia, ibid., "Proemium."

61. G. D. Cassini to Flamsteed, 8 Nov. 1673, in Flamsteed, *Lectures* (1975), 243, and *Meridiana* (1695), 16, resp.

62. Manfredi, *De gnomone* (1736), 56 (quote), 61.

63. Mengoli, *Anno* (1673), 258–9; Manfredi, *De gnomone* (1736), 21–3; Montanari to Cassini, 8 Feb. 1673, in ibid., 24–6.

64. G. D. Cassini, *Meridiana* (1695), 17–18.

65. Ibid., 18–20; Guglielmini, in ibid., preface (quote), 39–40.

66. G. D. Morgani, in Guglielmini, *Opera* (1719), *1*, 2 (quote); Raule, *Strenna stor. bol.*, *10* (1960), 224.

67. Cassini, *Meridiana* (1695), 20; Zanotti, *Meridiana* (1779), 40–2, giving φ = 44°29'39".

68. Manfredi, *De gnomone* (1736), 33–4, 37–40, and *CAS*, *1* (1731), 261, 589–91, 595–6 (text of ca. 1722).

69. Zanotti, *Meridiana* (1779), 6–8, 30 (quote); Fantuzzi, *Notizie* (1781), *8*, 266–7; *CAS*, *6* (1783), 110–11; Guarducci, AS, Bologna, *Mem.*, *2* (1905), 297, 300–1.

70. Zanotti, *Meridiana* (1779), 53; *CAS*, *6* (1783), 110–11.

71. Guarducci, AS, Bologna, *Mem.*, *2* (1905), 321.

72. Fabri, *Efemeridi* (1675), 6, 20; Maccaferri, *Il carobbio*, *7* (1981), 255–6; Paltrinieri, *Meridiane* (1995), 322–6, 332.

73. Raggi, *Cielo* (1991), 29, 31; Braccesi and Morigi, *Cielo in Bologna*, no. 5 (n.d.), 21–2; Paltrinieri, *Meridiane* (1995), 359–60.

74. Raggi, *Cielo* (1991), 42; Delitalia, AS, Bologna, Classe sci. fis., *Rendiconti*, *4:1* (1977), 29; Paltrinieri, *Meridiane* (1995), 368–71.

75. Riccioli, *Alm. nov.* (1651), *1:1*, 46.

76. Ibid., 59–60.

77. Riccioli, *Geogr.* (1661), 117, 144, 176.

78. Ibid., 172; *r* comes from sin Δφ = cos 35'28" = *r*/(*r* + 0.1955).

79. Riccioli, *Geogr.* (1661), 173–6.

80. Ibid., 163–4.

81. Malvasia to Cassini, 15 Oct. 1661, in Malvasia, *Ephem.* (1663), 152; in his response, 1 Nov. 1661, ibid., 153, Cassini gave a degree of latitude at 45° as 44 miles, implying a value for a degree of longitude of 62 miles.

82. Cassini to *fabbricieri*, 27 Nov. 1670, in Zanotti, *Meridiana* (1779), 12; Zanotti, ibid., 9–15. Delambre, *Hist.* (1821), *2*, 727, 739, indignantly and rightly objected to Cassini's insinuation that he, and not Picard, had found *C*.

83. G. D. Cassini, *Meridiana* (1695), 12, 20.

84. Guglielmini, in ibid., 45.

85. Zanotti, *Meridiana* (1779), 9–10.

86. Fontenelle, *Oeuvres* (1729), *3*, 171; Heilbron, *Weighing* (1993), 188–9.

5. The Pope's Gnomon

1. Newton was styled the "Light of the World" by the Spaulding Philosophical Society; Weld, *History* (1848), *1*, 422–3.

2. Pell, *Easter* (1664), iv–vii.

3. [Cassini], *HAS*, 1701, 108; Bianchini, "Adhortio," in *Solutio* (1703), 3–11, and in *Opuscula varia* (1753), *2*, 139–40; Celani, *Arch. ven.*, *36* (1888), 175. A good overview of calendrical questions in the early eighteenth century is Manfredi, *Elementi* (1744), 321–54 (text of 1732).

4. Leibniz to Rømer, 5 Mar. 1700, in Leibniz, *Epistolae* (1734), *1*, 213; Celani, *Arch. ven.*, *36* (1888), 172.

5. Wallis to Thomas, Archbishop of Canterbury, 13 June 1699, in *PT*, *21:257* (1699), 343–6, and to William, Bishop of Worcester, ibid., 350–4.

6. *HAS*, 1700, 128–9; Leibniz to Rømer, 5 and 18 Mar. 1700, in Leibniz, *Epistolae* (1734), *1*, 212, *2*, 3–4, and to Bianchini, 5 Mar. 1700, in Celani, *Arch. ven.*, *36* (1888), 176.

7. Gingerich, in Coyne et al., *Greg. ref.* (1983), 267; Jacobus Heerbrand, professor of theology, Tübingen, writing in 1584, quoted by Hoskin, in ibid., 260.

8. Petit, *Lettre* (1666), 12–13; Levera, in ibid., 24–8.

9. Desiderio, *Tavole* (1703), 55–6; Stein, Spec. Vat., *Misc. astr.*, *3* (1950), 42–3.

10. Manfredi, *Elementi* (1744), 190.

11. Fabri, *Synopsis* (1667), 227–8.

12. J. Cassini, *Elémens* (1740), 81; Leibniz to Rømer, 18 Mar. 1700, in Leibniz, *Epistolae* (1734), *2*, 4. Cf. Ginzel, *Handbuch* (1906), *3*, 273.

13. G. D. Cassini, *Meridiana* (1695), 1–2.

14. Leibniz to Rømer, 27 Jan. 1703, in Leibniz, *Epistolae* (1734), *2*, 8.

15. Bianchini, "Expositio," in *Solutio* (1703), 1; Bianchini, *GL*, *4* (1709), 65–6, 72; Fontenelle, *Oeuvres* (1728), *3*, 167, 171.

16. AS, Paris, *Hist.*, 1700, 128–9; G. D. Cassini, *MAS*, 1703, 42, 50; Bianchini, *Opuscula* (1753), *2*, 124.

17. Fontenelle, *HAS*, 1729, 116–20.

18. Bianchini, *De nummo* (1703), 4 (quote), and in Crescimbeni, *Vite* (1708), *1*, 209–10, 213; Noris, *Dissertationes*, in *Annus* (1691); Arck-

enholtz, *Mém.* (1751), *2*, 131; Cardella, *Mem.* (1792), *8*, 46–9.

19. *HAS*, 1701, 106; G. D. Cassini, *MAS*, 1703, 48–50, and *MAS*, 1704, 149, 151.

20. Desiderio, *Tavole* (1703), 5.

21. Piazza, *Gerarchia* (1703), 626–7.

22. Bianchini, *De nummo* (1703), 5–6, and *Opuscula* (1753), *2*, 129.

23. Marini, *Inscrizioni* (1785), viii.

24. Quoted by Alberti-Poja, *Meridiana* (1949), 19; Bianchini so places himself in *Camera* (1727), pl. 2 after p. 87.

25. Bianchini to Ciampini, 16 Sept. 1687, 23 Sept. and 21 Oct. 1688, in Carini, *Il muratori*, 1892, 167, 170, 172; Middleton, *BJHS*, *8* (1975), 141–2.

26. Bianchini, *Istoria* (1697), 36–8, 409, 519–20, and (1747), on the history of astronomy, 101, 107–9, 112–14, 147–50, 199–200; Croce, *Conv. critiche* (1950), *2*, 101–9; Lanza, *Lett. ital.*, *10* (1958), 339–48; Johns, *Papal art* (1993), 36.

27. Bianchini, *Carte* (1871), 43–77.

28. Liebniz to Tidius, 16 Nov. 1701, and to Rømer, 3 July 1703 (quote), in Leibniz, *Epistolae* (1734), *1*, 213, and *2*, 13.

29. Fontenelle, *Eloges* (1740), *2*, 381–2.

30. Respectively, Gabbrielli, *Heliometro* (1705), 29, and Leibniz to Rømer, n.d., in Celani, *Arch. ven.*, *36* (1888), 172.

31. Mazzoleni, *Vita* (1735), 2–6.

32. On Bianchini, M. di Livry, in Mazzoleni, *Vita* (1735), 11 (first quote), 29–30, and Barbazza, in Bianchini, *Istoria* (1747), f. a.6r; on Montanari, Bianchini, in Montanari, *Forze* (1694), ff. a.6v. (second quote), a.11, b.2r, b.7r; Bianchini, in Montanari, *Astrologia* (1685), vi–viii (third quote).

33. Ferrone, *Scienza* (1982), 42, 57, 375–8, and *Roots* (1995), 16, 22, 150–1. Cf. Montanari, *Cometes* (1665), 34, *Lettera* (1667), 65, and *Forze* (1694), 156–8; and Lopicolli, in Magrini, *Scienza* (1990), 25–31.

34. Bianchini, in Montanari, *Forze* (1694), ff. a.12r (quote), b.8–9.

35. Rotta, *Misc. seic.*, *2* (1971), 96–7, 165n186; Gomez Lopez, *Passioni* (1997), 16–17; Cavazza, *SMSUB*, *3* (1983), 44–5.

36. Bianchini, in Montanari, *Forze* (1694), ff. a.9–11, f. a.12r (quote).

37. Noris to Magliabecchi, 25 Jan. 1681, in Magliabecchi, *Epist. ven.* (1745), *1*, 133; Bianchini, in

Montanari, *Forze* (1694), f. a.12r; Rotta, *Misc. seic.*, *2* (1971), 131–2; Serena, *Barbarigo* (1963), *1*, 61; Bellinati, *Barbarigo* (1960), 165–6.

38. Mazzoleni, *Vita* (1735), 7–10; Cardella, *Mem.* (1792), *8*, 93–4 (quote); Ruggieri, in *Memorie* (1825), 40–50; Alberti-Poja, *Meridiana* (1949), 15; Carini, *Il muratori*, 1892, 156n8.

39. Eschinardi, *Cursus* (1689), "Ad lectorem"; Spoletino, in Crescimbeni, *Vite* (1708), *2*, 217–18, 222–6, 232–3, 240–1; Montalto, *Stud. rom.*, *10* (1962), 662–70; Schiavo, *Meridiana* (1993), 63; Leibniz to Magliabecchi, 8 Nov. 1691 and 12 Aug. 1704, in Magliabecchi, *Epist. germ.* (1746), *1*, 35, 102, and letters to Bianchini, in Celani, *Arch. ven.*, *36* (1888), 176 ff.; Johns, *Papal art* (1993), 23.

40. Spoletino, in Crescimbeni, *Vite* (1708), *2*, 200–1, 207–13, 238; Stephan, in Platen, *Queen Christina* (1966), 369; Masson, *Queen Christina* (1974), 347, 361; *DBI*, *25*, 136–9 (S. Grassi Fiorentini).

41. Eschinardi, *Ragguaglio* (1680), 2–3; Ferrone, *Scienza* (1982), 12–15, 20–2, 28–31, and *Roots* (1995), 5–8, 11; Maugain, *Etude* (1909), 27–8.

42. Crescimbeni, *Vite* (1708), *2*, 216, 241–4; Reusch, *Index* (1883), *2*, 14; Middleton, *BJHS*, *8* (1975), 140–1, 144, 147–8.

43. *GL*, *7* (1711), 110–12; Mazzoleni, *Vita* (1735), 14 (quote)–17, 24–25, 27–8; Mazzuchelli, *Scrittori* (1760), 1167–8; Schneider, in Jedin, *Hist.*, *6* (1981), 126–7; Pastor, *Hist.* (1891), *32*, 532–7.

44. Schneider, in Jedin, *Hist.*, *6* (1981), 127–34; Pastor, *Hist.* (1891), *32*, 572–6, and *33*, 8–14, 59–65, 90–109, 512; Johns, *Papal art* (1993), 1–4, 20–1, 33 (quote).

45. Cardella, *Mem.* (1792), *8*, 1–3; Johns, *Papal art* (1993), 33.

46. Mazzoleni, *Vita* (1735), 24, 28, 31, 35–40; Manfredi, in Bianchini, *Observationes* (1737), ii, dating the commission to 1701; Alberti-Poja, *Meridiana* (1949), 16–17.

47. Boscovich and Maire, *Voyage* (1770), 87.

48. Bianchini, *AE*, Oct. 1785, 471, in *Observationes* (1737), 13 (quote); Bianchini to Cassini, 29 Jun. 1699, Cassini Papers (Paris), A. B. 4.9.

49. Bianchini, *Opuscula* (1753), *2*, 128 (quote); Schiavo, *Meridiana* (1993), 44, 75.

50. Cassini to Manfredi, 5 Feb. 1703, Cassini Papers (Bologna), 4717; Manfredi, *Elementi* (1744), 207; Mazzoleni, *Vita* (1735), 38; Lalande, *Voyage* (1786), 311, and *Encl. meth.*,

Math. (1784), *2*, 282–3; Alberti-Poja, *Meridiana* (1949), 18.

51.   Bianchini to his brother, 18 Nov. 1702, in Bianchini, *Opuscula* (1753), *2*, 136–7.

52.   Johns, *Papal art* (1993), 36–7.

53.   Alberti-Poja, *Meridiana* (1949), 19–22; Mazzoleni, *Vita* (1735), 39–44.

54.   Noris to Magliabecchi, 7 June 1698, in Magliabecchi, *Epist. ven.* (1745), *1*, 179; Johns, *Papal art* (1993), 35.

55.   Cf. Ferrone, *Storia* (1982), 58, and *Roots* (1995), 22.

56.   Mazzoleni, *Vita* (1735), 52–9, 68; Bianchini to Cassini, 7 Apr 1705, Cassini Papers (Paris), A. B. 4.9; Bianchini, draft of letters to Louis XV, in Bianchini Papers (Bologna), B.211, f. 118–19; Sorbelli, *Inventari* (1939), 19–20; Maffei, in Bianchini, *Istoria* (1747), f. a.5v (quote).

57.   Alberti-Poja, *Meridiana* (1949), 22 (quote)-4; Ferrone, *Storia* (1982), 47, 59–64, and *Roots* (1995), 18, 23–4; Manfredi, in Bianchini, *Observationes* (1737), viii–xiii.

58.   Alberti-Poja, *Meridiana* (1942), 26–8; Mazzoleni, *Vita* (1735), 104–19; Maffei, in Bianchini, *Istoria* (1747), f. a.3r; Bianchini, *Phaenomena* (1728), 5 (the fall); Schiavo, *Meridiana* (1993), 48, 50.

59.   Moroni, *Dizionario* (1840), *12*, 91 (quote); Schiavo, *Meridiana* (1993), 177–88; Siebenhüner, *Münch. Jahrb. bild. Kunst, 6* (1955), 190–1, 202.

60.   Siebenhüner, *Münch. Jahrb. bild. Krunst, 6* (1955), 187–9, 193–5, 202–6; Piazza, *Gerarchia* (1703), 630.

61.   Schiavo, *Meridiana* (1993), 82–4; Filippo Titi, *Nuovo studio di pittura, scoltura ed architettura nelle chiese di Roma* (1763), cited by Johns, *Papal art* (1993), 34, 220n58.

62.   Alberti-Poja, *Meridiana* (1949), 37–40; Schiavo, *Meridiana* (1993), 70; Rotta, *Misc. seic., 2* (1971), 76.

63.   Bianchini, *De nummo* (1703), 20–4, 22 (quote).

64.   Schiavo, *Meridiana* (1993), 75.

65.   Ibid., 71–4; Bianchini, *De nummo* (1703), 3, 29–31.

66.   Bianchini, *De nummo* (1703), 1–2. Cf. Alberti-Poja, *Meridiana* (1949), 154.

67.   Schiavo, *Meridiana* (1993), 116–18, 125, 133–7, 141–6; Piazza, *Gerarchia* (1703), 628.

68.   Bianchini to Magliabecchi, 10 Aug. 1709, in Magliabecchi, *Epist. ven.* (1745), *1*, 326; Schiavo, *Meridiana* (1993), 63–4.

69.   Bianchini, *De nummo* (1703), 32–3, 37, and *Opuscula* (1753), *2*, 132 (quote); Buchowiecki, *Handbuch* (1970), 404.

70.   Bianchini, *De nummo* (1703), 29 (quote); Schiavo, *Meridiana* (1993), 99–100.

71.   Bianchini to P. M. Gabbrielli, [19] Aug. 1703, Gabbrielli Papers (Siena).

72.   Bianchini, *De nummo* (1703), 26–8, and *Opuscula* (1753), *2*, 133; *GL, 4* (1709), 83.

73.   Cotter, *History* (1968), 134–5.

74.   Cf. Schiavo, *Meridiana* (1993), 105. The axes of the ellipse are $h'\sigma\csc^2\phi$ and $h'\sigma\csc\phi$; the displacement of its center, $h'\sigma^2\ctn\phi\csc^2\phi$.

75.   Schiaro, *Meridiana* (1993), 100–4; cf. Alberti-Poja, *Meridiana* (1949), 121.

76.   Piazza, *Gerarchia* (1702), f. a.3v, 629–30.

77.   Bianchini, *De nummo* (1703), 52–9, 63, and *Solutio* (1703), xxxvii–lxi.

78.   G. D. Cassini, *MAS*, 1705, 44, 48–9 (quote).

79.   Leibniz to Bianchini, 28 Dec. 1705, in Celani, *Arch. ven., 36* (1888), 183.

80.   Bianchini, *De kalendario* (1703), esp., 91–6, on an ancient cycle of 112 years, and *Solutio* (1703), xii.

81.   Bianchini, *Opuscula* (1753), *2*, 139; Leibniz to Bianchini, 27 and 30 Nov. 1701, and to Noris, 8 Mar. 1702, in Celani, *Arch. ven., 36* (1888), 179, 186.

82.   G. D. Cassini, *MAS*, 1704, 145. Cf. Ferroni, *Storia* (1982), 45–6, and *Roots* (1995), 284n68.

83.   Bianchini to Magliabecchi, 19 July 1702, in Magliabecchi, *Epist. ven.* (1745), *1*, 320. Cf. Schiavo, *Meridiana* (1993), 26.

84.   Monaco, *Physis, 25* (1983), 418; Bianchini, *Observationes* (1737), 17–252.

85.   Bianchini, *Observationes* (1737), 54, 59 (1703–4, with Gabriele Manfredi), 115–17 (1715, with Eustachio); Bianchini, *De nummo* (1703), 32–83, esp. 81–2; Lami, *Memorabilia* (1742), *1*, 366; *DBI, 10*, 104–5, 109 (A. Fabri).

86.   Bianchini, *Observationes* (1737), 48–9, 66, 91, 103, 158–9, 238, 245, 254, 260 (quote); Mazzoleni, *Vita* (1735), 69–70, 90–1; Lalande, *Voyage* (1786), *4*, 313.

87.   Nordenmark, *Celsius* (1936), 23, 26, 37, 40, 44n (quote), 46 (quote).

88.   Bianchi to Celsius, 9 May 1734, Celsius Papers

(Uppsala); Manfredi, *De gnomone* (1736), 19. Cf. Leibniz to Huygens, 15/25 July 1690, in Rotta, *Misc. seic., 2* (1971), 182n357: "I've found very few people [in Italy] with whom to discuss anything beyond the ordinary in physics and mathematics."

89.   Nordenmark, *Celsius* (1936), 41, 44; Boscovich, "Del meridiano," Boscovich Papers (Berkeley), Pt. 1:89; Lalande, *Voyage* (1786), *4*, 316.

90.   *GL, 6* (1711), 126 (quote); Bruin and Bruin, in Maeyama and Saltzer, *Prismata* (1977), 28–31.

91.   Mazzuchelli, *Scrittori* (1760), 1169.

92.   Picard, *MAS, 1666–99, 7* (1729), 165 (text of 1671).

93.   B. Bianchini (a nephew), dedicatory letter, in Bianchini, *Observationes* (1737).

94.   Bianchini, *Appendice chorographica* (1724), in Bianchini, *Observationes* (1737), and *Phaenomena* (1728), 7–8; Manfredi, in Bianchini, *Observationes* (1737), viii–ix; Cavazza, *Settecento* (1990), 176.

95.   Maffei, in Bianchini, *Istoria* (1747), f. a.3v.

96.   J. Cassini, *Grandeur* (1720), 1–5; Konvitz, *Cartography* (1987), 5–8.

97.   J. Cassini, *Grandeur* (1720), 37, 46, 188, 192–4, 246–7; Delambre, *Grandeur* (1912), 4–5, 10–15, 18, 22.

98.   G. D. Cassini, *MAS*, 1713, 190–1; J. Cassini, *Grandeur* (1720), 9; Levallois, in Lacombe and Costabel, *Figure* (1988), 45–52; Costabel, in ibid., 98–103.

99.   Heilbron, *Weighing* (1993), 218–25, for this and the next three paragraphs. The radius of curvature at a point at latitude ϕ on an ellipsoidal earth of eccentricity e is $a(1 - 2e + 3e\sin^2\phi)$.

100.   Boscovich and Maire, *Voyage* (1770), 31.

101.   Ibid., 29; Cardella, *Mem.* (1792), *8*, 291–2.

102.   Rotta, *Illuminismo* (1974), 38.

103.   Boscovich, *CAS, 4* (1757), 354 (quote), 355; for Benedict and science, see Chapter 6 below.

104.   Boscovich, *CAS, 4*, 353, 361; Boscovich and Maire, *Voyage* (1770), 31–2.

105.   Boscovich, *CAS, 4* (1757), 361–2; Boscovich and Maire, *Voyage* (1770), 81–2; Nikolic, *AIHS, 14* (1961), 320–3.

106.   Boscovich and Maire, *Voyage* (1770), 48–49 (last quote), 52, 57–63, 74–75 (first quote).

107.   Boscovich and Maire, *Voyage* (1770), 32–3,

40–4, 117, 262–340, 489 (quote), 492; Nikolic, *AIHS, 14* (1961), 318–19; Westphal, *Zs. für Instr., 4* (1884), 161–2, and *5* (1885), 334. The remeasurement, commissioned by Napoleon on the urging of Laplace, was done by Barnaba Oriani; Tagliaferri and Tucci, *Giorn. fis., 34:4* (1993), 260–1, 264.

108.   Boscovich and Maire, *Voyage* (1770), 36; Nikolic, *AIHS, 14* (1961), 315–316.

109.   Zach, *Monat. Corr., 8* (1803), 515–21 (first quote), and *Mémoire* (1811), 7–8, 25, 93 (second quote), 123, 136; Westphal, *Zs. für Instr., 5* (1885), 344. Boscovich and Maire, in contrast, measured quite well; Antonelli, *Alcuni studi* (1871), 52–9.

110.   Heilbron, *Weighing* (1993), 239–42.

111.   J. Cassini, *MAS*, 1732, 454.

112.   G. D. Cassini, Cassini Papers (Paris), D1.13; Wolf, *Histoire* (1902), 84–86.

113.   Wolf, *Histoire* (1902), 88; J. Cassini, *MAS*, 1732, 452 (quote); Heilbron, *Electricity* (1979), 117.

114.   J. Cassini, *MAS*, 1732, 454n, 457–8.

115.   Ibid., 456–7, 460–1.

116.   Ibid., 458, 462.

117.   Ibid., 464, 469–70.

6. The Accommodation of Copernicus

1.   Favaro, Dep. stor. patria Romagna, *Atti, 22* (1904), 427–8, 440–4, and Ist. ven. sci., *Atti, 80:2* (1920/1), 16; Galileo to Marsili, 29 Nov. 1631, in Galileo, *Opere* (1890), *14*, 311–12.

2.   Marsili to Galileo, 17 Mar. 1631, and Cavalieri to Galileo, 21 May and 10 June 1631, in Galileo, *Opere* (1890), *14*, 225, 263, 275; Predieri, AS, Bol., *Mem., 3* (1851), 118–19, 123–4, 136.

3.   Galileo to Marsili, 5 Apr. 1631, in Galileo, *Opere* (1890), *14*, 240.

4.   The standard formulas give $\sin\angle QFW = \sin\varepsilon/\cos\phi$.

5.   Galileo to Marsili, 5 Apr. and 5 July 1631, in Galileo, *Opere* (1890), *14*, 240, 280–1.

6.   Marsili to Galileo, 8 July and 11 Oct. 1631, and Cavalieri to Galileo, 1 July 1631, in ibid., 279, 283, 300–1.

7.   Galileo, *Dialogo*, in ibid., *7*, 487.

8.   Fulgentio Micanzio to Galileo, 14 Oct. and 9 Dec. 1634, in Galileo, *Opere* (1890), *16*, 141, 172 (quote); cf. Giovanni Peroni to Galileo, 4 Jan. 1635, in ibid., 190.

9. Predieri, AS, Bol., *Mem., 3* (1851), 131–2.

10. Caramuel, *Perpendicularum inconstantia* (1643), in Monchamp, *Galilée* (1892), 143; Heilbron, *Electricity* (1979), 103–11.

11. Riccioli, *Alm. nov.* (1651), *1:2*, 348.

12. Galileo, *Dialogo*, in *Opere* (1890), *7*, 190; Koyré, APS, *Trans., 45* (1955), 334–5.

13. The velocity along the tower is $v\cos\alpha = v\sin\omega t = \omega(r+h)\sin\omega t \approx \omega^2(r+h)t$, where v is the velocity around the center D (Figure 6.2).

14. Hence it would have a period of 2 seconds. The quoted evaluation is Koyré's, in APS, *Proc., 97* (1953), 223.

15. Ibid., 229–30; Riccioli, *Alm. nov.* (1651), *1:1*, 85–7.

16. Koyré, APS, *Proc., 97* (1953), 230–2, and APS, *Trans., 45* (1955), 351–3, a translation of Riccioli, *Alm. nov.* (1651), *1:2*, 399–400.

17. Riccioli, *Alm. nov.* (1651), *1:2*, 400–1; Koyré, APS, *Trans., 45* (1955), 354.

18. Riccioli, *Alm. nov.* (1651), *1:2*, 479–500, *Astr. ref.* (1665), 81–4, and *Apologia* (1669), 40, 44, 52; Koyré, APS, *Trans., 45* (1955), 366–7, 390–5.

19. Koyré, APS, *Trans., 45* (1955), 355–95; Galluzzi, IMSS, *Ann., 2:2* (1977), 107, 111–12; degli Angeli to Magliabecchi, 6 Nov. 1671, in Magliabecchi, *Epist. ven.* (1745), *2*, 76.

20. Examples in Lattis, *Clavius* (1994), 161–4; Proverbio, in Bucciantini and Torroni, *Diffusione* (1997), 42–5, 52–5; and Heilbron, in Shumaker, *Dee* (1978), 56–7. Planetary distances were important for some astrological computations.

21. $r_i = s \tan\beta$.

22. The law of sines gives $r_s = r \sin\gamma/\sin(\gamma + \theta_E - \theta_P)$.

23. Gassendi, *Inst. astr.* (1653), 143; Thiriou, in Centre, *Copernic* (1975), 258–9.

24. Lenoble, *Mersenne* (1943), 392–410; Hine, *Isis, 64* (1973), 25–6, 30–1.

25. D'Addio, *Considerazioni* (1985), 113; Brandmüller, in Brandmüller and Greipl, *Copernico* (1992), 26.

26. Riccioli, *Alm. nov.* (1651), *1:2*, quotes on 309, 304, 309; cf. Stimson, *Grad. accept.* (1917), 82, 83.

27. Riccioli, *Alm. nov.* (1651), *1:1*, xxxiv.

28. For those who count, these divide into 20 for and 38 against the diurnal, and 29 for and 39 against the annual motion; ibid., *1:2*, 465–77.

29. Ibid., *1:1*, xix (quote), *1:2*, 500, 491 (reinterpretation). Cf. Stimson, *Grad. accept.* (1917), 79–80.

30. Letter of 26 Sept. 1671, in Galluzzi, IMSS, *Ann., 2* (1977), 94n.

31. Riccioli, *Apologia* (1669), 4, translation slightly modified from Stimson, *Grad. accept.* (1917), 83–4. Cf. Riccioli, *Alm. novum* (1651), *1:2*, 478–80, 486 (because God so decreed), and *Astr. ref.* (1665), 86 ("the motion of the sun, and the immobility of the earth, must be asserted on the sole authority of holy scripture"), and D'Addio, *Considerazioni* (1985), 114.

32. Riccioli, *Alm. novum* (1651), *1:2*, 476; Wilson, *AHES, 6* (1970), 104.

33. Cf. Delambre, *Hist.* (1821), *2*, 690, 695, 784. Cassini's later planisphere, designed before 1699, is described in Cassini, AS, *Mach., 1* (1735), 132–43; it may be compared with Rømer's Copernican model, ibid., 81–3.

34. Flamsteed to Newton, 7 Mar. 1680/1, in Newton, *Corresp.* (1959), *2*, 348.

35. Cassini, *Lettere astr.* (1665), 6 (quote), and *JS*, 22 Apr 1686, 140–2; Huygens, *Oeuvres* (1888), *21*, 311, and *22*, 723.

36. *HAS*, 1709, 82, 88 (quotes); Cassini, *MAS*, 1709, 251–4, and figures after p. 256; *DSB, 3*, 103 (Taton).

37. *JS*, 22 Feb. 1666, 295, 298 (first quote); Cassini, *Lettere . . . Falconieri* (1665), 3 (second quote); Wallis, *PT, 1* (6 Aug 1666), 270. Cf. Wilson, *AHES, 6* (1970), 127. Among the doubting Jesuits was Fabri, *Dialogi* (1665), 90; Campani, *Lettere* (1665), 3–4.

38. Auzout, *Lettre* (1665), 17.

39. *PT, 1:1* (6 Mar. 1664/5), 3; *1:4* (5 June 1665), 74–5; *1:8* (8 Jan. 1665/6), 144; *1:10* (12 Mar. 1665/6), 173.

40. Flamsteed to Newton, 11 Oct. 1694, in Newton, *Corresp.* (1959), *4*, 29.

41. *JS*, 22 Apr. 1686, 149, 152. The Latin fragment, which appears on the title page of Riccioli's *Almagestum*, requires the preceding verse: "The heavens tell out the glory of God, heaven's vault makes known his handiwork, one day speaks to another, night to night im-

parts knowledge" (Ps. 19: 1–2). See the illustrations at the head of Chapters 1 and 8.

42. Duhamel, *Historia* (1698), 229.

43. E.g., in *Origine* (1693), 42, and *Elémens* (1684), 82.

44. Riccioli, *Apologia* (1669), 3, tr. after Koyré, APS, *Trans.*, *45* (1955), 393; Galluzzi, IMSS, *Ann.*, *2* (1977), 124–5.

45. Galluzzi, IMSS, *Ann.*, *2* (1977), 123 (letter of 17 Jan. 1668), 128–9 (Borelli to Leopold, summer 1668). On Borelli's self-censorship see Gomez Lopez, *Passioni* (1997), 19–20; on tough censorship at Bologna, Battistella, *S. officio* (1905), 152–63.

46. Bosmans, AS, Bruss., *Biog. nat.*, *24*, cols. 441–3, 450–1 (the General's letter of 22 June 1652), and *Compas d'or*, *3* (1925), 65, 83–4.

47. Bosmans, AS, Bruss., *Biog. nat.*, *24*, cols. 452–6, 460–1; Monchamp, *Galilée* (1892), 152–3.

48. Tacquet, *Astr.* (1669), 12.

49. The biblical verses are Eccl. 1:5, Psalms 93:1 and 104:5, and Josh. 10:12; there were others, e.g., Psalms 19:5–6 (the sun runs its course), 1 Chron. 16:30 ("He has established the earth immovably"), 2 Kings 20:9–11 (the sun goes backward), Job 26:7 (the earth stands suspended), and Isaiah 40:12 ("Who . . . has gauged the heavens?"). Cf. Grant, APS, *Trans.*, *74:4* (1984), 62–3.

50. Tacquet, *Astr.* (1669), 330–1.

51. Bosmans, AS, Bruss., *Biog. nat.*, *24*, cols. 444–5.

52. *BU*, *7* (1854), 440 (Pillet); Heilbron, *Electricity* (1979), 105, 144; Deschales, *Cursus* (1674), *3*, 636 (quote).

53. Deschales, *Cursus* (1674), *1*, preface, and *3*, 287 (quote), 486.

54. Ibid., 277 (quote), 278–85.

55. Reusch, *Index* (1883), *2*, 503–4; Costantini, *Baliani* (1969), 79; Heilbron, *Electricity* (1979), 110, 113–14 (quote); Galluzzi, IMSS, *Ann.*, *2* (1977), 132–7.

56. Divini, *Pro sua annotatione* (1661), 46–8. Fabri was understood to be the author of Divini's books; Huygens, *Brevis assertio* (1660), 25, and degli Angeli to Magliabecchi, 30 July 1661, in Magliabecchi, *Epist. ven.* (1745), *2*, 72. Cf. Divini, *Brevis annotatio* (1660), 3–6, 35–40, regarding Fabri, "a man closely connected with me" (35).

57. Divini, *Pro sua annotatione* (1661), 49; cf. Duhem, *Phenomena* (1969), 111, 113, 117.

58. Baldini, in Michele, *Scienza* (1980), 520–3. Cf. Schreiber, *Natur und Offen.*, *49* (1903), 133–7, 142–3, 212–20.

59. Kircher, *Ars magna* (1646), 766–9, and memo, 4 July 1660, in Kircher Papers (Rome); Eschinardi, *Microcosmi* (1668), *1*, and *Cursus* (1689), 172–3, quoted by Torrini, *Dopo Galileo* (1979), 89; Lopicolli, in Magrini, *Scienza* (1990), 68–9. From an earlier generation, Clavius, Scheiner, and even Grassi were thought to favor Copernicus; Gambaro, *Astr.* (1989), 16–17, from correspondence in Galileo, *Opere* (1890), *13*, 202–3, 208–11, and *15*, 254.

60. Fabri, *Dialogi* (1665), 2–3 (quote), 43–4 (quote), 91 (last two quotes).

61. Kochansky, *AE*, Jul 1685, 317–18.

62. Ibid., 318–22; cf. Kochansky to Leibniz, 1 Jul 1671, in Leibniz, *Schr.* (1926), ser. 2, *1*, 138

63. Kochansky, *AE*, July 1985, 323–5 (quote); Descartes to Mersenne, 13 July 1638, quoted by Borgato, in Pepe, *Copernico* (1996), 220–1. Mersenne had proposed looking for a slight deflection of the shot to the East as a test for the diurnal moton.

64. Duhamel, *Astr. phys.* (1660), "Praefatio"; *BU*, *11* (1855), 464–5 (Weiss).

65. Duhamel, *Astr. phys.* (1660), 158–60, 163–81 (Descartes).

66. Duhamel, *Phil. vetus nova* (1682), *2*, 356–8.

67. Ibid., 365.

68. *BU*, *11* (1855), 464–5 (Weiss).

69. Leibniz to Landgraf Ernst, 1681, in Leibniz, *Schr.* (1926), ser. 2, *1*, 513–14.

70. Leibniz to Landgraf Ernst, 29 June/9 July 1688, in Leibniz, *Schr.* (1926), ser. 2, *5*, 185–6; Müller and Krönert, *Leben* (1969), 83, 90; Hiltebrandt, *Quell. Forsch.*, *10* (1907), 239, 243.

71. Leibniz to Landgraf Ernst, 29 June/9 July 1689, as discussed and translated by Meli, *Stud. leibn.*, *20* (1988), 21–24. Cf. Duhamel, *Phil. vetus nova* (1682), *2*, 363: "And I do not know whether Copernicus could have spoken differently from Joshua when he commanded the sun to stand still."

72. Müller and Krönert, *Leben* (1969), 99, 101.

73. Quotes from, resp., Leibniz to Magliabecchi, 20/30 Oct. 1689, in Meli, *Stud. leibn.*, *20*

(1988), 28, and to M. Thévenot, 3 Sept. 1691, in Leibniz, *Schr.* (1926), 352–3; Müller and Krönert, *Leben* (1969), 95–6, 98.

74. Leibniz to Ciampini, 14/24 Sept. 1690, in Leibniz, *Schr.* (1926), ser. 1, *6*, 248–9; Müller and Krönert, *Leben* (1969), 97; Robinet, *Iter* (1988), 42–50, 95–9.

75. Leibniz to Magliabecchi, 20/30 Oct. 1699, in Magliabecchi, *Epist. germ.* (1746), *1*, 93.

76. Quotes from Leibniz to Huygens, 4/14 Sept. 1694, and from Leibniz' paper, are from the translation in Leibniz, *Phil. essays* (1989), 91–2, 309. Cf. Leibniz to Hermann Corning, 3/13 Jan 1678, in Leibniz, *Schr.* (1926), ser. 2, *1*, 386, and Robinet, *Iter* (1988), 100–18.

77. Leibniz, *Phil. essays* (1989), 92–3; cf. Meli, *Stud. leibn.*, *20* (1988), 25–8.

78. Müller and Krönert, *Leben* (1969), 99; Robinet, *Iter* (1988), 227–8, 232–3, 276–9, 292, 298.

79. Viviano to Baldigiani, 1690, in Meli, *Stud. leibn.*, *20* (1988), 35–6; Robinet, *Iter* (1988), 51, 100, 116.

80. Müller and Krönert, *Leben* (1969), 98; Cardella, *Mem.* (1792), *7*, 151–3; Moroni, *Dizionario* (1840), *4*, 99; Bellinati, *Barbarigo* (1960), 47, 60–1, 102–5, 136–46, 165–6, 175–82, 247, and in *Galileo* (1982), 221–4; Serena, *Barbarigo* (1963), *1*, 59–60, 79, 83–8, 97; Robinet, *Iter* (1988), 185.

81. Leibniz to R. P. B., in Leibniz, *Math. Schr.* (1849), *6*, 146.

82. Baldigiani to Viviani, 25 Jan. 1693, in Torrini, *Dopo Galileo* (1979), 28. Baldigiani admitted his allegiance to Galileo to Viviani in a letter of 18 July 1678; Middleton, *BJHS*, *8* (1975), 154.

83. Leibniz to Bianchini, 13 Oct. 1703, in Celani, *Arch. ven.*, *36* (1888), 181–2; Robinet, *Iter* (1988), 55, 58, 61–2.

84. Leibniz to Huygens, 4/14 Sept. 1694, in Leibniz, *Phil. papers* (1969), 419; and to Morell, 29 Sept. and 4/14 May 1698, in Leibniz, *Textes inédits* (1948), *1*, 137–8, 126.

85. Bianchini, *Nova phaen.* (1728), 1–2, and plate IV; Fontenelle, *HAS*, 1729, 114; Schiavo, *Meridiana* (1993), 46.

86. Quoted by Rotta, *Misc. seic.*, *2* (1971), 82–3.

87. Reusch, *Index* (1883), *2*, 1, 6–7, 15.

88. Ibid., *1*, 173–4, and *2*, 8–9, 11; Brandmüller, in Brandmüller and Greipl, *Copernico* (1992), 48–9. The guess, that the usual number of

consultors was about 10, assigns an average tenure of five or six years to each of the two hundred who served from 1577 to 1644.

89. Brandmüller, in Brandmüller and Greipl, *Copernico* (1952), 50–1; Lackmann, *Bücherzensur* (1962), 26.

90. Reusch, *Index* (1883), *2*, 10–11.

91. Ceranski, *Bassi* (1996), 97; Rotta, *Illuminismo* (1974), 43.

92. Reusch, *Index* (1883), *2*, 5–6; Hilgers, *Index* (1904), 60–1.

93. Simoncelli, *Riv. stor. ital.*, *100* (1988), 72–3; Fontenelle, *HAS*, 1729, 115. Cf. Moroni, *Dizionario* (1840), *48*, 103–4.

94. Reusch, *Index* (1883), *2*, 671; Francini, *Riv. stor. chiesa*, *37* (1983), 438; Stasiewski, in Jedin, *Hist.*, *6* (1981), 532.

95. Delumeau and Cottret, *Cath.* (1996), 210–18; Cognet, in Jedin, *Hist.*, *6* (1981), 26–7, 31, 34–5, 40, 44–45, 48–9. The five theses (ibid., 38) are concerned mainly with the question whether man's will can resist inner grace.

96. Pastor, *Hist.* (1891), *32*, 576, 621; Reusch, *Index* (1883), *2*, 672–6; Francini, *Riv. stor. chiesa*, *37* (1983), 444–54.

97. Noris, letter to Magliabecchi, in Reusch, *Index* (1883), *2*, 676–7. Serving the censorship system often conferred upward mobility; Simoncelli, *Riv. stor. ital.*, *100* (1988), 74–8, 84.

98. Cognet, in Jedin, *Hist.*, *6* (1981), 53–7, 381–404.

99. Francini, *Riv. stor. chiesa*, *37* (1983), 454–62; Anon., *Obs.* (1749), 2; Pastor, *Hist.* (1891), *33*, chap. 5.

100. Schneider, in Jedin, *Hist.*, *6* (1981), 570; Haynes, *Philosopher* (1970), 184–8; Del Re, in Cecchelli, *Benedetto XIV* (1981), *1*, 656–8.

101. Benedict XIV, *Bref* (1748), 3–5, 8–11 (quote); Reusch, *Index* (1883), *2*, 832–4. Cf. Battistella, *S. Officio* (1905), 157–8. The misuse of the censorship in academic squabbles was an old story: "The inquisitor reported that the accusations came from many jealous rivals of the professor in Bologna, but that nothing really could be said against him;" ibid., citing a case of 1624.

102. Benedict XIV, *Bref* (1748), 5–6; Falco, in Fubini, *Cultura* (1964), 23–6 (quote), 34–6; Passerin, in ibid., 211–12, 219. When Benedict's letter to the Spanish inquisitor was leaked, Muratori asked the Pope to point out

where he had erred and direct him how to clear his name. Benedict replied in a friendly way that Muratori's views about the Pope's rights as temporal ruler were clearly censurable; Reusch, *Index* (1883), 2, 840–1.

103. Noris to Magliabecchi, 29 Apr. 1676, in Magliabecchi, *Epist. ven.* (1745), *1*, 83–4. Cf. Maugain, *Etude* (1909), 134, and Fisch, in Underwood, *Science* (1953), *1*, 524–5, for other relevant cases of special influence.

104. Hilgers, *Index* (1904), 93, 134–7; Garrido, *Censor* (1988), 84–118.

105. Garzend, *Inquisition* (1912), 6–7, 14–23, 56–7, 103, 106–12, 429–73. Cf. Costanzi, *Chiesa* (1897), 368–72.

106. Gassendi, *De motu impresso* (1642), in Garzend, *Inquisition* (1912), 468; Riccioli, *Alm. nov.* (1651), *1:1*, 52, and *1:2*, 489, 495–6, 500, also in Costanzi, *Chiesa* (1897), 375; Mengoli to Magliabecchi, 19 Oct. 1675, in Mengoli, *Corresp.* (1986), 82; Baldigiani to Viviani, in Brandmüller and Greipl, *Copernico* (1992), 29.

107. Borelli to Leopold de' Medici, 22 Jan. 1665, in Rotta, *Misc. seic.*, 2 (1971), 72.

108. Quoted in Cavazza, *Settecento* (1990), 215.

109. Manfredi, *De annuis* (1729), t.p., 5, 59, 64–6, 68–9, 73–6, 80.

110. Ibid., 3; Montanari, *Forze* (1694), 2–4; Mazzoleni, *Vita* (1935), 11; Cardella, *Mem.* (1792), *8*, 119; *DBI*, *33*, 127–30 (G. P. Brizzi); Nicolini, Dep. stor. patria Romagna, *Atti*, *20* (1930), 105, 108–16; Baiada et al., *Museo* (1995), 65; Moroni, *Dizionario* (1840), *19*, 163.

111. Baiada et al., *Museo* (1995), 63–5, 147.

112. Letter of 1729, quoted by Bortolotti, *Storia* (1947), 157.

113. Manfredi, *CAS*, *1* (1731), 619 (quote), 628–31, 639.

114. Baiada et al., *Museo* (1995), 66 (quote).

115. Giovanni Poleni (professor at the University of Padua), *De vorticibus coelestibus dialogus* (Padua, 1712), "Praefatio," quoted by Ferrone, *Scienza* (1982), 97n, and *Roots* (1995), 292n4; Newton, *Principia*, ed. T. Le Seur and F. Jacquier (Geneva, 1739–42), quoted by Reusch, *Index* (1883), 2, 399. Cf. Poleni, *GL*, 8 (1711), 200, 201.

116. Baliani, letter of 20 Dec. 1641, in Costantini, *Baliani* (1969), 52: "I hold as true only what is not opposed to Holy Scripture and . . . there-

fore I judge the opinion of Copernicus to be false"; Reynaud, *Erotemata* (1653), 251–2.

117. Duhem, *Phenomena* (1969), 5–23, 66–9, 76–91; Heilbron, *Weighing* (1993), 16–23, and in Bernhard et al., *Science* (1982), 52–9.

118. Manfredi, *Istituzioni* (1749), vii, ix. Cf. Campanacci Magnami, in Tarozzi and van Vloten, *Radici* (1989), 342–6, 350–1.

119. Manfredi to Algarotti, 19 Feb. 1737, in Cavazza, *Settecento* (1990), 245; Bianchi to Celsius, 9 May 1734, Celsius Papers (Uppsala).

120. Zan, in Cremante and Tega, *Scienza* (1984), 140–6, 147n (Zenotti to A. Leprotti, 6 Feb. 1737, quote).

121. Ferrone, *Scienza* (1982), 92–3, and *Roots* (1995), 34–5; Riccardi, *Bibl.* (1893), *1:1*, col. 512. The book also had the protection of a fictitious editor, a make-believe dedicatee, and the talisman from Virgil, "provehimur portu . . ."; Galileo, *Dialogo* (1710), t.p., and pt. 2, p. 69.

122. According to Lucio Ferraris, *Bibliotheca canonica* (1759), *3*, 336–7, as cited by D'Addio, *Considerazioni* (1985), 116n.

123. Restiglian, in *Galileo* (1982), 235–6; Bellinati, in Galluzzi, *Novità* (1984), 127–8; Fantoli, *Galileo* (1993), 398–9, 416, and (1994), 471–2, 492–3; Rotta, *Misc. seic.*, 2 (1971), 89; Serena, *Barbarigo* (1963), *1*, 69 (quote).

124. Calmet, in Galileo, *Opere* (1744), *4*, 1–2; "A chi legge," ibid., a.2r; Serena, *Barbarigo* (1963), *1*, 77 (evaluation of Toaldo). To his other virtues Toaldo added that of meridian maker; Toaldo, *Meridiana* (1838).

125. Brandmüller, in Brandmüller and Greipl, *Copernico* (1992), 32–3; Reusch, *Prozess* (1879), 460, and *Index* (1883), 2, 399–400.

126. Brandmüller, in Brandmüller and Greipl, *Copernico* (1992), 31, quoting *De aestu maris* (1747); Nikolic, *Arch. int. hist. sci.*, 14 (1961), 331–2, citing *De cometis* (1746), of which Casini, *Dix-huit. siècle*, *10* (1978), 99, prints the relevant text.

127. Brandmüller, in Brandmüller and Greipl, *Copernico* (1992), 35–7; Reusch, *Prozess* (1879), 440–1; Galileo, *Opere* (1890), *19*, 419.

128. Borgato, in Pepe, *Copernico* (1996), 211–14.

129. Ibid., 28–33, 239, 247–50, 255–60; Meli, in Harman and Shapiro, *Investigation* (1992), 424–41. Cf. Borgato and Fiocca, in Guglielmini, *Carteggio* (1994), 3–10.

130. D'Addio, *Considerazioni* (1985), 117–19; the documents in the case are published by Greipl, in Brandmüller and Greipl, *Copernico* (1992), 133–484, and in Maffei, *Settele* (1987).

131. Simoncelli, *Storia* (1992), 21–2, 34–5, 38, 43, 60–2, 73.

132. Ibid., 81–6, 104; Fantoli, *Galileo* (1993), 406–9, and (1994), 481–3.

133. Simoncelli, *Storia* (1992), 107–9, 112–21, 128–36.

134. Fantoli, *Galileo* (1993), 408–10, and (1994), 482–6; John Paul II, in Poupard, *Galileo* (1983), 195–9.

135. John Paul II, in Poupard, *Galileo*, xvi, xxii; Fantoli, *Galileo* (1993), 411–13, and (1994), 486–8.

136. Philippe de la Trinité, *Divinitas, 25* (1959), 36, quoted by Poupard, *Galileo* (1983), xv; Galileo to Tuscan Secretary of State, 6 Mar. 1616, in Finocchiaro, *Affair* (1989), 151.

137. Reusch, *Index* (1883), 2, 175, 598–9.

138. Redondi, *Galileo* (1987), 14–19, 31–5, 49, 179–96, 240–9, 261–3, 328–9, 333–5; Costantini, *Baliani* (1969), 72–3, 92–3. Cf. Costabel, *Vie des sciences, 4* (1987), 356–60.

139. Letter of [1679], in Leibniz, *Schr.* (1926), ser. 2, 1, 503.

140. Riccioli to Kircher, 22 Dec. 1646, and Grassi to Baliani, 25 Aug. 1652, both in Costantini, *Baliani* (1969), 97, 108. The machinery of the Jesuits' internal censorship to 1650 is described by Baldini, *Legem* (1992), 75–119.

141. Pachtler, *Ratio studiorum* (1887), 3, 121–7; Sortais, *Cartésianisme* (1929), 20.

142. Reusch, *Index* (1883), 2, 601; cf. Costantini, *Baliani* (1969), 60.

143. Ziggelaar, *Pardies* (1971), 80, 115, 119 (quote); Sortais, *Cartésianisme* (1929), 89–92; Bouillier, *Hist.* (1854), 1, 557–80.

144. Stimson, *Grad. accept.* (1917), 85–6.

145. *DBI, 24* , 326–8 (G. Gronda); *GL, 7* (1711), 113; Ceva, *AE,* 1695, 290–4; Pascal, Ist. lomb. sci. lett., *Rend., 48* (1915), 174–8; Ferrone, *Roots* (1995), 33–4, 291n58; Ceva to Grandi, 30 June 1700, in Paoli, *Ann. univ. tosc., 29:3* (1910), 77.

146. Paoli, *Ann. univ. tosc., 29:3* (1910), 64–8, 81–2 (quote); Pascal, Ist. lomb. sci. lett., *Rend., 48* (1915), 180–1.

147. Ceva to Grandi, 26 June 1700, in Paoli, *Ann. univ. tosc., 29:3* (1910), 75. Leibniz esteemed Ceva's brother, Giovanni, as a mathematician; Robinet, *Iter* (1988), 308.

148. Grandi to Ceva, 5 Dec. 1699 and 29 Jan. 1700, in Paoli, *Ann. univ. tosc., 28:6* (1908), 5–6 (quote), 23–4 (quote).

149. Grandi to Ceva, 8 May 1800, in ibid., *29:3* (1910), 59(quote)–60, and 16 Jan. 1700, in ibid., *28:6* (1908), 12–13.

150. Ceva, *Phil.* (1718), f. A.3v, 9, 28–30, 48, 68, 88; *GL, 7* (1711), 114–25, 128–30.

151. *GL, 7* (1711), 117 (quote); cf. Maugain, *Etude* (1909), 71.

152. *GL, 35* (1724), and *38* (1727); Maugain, *Etude* (1909), 168–71; *DBI, 24* , 326–7 (G. Gronda). Cf. Lastri, *L'osservatore, 1* (1821), 177–80.

153. Cousin, *JS,* 1841, 6, 9(quote)–10; Robinet, in Centre, *Copernic* (1975), 271–2; Charma and Marcel, *André* (1857), 2, 19, 234, 338–9.

154. Charma and Marcel, *André* (1857), 2, 340–2; Sortais, *Cartésienisme* (1929), 23, 28–32; Reusch, *Index* (1883), 2, 398–9 (quote).

155. Cousin, *JS,* 1841, 13 (quote, letter of 25 Apr. 1713), 15–17, 20–4; Charma and Marcel, *André* (1857), 1, 422–6, 2, 314–16, 319 (quotes).

156. Charma and Marcel, *André* (1857), 2, 292, 344–6.

157. Fisch, in Underwood, *Science* (1953), 1, 543; Renaldo, *Bartoli* (1979), 123–7; Lopiccoli, in Magrini, *Scienza* (1990), 21–2, 65–72.

158. Lopiccoli, in Magrini, *Scienza* (1990), 25–31; Magrini, in ibid., 9–10; Montanari to Magliabecchi, 1676, quoted by Baroncini, in *Materiali* (1975), 78–9.

159. Cavazza, *Settecento* (1990), 240, referring to Manfredi's *Il paradiso* (1698).

160. Maugain, *Étude* (1909), 136–42.

161. Duhamel, *Historia* (1698), 187, 267–72; Cassini Papers (Paris), D1.13, quote.

162. Russell, in Dobrzycki, *Reception* (1972), 213–14, dates the Copernican conquest of England to the 1660s. Tycho had competed well, especially among churchmen, but had lost out before midcentury; Johnson, *Astr. thought* (1937), 249, 253–63, 274–5.

163. Reyher, *Mathesis regia* (1693), 6–7, 11–55 (the examples), 61–68 (the branches).

164. For example, Cellarius, *Harmonia* (1661), 5–6, dedicated to the stadholder of the Netherlands; *Description de l'univers, contenant les différens systèmes du monde* (5 vols., 1693),

dedicated to Louis XIV; Coronelli, *Epitome* (1693), 26, 32, commissioned by the overseers of the University of Padua; Brisbar, *Calendrier* (1697), 203, 228, addressed to readers of history.

165. Lopez, *Riforma* (1964), 149, 158–64, 176–7, 182–7, 190–7; Cardella, *Mem.* (1792), 8, 5–6; Maugain, *Étude* (1909), 153–61, 224–32; Davia to Galiani, 11 Oct., 6 and 27 Dec. 1730, in Nicolini, Dep. stor. patr. Romagna, *Atti, 20* (1930), 118–19; Silvestre, *PT,* 22:265 (1700), 629, cited by Fisch, in Underwood, *Science* (1953), 544.

166. Monchamp, *Galilée* (1892), 187, 191–210, 215–19, 230–4, 263–96, 320–4, 329–30.

167. Brizzi, *Formazione* (1976), 22–51, 235–55; Rétat, *Dix-huit. siècle,* no. 8 (1976), 172, 179–81, 207, 211, 297–8.

168. Riccardi, *Bibl.,* pt. 2 (1880), xv–xvii.

169. Fontenelle, *HAS,* 1723, 127.

7. The Last Cathedral Observatories

1. Nordenmark, *Celsius* (1936), 44; Celsius Papers (Uppsala), Ms. A533d, f. 145.

2. Lemesle, *Saint-Sulpice* (1931), 26–30; Boinet, *Eglises* (1958), 2, 290–4; *BU, 23,* 199–200 (Labouderie).

3. *HAS,* 1743, 146; Lemmonier, *Inst.* (1746), xxxvii.

4. Boinet, *Eglises* (1958), 2, 294, 318, 335; Bouillet, *Saint Sulpice* (1899), 12; Hamel, *Histoire* (1900), 179–80; Lemesle, *Saint-Sulpice* (1931), 32, 162.

5. Sully, *Mercure,* July 1728, 1591–2; Heilbron, *Electricity* (1979), 279; Camus et al., *Astr., 104* (1990), 196.

6. Camus et al., *Astr., 104* (1990), 195, 197; Sully, *Mercure,* July 1728, 1601–5. Danti had run the end of the *meridiana* in San Domenico, Bologna, up a column.

7. *HAS,* 1743, 143–4; Camus et al., *Astr., 104* (1990), 199; *BU, 24,* 95–7 (Benchot); *DSB, 8,* 178–80 (T. Hankins).

8. Lemonnier, *HAS,* 1743, 365–6.

9. Delambre, *Histoire* (1827), xxxiii, dismisses Lemonnier as "always behind his century"; cf. ibid., 225, 228, 230, 406.

10. *DSB, 8,* 21–2 (Daumas); *Encycl., 8,* 386 (quote), perhaps Lalande, who uses the same phrase in *Ency. met., Math.* (1784), 2, 383–4.

11. Buchner, *Sonnenuhr* (1982), 7, 10–2, 34–37, 43, 72; Augustus reckoned that he was conceived at the autumnal equinox. Cf. Zanker, *Power* (1988), 144–5.

12. The text is given in Michaux, *Inventaire* (1885), 263; cf. Camus et al., *Astr., 104* (1990), 206–7.

13. Lemesle, *Saint-Sulpice* (1931), 59–62; Boinet, *Eglises* (1958), 2, 297–301, plate after p. 368 (Languet's monument); Malbois, *Gaz. beaux-arts, 75* (1933:1), 42, 46; Camus et al., *Astr., 104* (1990), 208.

14. KM = $2wh \cdot h\csc^2\alpha(\sigma/2)/h^2\ctn^2\alpha = w\sigma\sec^2\alpha$.

15. Dimensions from Gotteland, Soc. ast. France, *Obs. trav.,* no. 12 (1987), 26–7 (h = 24.55 m, w = 51.64 m), which differ from those in *Encylopédie, 8* (1765), 385, in Boucher d'Argis, *Variétés* (1752), 2, 346–7, 350, and in the descriptive sheet available at the church (h = 25.54 m, w = 51.64 m, XY = 10.72 m).

16. IJ = 2OP (tan $\sigma/2$)(w/x) ≈ $w\sigma\sec\alpha$; IJ:KM = cos α = 0.95.

17. Lemonnier, *MAS,* 1743, 363–4 (quote), repeated in *Encyclopédie, 8* (1765), 385, and Lemonnier, *Mercure,* Jan. 1744, 178–9.

18. Ibid.; Boucher d'Argis, *Variétés* (1752), 2, 330–1; Delisle, *MAS,* 1714, 241–2; Lalande, *Voyage* (1786), 2, 251. The midsummer sun's image had a diameter half that of the midwinter sun.

19. *HAS,* 1743, 145–6. The focal length wanted was 75csc65°40' = 82.3 feet.

20. Lemonnier, *MAS,* 1762, 265–6; Camus et al., *Astr., 104* (1990), 207–8.

21. Camus et al., *Astr., 104* (1990), 201–2, 209–14; Gotteland, Soc. astr. France, *Obs. trav.,* no. 12 (1987), 33–6.

22. Lalande to Ximenes, 6 Aug. 1755, in Mori, *Arch. stor. ital., 35* (1905), 378, and in Lalande, *Voyage* (1786), 2, 422; Algarotti to Zanotti, 5 Nov. 1763, in Algarotti, *Opere, 10* (1784), 367.

23. Piazzi, *Della specola* (1792), xxx; for his *meridiana* at Palermo see Chapter 8 below.

24. Barsanti and Rombai, *Ximenes* (1987), 27–30.

25. Ibid., 30, 107–9; *BU, 23,* 92–4 (Salfi); Ximenes later made an important library of mathematical and geographical books; Rombai, *Ximenes* (1987), 33–7, 201–25.

26. Rombai, *Ximenes* (1987), 38–40.

27. Ibid., 43; Mori, *Arch. stor. ital., 35* (1905),

376–80, 391–5, 417 (report by Ximenes, 26 Dec. 1777, quote); "contractor general," a contemporary tag, quoted by Rotta, *Illuminismo* (1974), 39.

28. Barsanti and Rombai, *Ximenes* (1987), 31–2, 109–11; Ximenes, *Notizie* (1752), 4–5 (quote), 17.

29. Ximenes, *Gnomone* (1757), 72 (quote), 75–82; Barsanti and Rombai, *Ximenes* (1987), 32, 113–14.

30. Barsanti and Rombai, *Ximenes* (1987), 14, 45, 53–4; Palcani, Soc. ital., *Mem., 5* (1790), xxvi.

31. Palcani, Soc. ital., *Mem., 5* (1709), xii–ix; Boscovich to Arnolfini, Feb. and Mar. 1781, in Arrighi, *Carteggi,* (1965), 57–62; Barsanti and Rombai, *Ximenes* (1987), 46–57.

32. Barsanti and Rombai, *Ximenes* (1987), 99–100. The observatory still exists.

33. *Novelle letterarie, 19* (1758), cols. 49–52, 82–5, 91–9, 114–19, 258–62, 273–7.

34. Ximenes, *Gnomone* (1757), xx–xxx, xxxvi–xxxvii.

35. Ibid., xxxv, xl–xli; Guasti, *Cupola* (1857), 211–12; Algarotti to Zanotti, 5 Nov. 1763, in Algarotti, *Opere, 10* (1784), 369.

36. Guasti, *Cupola* (1857), 185–6; Lalande, *Voyage* (1786), 2, 423.

37. Ximenes, *Gnomone* (1757), 11–13 (quote), 14–15.

38. Ximenes, *Gnomone* (1757), 16–20 (quote).

39. Ximenes, *Gnomone* (1757), 219 (quote), 220–3.

40. Ximenes, *Gnomone* (1757), 31(quote)–42, 235–84; Heilbron, *Weighing* (1993), 31–2, 96–8; Feldman, *HSPS, 15:2* (1985), 127–97.

41. Ximenes, *Gnomone* (1757), 52–7; Suter, *Isis, 55* (1964), 81–2.

42. Ximenes, *Gnomone* (1757), 87, 103, 233.

43. Ibid., xlii, 1, 41, 85; Richa, *Notizie* (1754), 6, 169–70; Zach, *Corr. astr., 1* (1818), 3.

44. Richa, *Notizie* (1754), 6, 152; Suter, *Isis, 55* (1964), 79.

45. Uzielli, *Toscanelli* (1894), 605–6; Paatz and Paatz, *Kirchen* (1940), 3, 381, 384, 536n385; Busignani and Bencini, *Chiese* (1993), 80.

46. Ximenes, *Notizie* (1752), 8–9, and *Gnomone* (1757), 316; Lemonnier, *MAS*, 1738, 209.

47. Rigaud in Bradley, *Misc. works* (1832), vii–viii, l–li; Delambre, *Hist.* (1827), 420, rated Bradley just after Hipparchus and Kepler.

48. Bradley, *PT, 35* (1729), in Bradley, *Misc. works* (1832), 5, 11, 13; Rigaud, in ibid., xii–xiii, lxii–lxx.

49. Rigaud, in Bradley, *Misc. works* (1832), xxx–xxxiii; Wolf, *Hist.* (1902), 153.

50. Bradley, *PT, 45* (1748), in Bradley, *Misc. works* (1832), 21–31; Rigaud, in ibid., lxii–lxix; Forbes, in *Greenw. Obs.* (1975), *1,* 91–6.

51. Lemonnier, *Inst.* (1746), iii–iv; D'Addio, *Considerazioni* (1985), 116–17; Zan, in Cremante and Tega, *Scienza* (1984), 146.

52. Manfredi, *De gnomone* (1736), 48–9.

53. *HAS*, 1716, 48–54 (quote); *BU, 25,* 356–7 (Weiss); Louville, *AE,* 1719, 281–3.

54. *HAS*, 1714, 68–9.

55. Louville, *AE,* 1719, 281–91, 292 (quote).

56. Ibid., 293 (quote); Louville, *MAS,* 1721, 173; *HAS,* 1721, 65–6; cf. Le Gentil, *MAS,* 1757, 180, 189, and Lalande, *MAS,* 1780, 287–9.

57. Cf. Lemonnier, *Histoire* (1741), "Projet," and lxxxvi–xc, and *Zodiaque* (1745), viii; Long, *Astr.* (1742), *1,* 274–87.

58. Malezieu, *MAS,* 1714, 332–3 (quotes), 324–7; Manfredi, *De gnomone* (1736), 58–9; *BU, 26,* 240–1 (Landrieux). Cf. *MAS,* 1715, 132–4.

59. Manfredi, *CAS, 1* (1731), 596; ibid., 262–4 (quote).

60. *CAS, 2:1* (1745), 439–40; Manfredi, *De gnomone* (1736), 79.

61. Ibid., 76–7.

62. Manfredi, *De gnomone* (1736), 78–82; *CAS, 1* (1731), 618.

63. Rigaud, in Bradley, *Misc. works* (1832), xxxiii.

64. Lemonnier, *MAS,* 1738, 361–2, and *MAS,* 1743, 68–9; Cassini de Thury, *MAS,* 1741, 128.

65. Cassini de Thury, *MAS,* 1741, 147–8, and *MAS,* 1748, 259–61. Cf. J. Cassini, *Elémens* (1740), xi.

66. *MAS,* 1743, 68–9, 365.

67. Lemonnier, *Zodiaque* (1745), ix, and *Inst.* (1746), xxvii–xxviii.

68. Lemonnier, *Zodiaque,* xxxiv–xxv, xliii–xliv, xlvii–xlix.

69. The Paris astronomers had advance notice of Bradley's discovery to help them reduce the observations made in Lapland. They therefore perceived before others the need to observe the obliquity at 19-year intervals, that is, at equal phases in the mutation cycle, to have any chance of measuring a small secular

change in ε. Maupertuis to Bradley, 27 Sept. 1737, and reply, 27 Oct. 1737, in Bradley, *Misc. works* (1832), 404–6, 409–10; Rigaud, in ibid., lxv; Lemmonier, *Inst.* (1746), xlvii.

70. Lemonnier, *MAS*, 1762, 263–6, confirmed in *MAS*, 1767, 422; Lalande, *MAS*, 1762, 267–8; *HAS*, 1762, 129–31.

71. Lemonnier, *MAS*, 1774, 252–3; Lalande, *MAS*, 1780, 293–4; *HAS*, 1774, 46.

72. Ximenes, AS, Siena, *Atti, 5* (1774), 39–50 (quote), and *Diss.* (1776), 11–14.

73. Ximenes, *Gnomone* (1757), 121–2, 179–80; Smart, *Spherical astronomy* (1962), 57.

74. Ximenes, *Gnomone* (1757), 179–80, 186, 193; Algarotti to Zanotti, 5 Nov. 1763, in Algarotti, *Opere, 10* (1784), 369–70.

75. Ximenes, *Diss.* (1776), 19–26 (the other astronomers were Bernoulli, Fontana, and Slop), 38, 43, 48, 53, 56, 63, 66, 73, 87–99 (how to separate the periodic from the secular), 100 (quote); Boscovich to Conti, 27 Aug. 1780, in Boscovich, *Lettere* (1980), 7.

76. Euler, AS, Berlin, *Mém.*, 1754, 796–9, 800 (quote).

77. Ibid., 311, 318, 322.

78. Euler to Mayer, 27 May 1755, in Forbes, *Corresp.* (1971), 98 (E = –45''); *HAS*, 1780, 38–9; Lalande, *MAS*, 1780, 289–91.

79. Lalande, *MAS*, 1758, 359–61, 369.

80. Cassini de Thury, *MAS*, 1780, 472–3.

81. Ximenes, AS, Siena, *Atti, 5* (1774), 36–7, 51–4, and *Diss.* (1776), 14–17; cf. Ximenes' *Notizie* (1752), 10–11, criticizing overreliance on theory.

82. Lalande, *MAS*, 1762, 268, and *MAS*, 1780, 291–3; *HAS*, 1780, 39–40.

83. Lalande, *MAS*, 1780, 295–8, 303; Cassini de Thury, *MAS*, 1780, 471–4 (quote). Cf. Cesaris, in *EA*, 1775, 85–6, and *EA*, 1776, 86–7, inclining toward Lacaille's –44''.

84. J. D. Cassini, *MAS*, 1778, 484–90, 494–5, 502.

85. Ibid., 504; *HAS*, 1778, 30–1.

86. Lalande, *Astronomie* (1792), 2, 58. Laplace's expression for ε(t), the variable part of ε: $\varepsilon(t) = 933''\cos 18''t - 3140''\sin 33''t$, where t is measured in years from 1700.

87. Manzini, *L'Occhiale* (1660), f.*[3]rv.

88. Grillot, *Nuncius*, 2:2 (1987), 148; Wolf, *Hist.* (1902), 161–2. This ratio (diameter of objective/focal distance = 1/300) held for the longest lenses; for a focal distance f of 35', the

ratio was about 1/150. Cf. Smith, *Opticks* (1738), *1*, 135–9. From Appendix J, equation 6, $GF_0/2 = a(n/2(n - 1))^2(a/f)^2$, where $2a$ is the diameter, and n the index of refraction, of the lense.

89. $\beta/\alpha \approx \tan\beta/\tan\alpha = (FF'/f_e)/(FF'/f_0)$.

90. Wolf, *Hist.* (1902), 161–2.

91. Auzout, *Lettre* (1665), 25–6, gives $f_0 = 300$ feet, $f_e = 6$ inches, m (magnification) = 600.

92. Smith, *Opticks* (1738), *1*, 134–5, 140; King, *History* (1955), 43–4, 49–50.

93. Wolf, *Hist.* (1902), 162–3, 167–8; G. D. Cassini, *Origine* (1693), 35–6.

94. Bianchini, *Nova phaen.* (1728), 58–9, and figures VII, VIII; Monaco, *Physis, 25* (1983), 419, 426–31; Bedini, in Anderson et al., *Instr.* (1993), 114.

95. Napoli, *Nuove invenzione* (1686), 10–11 (with plates); Sebastiani, in AS, Paris, *Machines* (1735), *1*, 93–5, with illustration, dating from before 1699.

96. *Biog. nat. de Belgique, 8* (1885), cols. 154–6 (A. Siret); Bosmans, *Rev. quest. sci., 13* (1928), 221, quoting an eyewitness of 1683; Gottignies, *Logistica* (1687), "Praefatio," sig. a.4v (quote); *PT, 1:12* (1666), 209. Cf. Petit to Huygens, 2 Sept. 1662, in Huygens, *Oeuvres* (1888), *4*, 236.

97. Monaco, *Physis, 25* (1983), 427–8.

98. Smith, *Opticks* (1738), *2*, 355–62, gives a translation of Huygens' *Astroscopia* (1684), also in Huygens, *Oeuvres* (1888), *21*, 202–22; the foreign objective (no. 616 of the figure) is from La Hire, *MAS*, 1715, 7–9. Cf. Huygens, *Oeuvres* (1888), *21*, 19–20, and *22*, 730.

99. Grillot, *Nuncius*, 2:2 (1987), 151; Wolf, *Hist.* (1902), 156–7; Alexander, *Saturn* (1962), 111–12, 278. Iapetus has a diameter of 750 miles.

100. Colbert to Bishop of Laon, 10 Sept. 1671 and 15 Jan. 1672, and reply, 13 July 1672, in Wolf, *Hist.* (1902), 157–8; Cardella, *Mem.* (1792), *7*, 208–10. The cost was 4400 livres. Cf. King, *Hist.* (1955), 56–60.

101. Delambre, *Hist.* (1821), *2*, 785; Wolf, *Hist.* (1902), 159; G. D. Cassini, in ibid., 164–6. By 1686 Cassini had managed to see all five moons of Saturn with an 84-foot Campani glass; Cassini, *JS*, 1686, 150–1.

102. Wolf, *Hist.* (1902), 172–3; Grillot, *Nuncius*, 2:2 (1987), 153–4.

103. Flamsteed to Newton, 27 Dec. 1684, and 9 Sept. 1686 (quote), in Newton, *Corresp.* (1959), 2, 405, 449, 450n2; Rigaud, in Bradley, *Misc. works* (1832), ii–iv, ix. Flamsteed remained skeptical. Even Huygens had had trouble finding Cassini's satellites; Smith, *Opticks* (1738), 2, 361–2.

104. Réaumur, *MAS*, 1713, 300–3, describing Bianchini's machine.

105. J. Cassini, *MAS*, 1714, 363–4.

106. Ibid., 365–7 (first quote), 367, 369.

107. Bianchini, *Nova phaen.* (1728), 87–91; Fontenelle, *HAS*, 1729, 109, 111–13; Introduction, above, for the Portuguese connection.

108. Melchiore a Briga to Bianchini, 3 Sept. 1626, in Bianchini, *Nova phaen.* (1728), 86 (quote); Hunt and Moore, *Venus* (1982), 36–9, 82, 128; Bianchini to Maraldi, 17 Apr. 1727, Cassini Papers (Paris), Ms A. B.49; Bianchini to James Jurin (secretary of the Royal Society), 28 Dec. 1726, in Rusnock, *Correspondence* (1996), 330. This last reports Bianchini's visit to Bologna and an interruption in the observations caused by the discovery of ancient ruins that needed his attention.

109. Cassini de Thury, *MAS*, 1741, 114; J. D. Cassini, *MAS*, 1780, 39.

110. Lalande, *MAS*, 1762, 268. Cf. Zach, *Corr. astr.*, 3 (1819), 276–7.

111. Wolf, *Hist.* (1902), 136–8; Olmsted, *Isis, 40* (1949), 224–5; McKeon, *Physis, 14* (1972), 228–30. The following account is adapted from Heilbron, *Weighing* (1993), 38–46.

112. McKeon, *Physis, 13* (1971), 246–66; Repsold, *Astr. Messw.* (1908), 1, 40–3; Chapman, *Dividing the circle* (1990), 26–30, 40–4.

113. Delambre, *Hist.* (1821), 2, 598–613; Débarbet, in Picolet, *Picard* (1987), 161–8; cf. McKeon, *Physis, 13* (1971), 276–7.

114. G. D. Cassini, *Meridiana* (1695), 15–17.

115. Picard, *Mesure* (1671), 165 (quote), 172; Repsold, *Astr. Messw.* (1908), 1, 44, 65.

116. Taton, in Picolet, *Picard* (1987), 209, 218; Levallois, in ibid., 234–5, 240–3.

117. Delambre, *Hist.* (1821), 602–3, 612–17, 621–4, 627.

118. Repsold, *Astr. Messw.* (1908), 1, 48–54; Lévy, in Picolet, *Picard* (1987), 137; Chapman, *Dividing the circle* (1990), 63–5.

119. Lemonnier, *Histoire* (1741), "Projet"; Wolf, *Hist.* (1902), 138, 140, 148–9.

120. Smeaton, *PT, 76* (1786), 6 (quote); Chapman, *Ann. sci., 40* (1983), 463–8, and *Dividing the circle* (1990), 54–60; Howse, *Greenw. Obs.* (1975), 3, 17–21.

121. *DSB, 5,* 490–2 (E. A. Battison).

122. Smith, *Opticks* (1738), 2, 332(quote)–6.

123. Shuckburgh, *PT, 83* (1793), 75; King, *Hist.* (1955), 109–18; Daumas, *Instruments* (1953), 234–5, 258; Chapman, *Ann. sci., 40* (1983), 470, and *Dividing the circle* (1990), 70–3.

124. Chapman, *Dividing the circle* (1990), 67–70.

125. Ibid., 71–6.

126. Rigaud, in Bradley, *Misc. works* (1832), xiii–xv, xxvii–xxviii.

127. Chapman, *Dividing the circle* (1990), 83–5, 88–9.

128. Daumas, *Instruments* (1953), 230, 238, 255; Chapman, *Jl hist. astr., 14* (1983), 134, and *Dividing the circle* (1990), 109.

129. Shuckburgh, *PT, 83* (1793), 93–4, 98; Chapman, *Dividing the circle* (1990), 113–16.

130. Lemonnier, *Description* (1774), 1–16, quote on 3–4.

131. Louville, *MAS*, 1714, 65 (quote), 73–7; Cassini de Thury, *MAS*, 1741, 114.

132. Wolf, *Hist.* (1902), 176–9, 181–5; Chaulnes, *MAS*, 1765, 411–15.

133. Oriani to Cesaris, 28 Sept. 1786, in Mandrino et al., *Viaggio* (1994), 199 (second quote); Wolf, *Hist.* (1902), 254–5 (other quotes), 265, 288, 292.

134. David, *Dix-huit. siècle, 14* (1982), 279–84; Lalande to Oriani, 27 June 1786, in Mandrino et al., *Viaggio* (1994), 182. Astronomy in Lombardy is discussed in Chapter 8 below.

135. Chapman, *Dividing the circle* (1990), 96, citing Bird, *The method of dividing mathematical instruments* (London, 1767), 13–14.

136. Cacciatore, *Libri* (1826), 86, 115–27, 132–5. Cf. Piazzi, *Soc. ital. sci., Mem., 11* (1804), 426, 436–7, 443.

137. Delambre, *Hist.* (1827), 180–1, 403–6; Zach, *Corr. astr., 3* (1819), 269–70, 275.

138. Bianchi, *Soc. ital. sci., Mem., 22* (1839), cxxx–cxxxi; cf. Bianchi, in Modena, Obs., *Atti, 1* (1834), t.p., [vi–viii], xv–xvii, and Zach, *Corr. astr., 1* (1818), 6.

139. Calendrelli and Conti, *Opuscoli astr.*, *1* (1803), vi–xi (quote); *DBI, 16,* 440–2 (U. Baldini).

8. Time Telling

1. Proverbio, *Nuncius, 2:2* (1987), 123–4; Ximenes, *Diss.* (1776), dedication.

2. Reggio, Cesaris, and Oriani to Wilczek, 1785/6, in Mandrino et al., *Viaggio* (1994), 174; Wilceck to Oriani, 18 Apr., and Oriani to Cesaris, 5 Aug. 1786, in ibid., 165, 190.

3. Ibid., 20–1.

4. Cesaris, *EA,* 1792, App., 91 (quote); Gabba, Soc. astr. ital., *Mem., 28* (1957), 154; *DBI, 24,* 213–14 (U. Baldini).

5. Bianchi, Soc. ital. sci., *Mem., 22* (1839), cxix, cxxii, cxxvi–cxxix (quote).

6. Gilii, *Memoria* (1805), 19, 27–8; Passano et al., *Meridiana* (1977), 9.

7. Quoted by Tenard, *Dix-huit. siècle,* no. 8 (1976), 96.

8. Toaldo, *Raccolta* (1802), *2,* 260n (quote); Fabroni, in ibid., *1,* xxx–xxxi.

9. Fabroni, in ibid. *1,* 28, and *2,* 252–3; Piazzi, *Gior. sci. lett. arti Sicilia, 7* (1824), 145–7 (quote), 152, 157–60, 171–2.

10. Piazzi, *Gior. sci. lett. arti Sicilia, 7* (1824), 148; Toaldo, *Raccolta* (1802), *2,* 256–9. Cf. Walker, *Phil. mag., 25* (1806), 173.

11. Wilczek, Order of 23 Oct. 1786, reproduced in Passano et al., *Meridiana* (1977), 15.

12. Gabba, Ist. lomb. sci. lett., *Rend., 54* (1921), 447 (quote); Cesaris, *EA,* 1787, App., 123.

13. Cesaris to Oriani, 3 June 1786, in Mandrino et al., *Viaggio* (1994), 194.

14. Catteneo, *Duomo* (1985), 126.

15. Cesaris, *EA,* 1787, App., 124, 127, 133–7, 141–4, quote on 142; Gabba, Ist. lomb. sci. lett., *Rend., 54* (1921), 448–9; Passano et al., *Meridiana* (1877), 17–20, 42; Schiavo, *Meridiana* (1993), 70.

16. Ackerman, *Distance points* (1991), 215–20.

17. Ibid., 216, 221–6; D'Amico and Grandi, *Tramonto* (1987), 28–9, 37.

18. Passano et al., *Meridiana* (1977), 26, 44, 46.

19. Quotes from, resp., Bemporad, Soc. astr. ital., *Mem., 3* (1926), 396, and Piazzi to Oriani, 2 Mar. 1795, in Piazzi, *Corrispondenza* (1874), 35.

20. Angelitti, Soc. astr. ital., *Mem., 3* (1925), 370–9; Bemporad, ibid., 396–7, 400; Piazzi, *Risultati* (1801), 18–19, and *Della scoperta* (1802), 58–9 (quote).

21. Cacciatore, *Gior. sci. lett. art. Sicilia, 7* (1824), 172.

22. Piazzi to Oriani, 2 Mar 1795, and reply, 3 June 1795, in Piazzi, *Corrispondenza* (1874), 35, 37 (first and third quotes); Piazzi, *Discorso* (1790), 5 (second quote).

23. *DBI, 16,* 14–15 (U. Baldini); Piazzi, in *Reale osserv.* (1806), 1; Cacciatore, in *Libri* (1826), *1,* xviii, xxiv–xxv, and in Müller, *Biografie* (1853), 86–8.

24. Oriani to Piazzi, 3 June 95, and Piazzi to Oriani, 10 Dec. 1795 and 2 Sept. 1796, in Piazzi, *Corrispondenza* (1874), 37, 40, 42; Cacciatore, *Gior. sci. lett. art. Sicilia, 7* (1824), 173–5.

25. Cacciatore, *Gior. sci. lett. art. Sicilia, 7* (1824), 173–6.

26. Piazzi, *Gior. sci. lett. arti Sicilia, 7* (1824), 161–2.

27. Zanotti, *Meridiana* (1779), 8; Villa, *SMSUB, 1* (1956), 479.

28. Cf. Schiavo, Centro di studi stor. arch., *Boll., 8* (1954), 22n.

29. Gilii, *Memoria* (1805), 7, 9 (quote), 18, 26–7; Lais, *Memorie, 7* (1890), 49–62.

30. E. P., *Nuovo orologio* (1859), 2–3.

31. Lalande, *Ency. meth., Math.* (1784), *2,* 385; Camus et al., *Astr., 104* (1980), 196.

32. Howse, *Greenw. time* (1980), 82; Dohrn–van Rossum, *Gesch.* (1992), 317, and *Hist.* (1996), 342–6; Lalande, *Ency. meth., Math.* (1784), *2,* 385.

33. Gabbrielli, *Heliometro* (1705), 1–2; Ricci, Acc. fisiocritici, *Mem., 2* (1985), 319, 321, 323.

34. Alessandretti, *Meridiana* (1990), 6.

35. Sergardi to Gabbrielli, 4 and 11 Nov., 30 Dec. 1702, and Bianchini to same, 18 Aug. 1703 and 15 Mar. 1704, in Ricci, Acc. fisiocritici, *Mem., 2* (1985), 322–4.

36. Vaselli, in Crescimbeni, *Vite* (1708), *2,* 36; *GL, 6* (1711), 129; Procès verbaux, 1705, in Rossi, Acc. fisiocritici, *Mem., 5* (1985), 326.

37. Vaselli, in Crescimbeni, *Vite* (1708), *2,* 32–3, 38–41; *GL 6* (1711), 121–4.

38. *GL 6* (1721), 130–5; Gabbrielli, *Heliometro* (1705), 3, 8–18, 31–4.

39. Gabbrielli, *Heliometro* (1705), 39–44; *GL, 6* (1711), 137–48; Ricci, Acc. fisiocritici, *Mem., 2* (1985), 329–32.

40. Ricci, Acc. fisiocritici *Mem., 2* (1985), 334–6; Bianchini, *Observationes* (1737), 239–40.

41. Ricci, Acc. fisiocritici, *Mem., 2* (1985), 336–42, 352–5.

42. Ibid., 344–7.

43. Ibid., 348–51.

44. Alessandretti, *Meridiana* (1990), 4–7, 10–15 (quote), 34.

45. Ibid., 15–21, 38–40.

46. J. Cassini, *MAS*, 1732, 452–3.

47. Zach to David, 22 June 1796, in Zach, *Briefe* (1938), 120.

48. Quetelet, Bruss. Obs., *Ann.*, 1837, 226–7.

49. Ibid., 216 (first quote); Piazzi, *Gior. sci. lett. arti Sicilia, 7* (1824), 166; Quetelet, *Histoire* (1864), 415, 417 (third quote).

50. Quetelet, *Histoire* (1864), 418–20, and Quetelet, Bruss., Obs., *Ann.*, 1837, 218–19.

51. Quetelet, *Histoire* (1864), 421–4; Mailly, *Essai* (1875), 121–2.

52. Giuffrida d'Angelo, *Nuovo metodo* (1838), [1]-[3], and *Brevi cenni* (1841), 3–4; Gemmelaro, Acc. gioenia, *Gior., 6:6* (1841), 24–5; Prestinenza, *Astr.*, no. 164, Apr. 1966, 15; Grassi, *Vita* (n.d.), 16(quote)–17.

53. *Deut. allg. Biog., 30*, 284–5 (v. Gümbal); Listing, Akad. Wiss., Gött., *Nachr.*, 1876, 547–8.

54. Roy. Astr. Soc., *Monthly notices, 51* (1891), 199–202; Hall, in *Peters* (1890), 5–7 (on Gauss); Sartorius, *Aetna* (1880), *1*, 69 (quote), 70.

55. Sartorius, *Aetna* (1880), *1*, 77, 80 (quote), 82, 105 (quote), 108–11, 114–24, 128; Listing, Akad. Wiss., Gött., *Nachr.*, 1876, 548–54.

56. Giuffrida d'Angelo, *Brevi cenni* (1841), [3]; Gemmelaro, Acc. gioenia, *Gior., 6:6* (1841), 23; Bertucci, *Guida* (1846), 52–5; Prestinenza, *Astr.*, no. 164, Apr. 1966, 15.

57. Sartorius, *Aetna* (1880), *1*, 139; Cacciatore, *Riflessioni* (1835); Hall, in *Peters* (1890), 8.

58. Hall, in *Peters* (1890), quoting Alexander von Humboldt's opinion of Peters; *Dict. Am. Biog., 14*, 503–4 (R. S. Duggan); Frost, in Peters, *Positions* (1907), xiii; Knobel, in Peters and Knobel, *Catalogue* (1915), 7–8.

59. Quotes from, resp., Gravagno, *Storia* (1992), 329, and an editorial in the *Gazzetta di Acireale, 4:13* (23 Aug. 1882), [4].

60. Gravagno, *Storia* (1992), 330; Ronsisvalle, *Astronomi* (1989), 22 (quote), 84 (dimensions); Anon., Acc. sci. lett. arti, Acireale, *Atti, 2* (1890), xiii.

61. Gravagno, *Storia* (1992), 90; Touring Club Italiana, *Guida, Sicilia* (1958), 450.

62. Ronsisvalle, *Astronomi* (1989), 48.

63. Portoghese, *Borromini* (1966), 225–6.

64. Russell, *We will remember* (1991), 14, 16, 22, 62, 64.

65. Ibid., 16, 78, 80–1 (quote).

Appendices

1. Danti, *Primo volume* (1578), 286.

2. Danti, *Sfera* (1571), 26.

3. Vitruvius, *Architecture* (1914), bk. ix, chap. vii, pp. 270–2; Clavius, *Gnomonices* (1581), 8, 17 (quote).

4. Cf. Schütz, *Sterne und Weltr., 6* (1989), 364.

5. Cassini, *Meridiana* (1695), 14–15 (quote), and *Specimen* (1656), 31–2. Cf. Baiada, *Museo* (1995), 36: Cassini's work at San Petronio "destroyed Aristotelian physics in the skies."

6. Cf. Mercator, *PT, 5:57* (25 Mar. 1670), 1168–70.

7. Riccioli, *Astr. ref.* (1665), *1*, 3–4; the square of the time figures in $\Delta\lambda$ so 3.5" in 12 hours corresponds to 15" a day. Manfredi, *De gnomone* (1736), 14–15, gives ν as a function of $\Delta\lambda$; as expected, for Δt = 12 hours, $\Delta\lambda$ = 30'.

Credits

The illustrations opening each chapter are from Riccioli's *Almagestum novum* (1651). The Introduction shows the lower half of the frontispiece: Astronomy finds Tycho's system of the world to outweigh Copernicus' while the all-seeing Argos peers through a telescope from an eye conveniently placed on his knee. Chapters 1 and 8 show the upper part of the frontispiece: symbols of day and night enlivened by the verse "one day speaks to another, night to night imparts knowledge," the subject of their conversation being God's handiwork in the heavens (Psalm 19:1–2). Chapters 2 to 7 bear the world systems in the order in which they stand on pp. 101–3 of the *Almagestum:* 2, the Ptolemaic (all the planets and luminaries around the earth, Mercury [☿] and Venus [♀] beneath the sun [☉]); 3, the Egyptian, usually ascribed to Martianus Capella (Mercury and Venus around the sun, the sun around the earth); 4, the Platonic (as in the Ptolemaic, except that Mercury and Venus are above the sun); 5, the Copernican (all the planets, including the earth, around the sun, the moon [☽] around the earth); 6, the Tychonic (all the planets around the sun, the sun and moon around the earth); and 7, the Ricciolian (as in the Tychonic, except that Jupiter [♃] and Saturn [♄] circle the earth). Courtesy of the Istituto e Museo di Storia della Scienza, Florence, and the Houghton Library, Harvard University.

Figures

Plates